Classroom Manual for

# Automotive Engine Repair and Rebuilding

Second Edition

# Classroom Manual for

# Automotive Engine Repair and Rebuilding

## Second Edition

Barry Hollembeak
Technical Training Inc.

Don Knowles
Knowles Automotive Training

Jack Erjavec
Series Advisor
Professor Emeritus, Columbus State Community College

DELMAR
THOMSON LEARNING™

Australia Canada Mexico Singapore Spain United Kingdom United States

**Today's Technician: Automotive Engine Repair and Rebuilding Classroom Manual 2E**
Barry Hollembeak and Don Knowles

**Business Unit Director:**
Alar Elken

**Executive Editor:**
Sandy Clark

**Acquisitions Editor:**
Sanjeev Rao

**Developmental Editor:**
Allyson Powell

**Editorial Assistant:**
Matthew Seeley

**Executive Marketing Manager:**
Maura Theriault

**Marketing Coordinator:**
Brian McGrath

**Executive Production Manager:**
Mary Ellen Black

**Production Coordinator:**
Toni Hansen

**Project Editor:**
Ruth Fisher

**Art/Design Coordinator:**
Cheri Plasse

**Cover Illustration:**
Bruce Kaiser



Printed in the United States of America
2 3 4 5 QWD 05 04 03 02

For more information contact Delmar,
3 Columbia Circle, PO Box 15015,
Albany, NY 12212-5015.

Or find us on the World Wide
Web at http://www.delmar.com



***Library of Congress Cataloging-in-Publication Data:***
Hollembeak, Barry.
Automotive engine repair and rebuilding / Barry Hollembeak, Jack Erjavec. — 2nd ed.
p. cm. — (Today's technician)
Includes index.
Contents: [1] Classroom manual for — [2] Shop manual for.
ISBN 0-7668-1626-5 (alk. paper) — ISBN 0-7668-1627-3 (alk. paper) — ISBN 0-7668-1628-1 (alk. paper)
1. Automobiles—Motors—Maintenance and repair. I. Erjavec, Jack. II. Title. III. Series.

TL210.H644 2002
629.25′028′8—dc21 2001028773

**NOTICE TO THE READER**

Publisher does not warrant or guarantee any of the products described herein or perform any independent analysis in connection with any of the product information contained herein. Publisher does not assume, and expressly disclaims, any obligation to obtain and include information other than that provided to it by the manufacturer.

The reader is expressly warned to consider and adopt all safety precautions that might be indicated by the activities herein and to avoid all potential hazards. By following the instructions contained herein, the reader willingly assumes all risks in connection with such instructions.

The publisher makes no representation or warranties of any kind, including but not limited to, the warranties of fitness for particular purpose or merchantability, nor are any such representations implied with respect to the material set forth herein, and the publisher takes no responsibility with respect to such material. The publisher shall not be liable for any special, consequential, or exemplary damages resulting, in whole or part, from the readers' use of, or reliance upon, this material.

# CONTENTS

# PREFACE

Thanks to the support the *Today's Technician* series has received from those who teach automotive technology, Delmar Publishers is able to live up to its promise to provide new editions every three years. We have listened to our critics and our fans and present this new revised edition. By revising our series every three years, we can and will respond to changes in the industry, changes in the certification process, and to the ever-changing needs of those who teach automotive technology.

The *Today's Technician* series, by Delmar Publishers, features textbooks that cover all mechanical and electrical systems of automobiles and light trucks. Principal titles correspond with the eight major areas of ASE (National Institute for Automotive Service Excellence) certification. Additional titles include remedial skills and theories common to all of the certification areas and advanced or specialized subject areas that reflect the latest technological trends.

Each title is divided into two manuals: a Classroom Manual and a Shop Manual. Dividing the material into two manuals provides the reader with the information needed to begin a successful career as an automotive technician without interrupting the learning process by mixing cognitive and performance-based learning objectives.

Each Classroom Manual contains the principles of operation for each system and subsystem. It also discusses the design variations used by different manufacturers. The Classroom Manual is organized to build upon basic facts and theories. The primary objective of this manual is to allow the reader to gain an understanding of how each system and subsystem operates. This understanding is necessary to diagnose the complex automobile systems.

The understanding acquired by using the Classroom Manual is required for competence in the skill areas covered in the Shop Manual. All of the high-priority skills, as identified by ASE, are explained in the Shop Manual. The Shop Manual also includes step-by-step instructions for diagnostic and repair procedures. Photo Sequences are used to illustrate many of the common service procedures. Other common procedures are listed and are accompanied with fine-line drawings and photographs that allow the reader to visualize and conceptualize the finest details of the procedure. The Shop Manual also contains the reasons for performing the procedures, as well as when that particular service is appropriate.

The two manuals are designed to be used together and are arranged in corresponding chapters. Not only are the chapters in the manuals linked together, the contents of the chapters are also linked. Both manuals contain clear and thoughtfully selected illustrations. Many of the illustrations are original drawings or photos prepared for inclusion in this series. This means that the art is a vital part of each manual.

The page layout is designed to include information that would otherwise break up the flow of information presented to the reader. The main body of the text includes all of the "need-to-know" information and illustrations. In the side margins are many of the special features of the series. Items such as definitions of new terms, common trade jargon, tool lists, and cross-referencing are placed in the margin, out of the normal flow of information so as not to interrupt the thought process of the reader.

## Highlights of This Edition—Classroom Manual

The text has been updated and the chapters reorganized to improve flow and enhance learning. In this edition, more emphasis is placed on overhead cam (OHC) and dual overhead cam (DOHC). In addition, the Theory of Engine Operation chapter is expanded to include Alternate Power Systems such as hybrid vehicles and fuel cells, as well as information on balance shafts and engine vibration. Information on the latest electronic ignition (EI) systems, coil-on-spark-plug systems, reverse flow cooling systems, and message centers is provided, as is the latest information on intake systems for fuel-injected engines. Finally, a discussion of variable camshaft timing and lift can be found, as well as an expanded chapter on high-performance engines.

## Highlights of This Edition—Shop Manual

This text has been updated and the chapters reorganized to improve flow and enhance learning. In this edition, more emphasis is placed on OHC and DOHC engines. A discussion regarding United States Customary (USC) and metric measurements is included, as well as expanded information on all areas of engine and engine operating system diagnosis. Information on replacing timing belts on DOHC engines and servicing engines with variable camshaft timing is also provided. In addition, balance shaft and balance shaft bearing diagnosis and service are covered, as well as thread repair and gasket, seal, and sealant service. Finally, to provide additional assessment and learning opportunities, job sheets are included at the end of each chapter, and ASE Challenge questions can be found throughout.

# Classroom Manual

To stress the importance of safe work habits, the Classroom Manual dedicates one full chapter to safety. Included in this chapter are common safety practices, safety equipment, and safe handling of hazardous materials and wastes. This includes information on MSDS sheets and OSHA regulations. Other features of this manual include:

## Cognitive Objectives

These objectives define the contents of the chapter and define what the student should have learned upon completion of the chapter.

*Each topic is divided into small units to promote easier understanding and learning.*

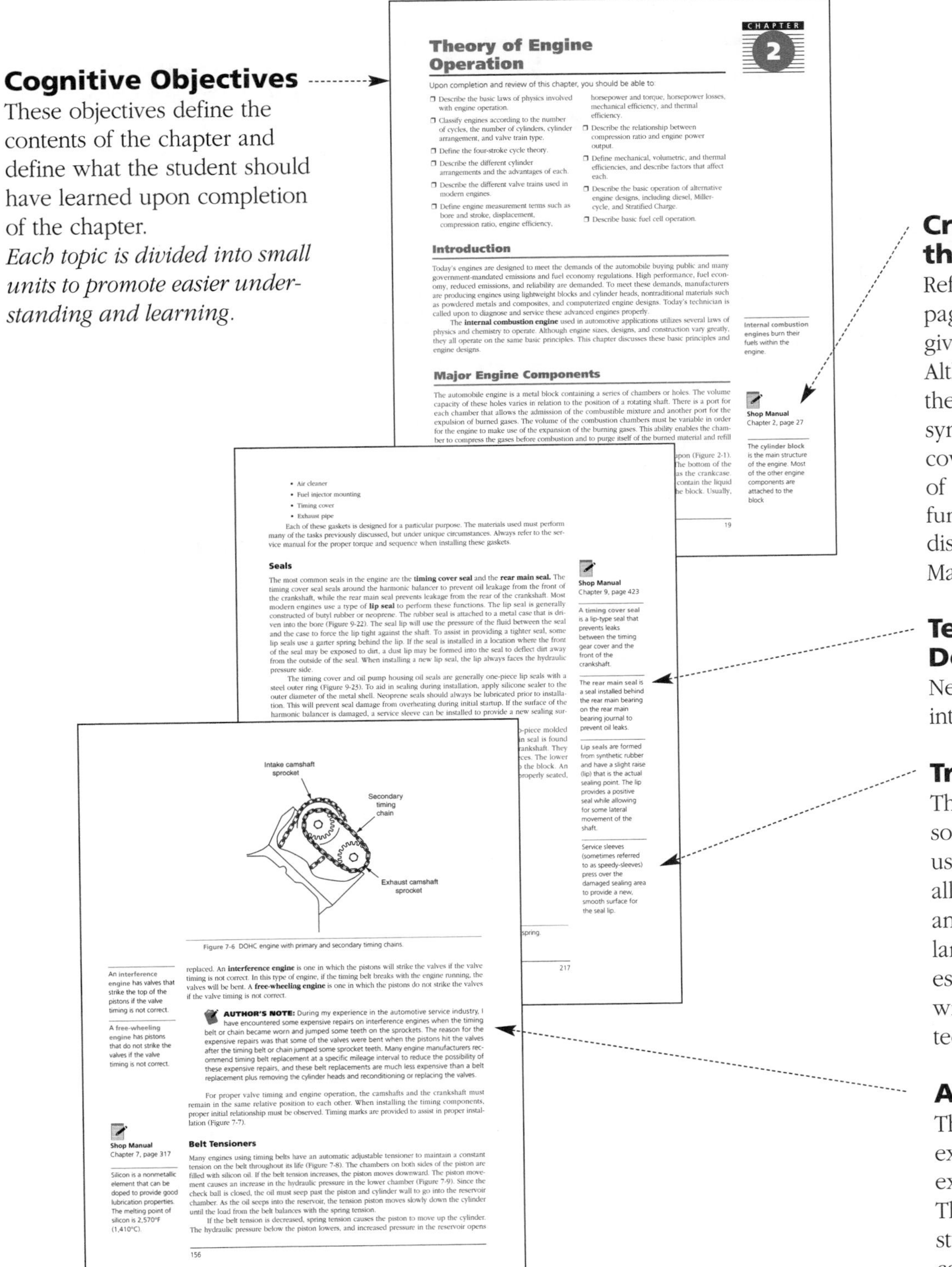

CHAPTER 2

### Theory of Engine Operation

Upon completion and review of this chapter, you should be able to:

- ❒ Describe the basic laws of physics involved with engine operation.
- ❒ Classify engines according to the number of cycles, the number of cylinders, cylinder arrangement, and valve train type.
- ❒ Define the four-stroke cycle theory.
- ❒ Describe the different cylinder arrangements and the advantages of each.
- ❒ Describe the different valve trains used in modern engines.
- ❒ Define engine measurement terms such as bore and stroke, displacement, compression ratio, engine efficiency, horsepower and torque, horsepower losses, mechanical efficiency, and thermal efficiency.
- ❒ Describe the relationship between compression ratio and engine power output.
- ❒ Define mechanical, volumetric, and thermal efficiencies, and describe factors that affect each.
- ❒ Describe the basic operation of alternative engine designs, including diesel, Miller-cycle, and Stratified Charge.
- ❒ Describe basic fuel cell operation.

#### Introduction

Today's engines are designed to meet the demands of the automobile buying public and many government-mandated emissions and fuel economy regulations. High performance, fuel economy, reduced emissions, and reliability are demanded. To meet these demands, manufacturers are producing engines using lightweight blocks and cylinder heads, nontraditional materials such as powdered metals and composites, and computerized engine designs. Today's technician is called upon to diagnose and service these advanced engines properly.

The **internal combustion engine** used in automotive applications utilizes several laws of physics and chemistry to operate. Although engine sizes, designs, and construction vary greatly, they all operate on the same basic principles. This chapter discusses these basic principles and engine designs.

**Internal combustion engines** burn their fuels within the engine.

#### Major Engine Components

The automobile engine is a metal block containing a series of chambers or holes. The volume capacity of these holes varies in relation to the position of a rotating shaft. There is a port for each chamber that allows the admission of the combustible mixture and another port for the expulsion of burned gases. The volume of the combustion chambers must be variable in order for the engine to make use of the expansion of the burning gases. This ability enables the chamber to compress the gases before combustion and to purge itself of the burned material and refill ... upon (Figure 2-1). The bottom of the ... as the crankcase. ... contain the liquid ... he block. Usually,

**Shop Manual**
Chapter 2, page 27

**The cylinder block** is the main structure of the engine. Most of the other engine components are attached to the block

19

- Air cleaner
- Fuel injector mounting
- Timing cover
- Exhaust pipe

Each of these gaskets is designed for a particular purpose. The materials used must perform many of the tasks previously discussed, but under unique circumstances. Always refer to the service manual for the proper torque and sequence when installing these gaskets.

#### Seals

The most common seals in the engine are the **timing cover seal** and the **rear main seal.** The timing cover seal seals around the harmonic balancer to prevent oil leakage from the front of the crankshaft, while the rear main seal prevents leakage from the rear of the crankshaft. Most modern engines use a type of **lip seal** to perform these functions. The lip seal is generally constructed of butyl rubber or neoprene. The rubber seal is attached to a metal case that is driven into the bore (Figure 9-22). The seal lip will use the pressure of the fluid between the seal and the case to force the lip tight against the shaft. To assist in providing a tighter seal, some lip seals use a garter spring behind the lip. If the seal is installed in a location where the front of the seal may be exposed to dirt, a dust lip may be formed into the seal to deflect dirt away from the outside of the seal. When installing a new lip seal, the lip always faces the hydraulic pressure side.

The timing cover and oil pump housing oil seals are generally one-piece lip seals with a steel outer ring (Figure 9-23). To aid in sealing during installation, apply silicone sealer to the outer diameter of the metal shell. Neoprene seals should always be lubricated prior to installation. This will prevent seal damage from overheating during initial startup. If the surface of the harmonic balancer is damaged, a service sleeve can be installed to provide a new sealing sur- ... -piece molded ... in seal is found ... rankshaft. They ... ces. The lower ... the block. An ... properly seated, ... spring.

**Shop Manual**
Chapter 9, page 423

**A timing cover seal** is a lip-type seal that prevents leaks between the timing gear cover and the front of the crankshaft.

**The rear main seal** is a seal installed behind the rear main bearing on the rear main bearing journal to prevent oil leaks.

**Lip seals** are formed from synthetic rubber and have a slight raise (lip) that is the actual sealing point. The lip provides a positive seal while allowing for some lateral movement of the shaft.

Service sleeves (sometimes referred to as speedy-sleeves) press over the damaged sealing area to provide a new, smooth surface for the seal lip.

217



Figure 7-6 DOHC engine with primary and secondary timing chains.

replaced. An **interference engine** is one in which the pistons will strike the valves if the valve timing is not correct. In this type of engine, if the timing belt breaks with the engine running, the valves will be bent. A **free-wheeling engine** is one in which the pistons do not strike the valves if the valve timing is not correct.

**AUTHOR'S NOTE:** During my experience in the automotive service industry, I have encountered some expensive repairs on interference engines when the timing belt or chain became worn and jumped some teeth on the sprockets. The reason for the expensive repairs was that some of the valves were bent when the pistons hit the valves after the timing belt or chain jumped some sprocket teeth. Many engine manufacturers recommend timing belt replacement at a specific mileage interval to reduce the possibility of these expensive repairs, and these belt replacements are much less expensive than a belt replacement plus removing the cylinder heads and reconditioning or replacing the valves.

For proper valve timing and engine operation, the camshafts and the crankshaft must remain in the same relative position to each other. When installing the timing components, proper initial relationship must be observed. Timing marks are provided to assist in proper installation (Figure 7-7).

#### Belt Tensioners

Many engines using timing belts have an automatic adjustable tensioner to maintain a constant tension on the belt throughout its life (Figure 7-8). The chambers on both sides of the piston are filled with silicon oil. If the belt tension increases, the piston moves downward. The piston movement causes an increase in the hydraulic pressure in the lower chamber (Figure 7-9). Since the check ball is closed, the oil must seep past the piston and cylinder wall to go into the reservoir chamber. As the oil seeps into the reservoir, the tension piston moves slowly down the cylinder until the load from the belt balances with the spring tension.

If the belt tension is decreased, spring tension causes the piston to move up the cylinder. The hydraulic pressure below the piston lowers, and increased pressure in the reservoir opens

**An interference engine** has valves that strike the top of the pistons if the valve timing is not correct.

**A free-wheeling engine** has pistons that do not strike the valves if the valve timing is not correct.

**Shop Manual**
Chapter 7, page 317

Silicon is a nonmetallic element that can be doped to provide good lubrication properties. The melting point of silicon is 2,570°F (1,410°C).

156

## Cross-References to the Shop Manual

Reference to the appropriate page in the Shop Manual is given whenever necessary. Although the chapters of the two manuals are synchronized, material covered in other chapters of the Shop Manual may be fundamental to the topic discussed in the Classroom Manual.

## Terms to Know Definitions

New terms are pulled out into the margin and defined.

## Trade Jargon

These marginal notes give some of the common terms used for components and allow the reader to speak and understand the language of the trade, especially when conversing with an experienced technician.

## Author's Notes

This feature includes simple explanations, stories, or examples of complex topics. These are included to help students understand difficult concepts.

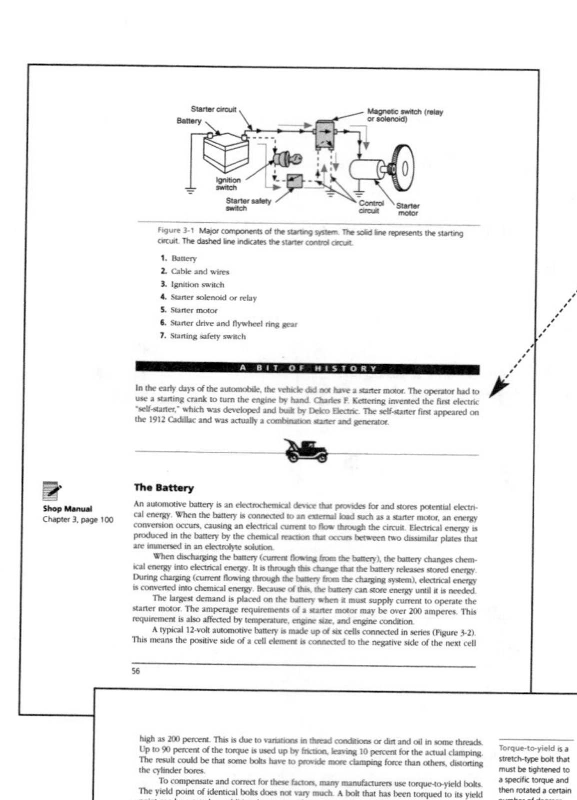


Figure 3-1 Major components of the starting system. The solid line represents the starting circuit. The dashed line indicates the starter control circuit.

1. Battery
2. Cable and wires
3. Ignition switch
4. Starter solenoid or relay
5. Starter motor
6. Starter drive and flywheel ring gear
7. Starting safety switch

A BIT OF HISTORY

In the early days of the automobile, the vehicle did not have a starter motor. The operator had to use a starting crank to turn the engine by hand. Charles F. Kettering invented the first electric "self-starter," which was developed and built by Delco Electric. The self-starter first appeared on the 1912 Cadillac and was actually a combination starter and generator.

Shop Manual
Chapter 3, page 100

**The Battery**

An automotive battery is an electrochemical device that provides for and stores potential electrical energy. When the battery is connected to an external load such as a starter motor, an energy conversion occurs, causing an electrical current to flow through the circuit. Electrical energy is produced in the battery by the chemical reaction that occurs between two dissimilar plates that are immersed in an electrolyte solution.

When discharging the battery (current flowing from the battery), the battery changes chemical energy into electrical energy. It is through this change that the battery releases stored energy. During charging (current flowing through the battery from the charging system), electrical energy is converted into chemical energy. Because of this, the battery can store energy until it is needed.

The largest demand is placed on the battery when it must supply current to operate the starter motor. The amperage requirements of a starter motor may be over 200 amperes. This requirement is also affected by temperature, engine size, and engine condition.

A typical 12-volt automotive battery is made up of six cells connected in series (Figure 3-2). This means the positive side of a cell element is connected to the negative side of the next cell

56

## A Bit of History

This feature gives the student a sense of the evolution of the automobile. This feature not only contains nice-to-know information, but also should spark some interest in the subject matter.

high as 200 percent. This is due to variations in thread conditions or dirt and oil in some threads. Up to 90 percent of the torque is used up by friction, leaving 10 percent for the actual clamping. The result could be that some bolts have to provide more clamping force than others, distorting the cylinder bores.

To compensate and correct for these factors, many manufacturers use torque-to-yield bolts. The yield point of identical bolts does not vary much. A bolt that has been torqued to its yield point can be rotated an additional amount without any increase in clamping force. When a set of **torque-to-yield** fasteners is used, the torque is actually set to a point above the yield point of the bolt. This assures the set of fasteners will have an even clamping force.

Manufacturers vary on specifications and procedures for securing torque-to-yield bolts. Always refer to the service manual for exact procedures. In most instances, a torque wrench is first used to tighten the bolts to their yield point. Next, the bolt is turned an additional amount as specified in the service manual.

The graph in Figure 4-10 indicates that a bolt can be elongated considerably at its yield point before it reaches its failure point. Also notice, the clamp load is consistent between the proof load and the failure point of the bolt. Bolts that are torqued to their yield points have been stretched beyond their elastic limit and require replacement whenever they are removed or loosened.

Torque-to-yield is a stretch-type bolt that must be tightened to a specific torque and then rotated a certain number of degrees.

Shop Manual
Chapter 5, page xx

A BIT OF HISTORY

In the 1940s Ralph H. Miller, an American, designed an engine with a compressor that forced more air-fuel mixture into the cylinder during the compression stroke. This design may be called a Miller-cycle engine. This type of engine is used in marine and industrial applications in addition to late-model Mazda Millenias.

**Summary**

- Metals are divided into two basic groups: ferrous and nonferrous. Ferrous metals are those containing iron. Nonferrous metals contain no iron.
- Alloys are mixtures of two or more metals.
- Iron containing very low carbon (between 0.05 and 1.7 percent) is called steel.

Terms to Know
Alloys
Annealing
Bolt diameter
Bolt length
Carbon steel
Case hardening
Ceramics
Coke
Composite
Ferrous metals
Grade
Grade marks
Gray cast iron
Metallurgy
Nodular iron
Nonferrous metals

## Terms to Know List

A list of new terms appears next to the Summary.

## Summaries

Each chapter concludes with a summary of key points from the chapter. These are designed to help the reader review the chapter contents.

**Review Questions**

**Short Answer Essays**

1. Explain how the intake manifold may be designed to improve airflow.
2. Explain the operation of a fuel system with dual intake runners and dual injectors for each cylinder.
3. Describe combustion chamber modifications that increase volumetric efficiency and engine power.
4. Explain camshaft modifications that increase volumetric efficiency.
5. Explain valve float and describe how it may be reduced.
6. Describe the design of a tuned exhaust manifold.
7. Explain how a turbocharger or supercharger supplies more engine power.
8. Describe basic turbocharger operation.
9. What is the purpose of the intercooler?
10. Explain basic supercharger operation.

**Fill-in-the-Blanks**

1. At low engine speeds, ____________ ____________, ____________ intake runners improve volumetric efficiency.
2. A tuned exhaust manifold has ____________ ____________ ____________ connected to each exhaust port in the cylinder head.
3. At high engine speeds ____________ ____________, ____________ intake manifold runners improve volumetric efficiency.
4. The exhaust flows past the ____________ wheel in a turbocharger.
5. The intake airflow is forced into the intake manifold by the ____________ wheel in a turbocharger.
6. The wastegate diaphragm is moved by ____________ pressure from the intake manifold.
7. Turbocharged or supercharged engines have ____________ compression ratios compared to normally aspirated engines.
8. A supercharger is ____________-driven from the engine.
9. The air flows through the intercooler ____________ it flows through the supercharger.
10. Supercharger rotor speed is limited by ____________ ____________.

242

## Review Questions

Short answer essay, fill-in-the-blank, and multiple-choice questions are found at the end of each chapter. These questions are designed to accurately assess the student's competence in the stated objectives at the beginning of the chapter.

# Shop Manual

To stress the importance of safe work habits, the Shop Manual also dedicates one full chapter to safety. Other important features of this manual include:

## Performance-Based Objectives

These objectives define the contents of the chapter and define what the student should have learned upon completion of the chapter. These objectives also correspond with the list of required tasks for ASE certification.

Although this textbook is not designed to simply prepare someone for the certification exams, it is organized around the ASE task list. These tasks are defined generically when the procedure is commonly followed and specifically when the procedure is unique for specific vehicle models. Imported and domestic model automobiles and light trucks are included in the procedures

## Basic Tools List

Each chapter begins with a list of the Basic Tools needed to perform the tasks included in the chapter.

## Terms to Know Definitions

New terms are pulled out into the margin and defined.

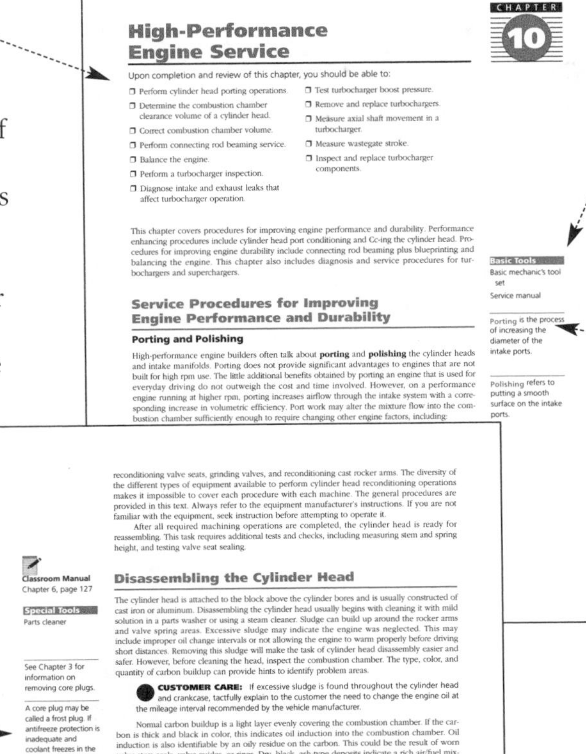

CHAPTER 10

High-Performance Engine Service

Upon completion and review of this chapter, you should be able to:

- ❒ Perform cylinder head porting operations.
- ❒ Determine the combustion chamber clearance volume of a cylinder head.
- ❒ Correct combustion chamber volume.
- ❒ Perform connecting rod beaming service.
- ❒ Balance the engine.
- ❒ Perform a turbocharger inspection.
- ❒ Diagnose intake and exhaust leaks that affect turbocharger operation.
- ❒ Test turbocharger boost pressure.
- ❒ Remove and replace turbochargers.
- ❒ Measure axial shaft movement in a turbocharger.
- ❒ Measure wastegate stroke.
- ❒ Inspect and replace turbocharger components.

This chapter covers procedures for improving engine performance and durability. Performance enhancing procedures include cylinder head port conditioning and Cc-ing the cylinder head. Procedures for improving engine durability include connecting rod beaming plus blueprinting and balancing the engine. This chapter also includes diagnosis and service procedures for turbochargers and superchargers.

Basic Tools
Basic mechanic's tool set
Service manual

Service Procedures for Improving Engine Performance and Durability

Porting and Polishing

High-performance engine builders often talk about **porting** and **polishing** the cylinder heads and intake manifolds. Porting does not provide significant advantages to engines that are not built for high rpm use. The little additional benefits obtained by porting an engine that is used for everyday driving do not outweigh the cost and time involved. However, on a performance engine running at higher rpm, porting increases airflow through the intake system with a corresponding increase in volumetric efficiency. Port work may alter the mixture flow into the combustion chamber sufficiently enough to require changing other engine factors, including:

Porting is the process of increasing the diameter of the intake ports.

Polishing refers to putting a smooth surface on the intake ports.

reconditioning valve seats, grinding valves, and reconditioning cast rocker arms. The diversity of the different types of equipment available to perform cylinder head reconditioning operations makes it impossible to cover each procedure with each machine. The general procedures are provided in this text. Always refer to the equipment manufacturer's instructions. If you are not familiar with the equipment, seek instruction before attempting to operate it.

After all required machining operations are completed, the cylinder head is ready for reassembling. This task requires additional tests and checks, including measuring stem and spring height, and testing valve seat sealing.

Classroom Manual
Chapter 6, page 127

Special Tools
Parts cleaner

Disassembling the Cylinder Head

The cylinder head is attached to the block above the cylinder bores and is usually constructed of cast iron or aluminum. Disassembling the cylinder head usually begins with cleaning it with mild solution in a parts washer or using a steam cleaner. Sludge can build up around the rocker arms and valve spring areas. Excessive sludge may indicate the engine was neglected. This may include improper oil change intervals or not allowing the engine to warm properly before driving short distances. Removing this sludge will make the task of cylinder head disassembly easier and safer. However, before cleaning the head, inspect the combustion chamber. The type, color, and quantity of carbon buildup can provide hints to identify problem areas.

See Chapter 3 for information on removing core plugs.

**CUSTOMER CARE:** If excessive sludge is found throughout the cylinder head and crankcase, tactfully explain to the customer the need to change the engine oil at the mileage interval recommended by the vehicle manufacturer.

A core plug may be called a frost plug. If antifreeze protection is inadequate and coolant freezes in the cylinder head, the core plugs are usually pushed out of the head.

Normal carbon buildup is a light layer evenly covering the combustion chamber. If the carbon is thick and black in color, this indicates oil induction into the combustion chamber. Oil induction is also identifiable by an oily residue on the carbon. This could be the result of worn valve stem seals, valve guides, or rings. Dry, black, ash-type deposits indicate a rich air/fuel mixture or ignition misfire. Some common causes include a misadjusted carburetor, excessive fuel pressure, leaking injectors, or faulty spark plug cables or spark plugs.

If the combustion chamber is clean and the metal is shiny, engine coolant may have leaked into it. This can be caused by a damaged head gasket, cracked cylinder head, or cracked engine block.

After the initial inspection of the cylinder head is complete, the head should be cleaned. Before cleaning the cylinder head, remove the core and galley plugs. Remove carbon deposits from the combustion chambers with a scraper or wire wheel, then clean the cylinder head in a parts cleaner or use a steam cleaner. This cleaning will make cylinder head inspection easier and disassembly safer. It will be necessary to clean the cylinder head again after the valves and valve train components are removed. Once the initial cleaning is complete, inspect the cylinder head

Classroom Manua
Chapter 6, page 13

Special Tools
Sheet of glass
Outside micromete
Bore gauge

## Trade Jargon

These marginal notes give some of the common terms used for components and allow the reader to speak and understand the language of the trade, especially when conversing with an experienced technician.

## Photo Sequences

Many procedures are illustrated in detailed Photo Sequences. These detailed photographs show the students what to expect when they perform particular procedures. They also can provide the student a familiarity with a system or type of equipment, which the school may not have.

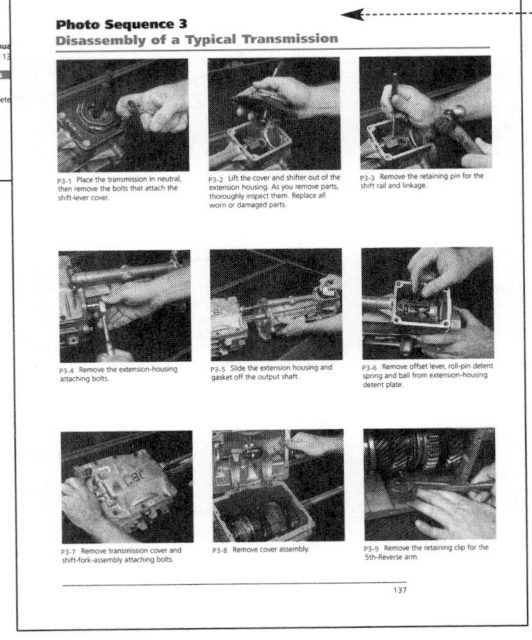

Photo Sequence 3
Disassembly of a Typical Transmission

P3-1 Place the transmission in neutral, then remove the bolts that attach the shift-lever cover.

P3-2 Lift the cover and shifter out of the extension housing. As you remove parts, thoroughly inspect them. Replace all worn or damaged parts.

P3-3 Remove the retaining pin for the shift rail and linkage.

P3-4 Remove the extension-housing attaching bolts.

P3-5 Slide the extension housing and gasket off the output shaft.

P3-6 Remove offset lever, roll-pin detent spring and ball from extension-housing detent plate.

P3-7 Remove transmission cover and shift-fork-assembly attaching bolts.

P3-8 Remove cover assembly.

P3-9 Remove the retaining clip for the 5th-Reverse arm.

137

## Special Tools List

Whenever a Special Tool is required to complete a task, it is listed in the margin next to the procedure.

## Service Tips

Whenever a short-cut or special procedure is appropriate, it is described in the text. These tips are generally those things commonly done by experienced technicians.

## Cautions and Warnings

Throughout the text, warnings are given to alert the reader to potentially hazardous materials or unsafe conditions. Cautions are given to advise the student of things that can go wrong if instructions are not followed or if a nonacceptable part or tool is used.

12. Installation of the battery. Before making connections, make sure all wires are properly connected. Connect the battery, positive terminal first, then the negative terminal.
13. Fill the radiator.

**SERVICE TIP:** Since water is easier to clean up than antifreeze, fill the radiator with pure water until it is determined there are no leaks.

14. Fill the crankcase with the proper oil type and quantity.
15. Fill the brake master cylinder or clutch cylinder as needed.
16. Fill the transmission with the proper fluid.
17. Follow the service manual procedures to fill the air-conditioning compressor with oil, then connect the hoses.

**CAUTION:** Make sure to install the proper oil into the system. Oil used in R-12 and R-134a systems is not interchangeable.

**WARNING:** Never wear jewelry when working around the vehicle. Gold, silver, and copper are excellent conductors of electricity. Your body is also a conductor of electricity. If the jewelry makes contact with a current-carrying wire or terminal, severe injury may result. Jewelry may also become caught on objects or in fans and other rotating parts.

**Rear-Wheel Drive Vehicles**

Special Tools
Engine hoist
Hoist

On most rear-wheel drive vehicles, it is possible to install the engine through the hood opening. Begin by attaching the engine lifting fixture to the engine. The following procedure is typical of most engine installations:

**CAUTION:** Make sure the engine hoist is capable of lifting the weight of the engine to a height sufficient to clear the engine compartment.

**WARNING:** Do not lift the engine by the intake manifold. This action may cause the intake manifold or retaining bolts to break, allowing the engine to drop suddenly, resulting in technician injury.

1. Lift the engine high enough to clear the vehicle and center it over the motor mount towers. If the transmission is attached, it will have to be tipped to clear the bulkhead and lowered slowly in order to center it.
2. Slowly lower the engine until the engine settles into the front mounts.
3. Use a transmission jack to support the transmission. Locate the jack in an area that will not interfere with installation of the rear motor mount.
4. Secure the front motor mounts and cross members.
5. Install the rear motor mount and cross member (Figure 9-60).

Complete under vehicle installation by raising the vehicle on the hoist. Install transmission and engine braces and connect the exhaust system to the exhaust manifold. Make sure to install any splash shields and/or heat shields.

Connect the clutch and shift lever linkages, and adjust as needed. Many clutch assemblies use hydraulic cylinders to actuate the clutch. These require the air to be bled for proper operation. On automatic transmissions, connect the cooling lines. In addition, connect all wiring harnesses that are routed to the underside of the engine. Finally, connect the drive shaft using the marks made during removal to align it properly.

Lower the vehicle and complete the uppe...

1. Connecting the fuel lines.
2. Connecting the heater hoses.

## Cross-References to the Classroom Manual

Reference to the appropriate page in the Classroom Manual is given whenever necessary. Although the chapters of the two manuals are synchronized, material covered in other chapters of the Classroom Manual may be fundamental to the topic discussed in the Shop Manual.

4. Install three supercharger mounting bolts, and tighten these bolts to the specified torque.
5. Install three bolts that retain the intercooler adapter to the intake manifold, and tighten these bolts to the specified torque.
6. Install the intercooler inlet and outlet tubes, and tighten all mounting bolts to the specified torque.
7. Install the coolant hoses on the throttle body, and tighten these hose clamps.
8. Install a new EGR valve gasket, and install the EGR valve. Tighten the EGR valve-mounting bolts to the specified torque.
9. Install the throttle linkage and bracket, and tighten the bracket bolts to the specified torque.
10. Connect all the vacuum hoses to the original locations on the air intake plenum, EGR valve, and EGR transducer.
11. Install the spark plug wires on the spark plugs in the right cylinder head, and connect the electrical connectors to the throttle position sensor, air charge temperature sensor, and idle air control valve.
12. Install the cowl covers.
13. Install and tighten the air inlet tube on the throttle body.
14. Install the supercharger drive belt.
15. Refill the cooling system to the proper level.
16. Connect the battery ground cable.
17. Start the engine, and check for air leaks in the supercharger system.

Classroom Manual
Chapter 10, page 238

**CUSTOMER CARE:** When talking to customers, always remember the two Ps: pleasant and polite. There may be many days when we do not feel like being pleasant and polite. Perhaps we have several problem vehicles in the shop with symptoms that are difficult to diagnose and correct. Some service work may be behind schedule, and customers may be irate because their vehicles are not ready on time; however, we should always remain pleasant and polite with customers. Our attitude does much to make the customer feel better and to realize their business is appreciated. A customer may not feel very happy about an expensive repair bill, but a pleasant attitude on our part may help to improve their feelings. When the two Ps are remembered by service personnel, customer relations are enhanced, and the customer feels like returning to the shop. Conversely, if service personnel have a grouchy, indifferent attitude, the customer may be turned off and take his or her business to another shop.

CASE STUDY

A customer was concerned about lack of power on a Dodge Spirit with a 2.5L turbocharged engine. The customer said the engine had much slower acceleration compared to what it had previously.

The technician performed a basic turbocharger inspection and found that all linkages, hoses, and lines were in normal condition. A check for intake and exhaust leaks did not reveal any problems, and the turbocharger did not have an excessive noise level. The technician checked the turbocharger boost pressure during a road test, and discovered it...

## Customer Care

This feature highlights those little things a technician can do or say to enhance customer relations.

## Job Sheets

Located at the end of each chapter, the Job Sheets provide a format for students to perform procedures covered in the chapter. A reference to the ASE Task addressed by the procedure is listed on the Job Sheet.

**JOB SHEET 14**

14

Name ______________________ Date __________

**Engine Removal**

Upon completion of this job sheet, you should be able to properly prepare and remove the engine from a vehicle.

**Tools and Materials**

Engine lift
Engine stand
Technician's tool set

**Describe the vehicle being worked on:**

Year ______ Make ______________ VIN ______________

Model ______________________

**Procedure** — Task Completed

Your instructor will assign you (or your group) to a vehicle that needs to have its engine removed. Using the proper service manual, you will identify and perform the required steps to remove the engine.

1. Is the vehicle FWD or RWD? ☐ FWD ☐ RWD
   If FWD, according to the service manual does the engine come out from the top or the bottom? ☐ Top ☐ Bottom
2. Following all safety cautions and warnings, clean the engine and compartment. ☐
3. On a separate sheet of paper, make a list of any special tools and equipment that are required to remove the engine. After making the list, gather the tools and organize them in your area. ☐
4. Center the vehicle over the frame contact hoist. ☐
5. Disconnect and remove the battery. ☐
6. Lift the vehicle and drain all necessary fluids. ☐
7. Is the transmission coming out with the engine? ☐ Yes ☐ No
8. Make a list of all the components that can be removed easily while the vehicle is on the hoist.

211

## Case Studies

Case Studies concentrate on the ability to properly diagnose the systems. Beginning with Chapter 3, each chapter ends with a case study in which a vehicle has a problem, and the logic used by a technician to solve the problem is explained.

the specified cylinder is at TDC on the comprssion stroke. With the feeler gauge installed between the rocker arm and valve stem, adjust the rocker arm until a slight drag is felt as the gauge is pulled out. Next, rotate the engine until another specified cylinder is at TDC on the compression stroke, and adjust the remaining valves.

**CUSTOMER CARE:** Always concentrate on quality workmanship and customer satisfaction. Most customers do not mind paying for vehicle repairs if the work is done properly and their vehicle problem is corrected. A follow-up phone call to determine customer satisfaction a few days after servicing a vehicle indicates that you consider quality work and customer satisfaction a priority.

**CASE STUDY**

A vehicle was towed to the shop because it would not start. The technician learned from the customer that the vehicle was running fine, then the customer turned off the engine and left the vehicle for a few hours. When he attempted to restart the engine, it would turn over, but it would not "fire." The technician located the timing marks on TDC number one cylinder, then removed the distributor cap. The rotor was not pointing to the number one spark plug terminal. The technician knew the problem was a slipped timing belt, and prepared a written estimate for the customer.

### Terms to Know

| | | |
|---|---|---|
| Base circle | Leak-down | Overlap |
| Camshaft | Lifter | Pushrods |
| Duration | Lobe lift | Rocker arm |
| Heel | Nose | Valve train |

### ASE-Style Review Questions

1. *Technician A* says most engines are valve timed at bottom dead center (BDC) number one piston.
   *Technician B* says valve train timing is only required on engines with balance shafts.
   Who is correct?
   **A.** A only **C.** Both A and B
   **B.** B only **D.** Neither A nor B

2. Engine valve timing is being discussed.
   *Technician A* says being off on valve timing as little as three teeth can result in piston and valve contact.
   *Technician B* says proper alignment of the timing marks sets the correct relationship between piston position and valve opening.
   Who is correct?
   **A.** A only **C.** Both A and B
   **B.** B only **D.** Neither A nor B

Who is correct?
**A.** A only **C.** Both A and B
**B.** B only **D.** Neither A nor B

4. Camshaft inspection is being discussed.
   *Technician A* says worn journals can cause changes in valve timing.
   *Technician B* says wear on the ramps of the lobes has little effect on valve operation.
   Who is correct?
   **A.** A only **C.** Both A and B
   **B.** B only **D.** Neither A nor B

5. *Technician A* says after the timing belt is removed, do not rotate the crankshaft.
   *Technician B* says to mark or organize the lifters so their original positions are maintained.
   A and B
   her A nor B

## Terms to Know List

A list of new terms appears after the case study.

## ASE-Style Review Questions

Each chapter contains ASE-style review questions that reflect the performance-based objectives listed at the beginning of the chapter. These questions can be used to review the chapter as well as to prepare for the ASE certification exam.

### ASE-Style Review Questions

1. The camshaft is being installed in the block of an OHV engine.
   *Technician A* says the camshaft should rotate freely in the bearings.
   *Technician B* says bearing high spots can be removed by scraping the bearing surface.
   Who is correct?
   **A.** A only **C.** Both A and B
   **B.** B only **D.** Neither A nor B

2. *Technician A* says excessive camshaft end play may cause timing chain noise.
   *Technician B* says camshaft end play is checked using a feeler gauge.
   Who is correct?
   **A.** A only **C.** Both A and B
   **B.** B only **D.** Neither A nor B

3. Main bearing installation is being discussed.
   *Technician A* says many manufacturers use a select fit bearing selection procedure.
   *Technician B* says most manufacturers identify the use of select bearings by codes on the engine block and the crankshaft.
   Who is correct?
   **A.** A only **C.** Both A and B
   **B.** B only **D.** Neither A nor B

4. *Technician A* says Plastigage strips should be placed lengthwise on the journal.
   *Technician B* says Plastigage measures total bearing clearance.
   Who is correct?
   **A.** A only **C.** Both A and B
   **B.** B only **D.** Neither A nor B

5. *Technician A* says the pilot bushing supports the front of the transmission input shaft and maintains centering.
   *Technician B* says the thrust bearing is aligned by moving the crankshaft back and forth.
   Who is correct?
   **A.** A only **C.** Both A and B
   **B.** B only **D.** Neither A nor B

6. *Technician A* says most piston assemblies are nondirectional.
   *Technician B* says the notch on the piston head usually faces toward the front of the engine block.
   Who is correct?
   **A.** A only **C.** Both A and B
   **B.** B only **D.** Neither A nor B

7. *Technician A* says to rotate the crankshaft after each main bearing journal is torqued.
   *Technician B* says the maximum rotating torque should not exceed 5 lb. ft.
   Who is correct?
   **A.** A only **C.** Both A and B
   **B.** B only **D.** Neither A nor B

8. If a harmonic balancer outer ring has slipped on the rubber mounting, the technician should:
   **A.** Twist the outer ring back to the original position.
   **B.** Secure the outer ring by welding it in place.
   **C.** Replace the harmonic balancer.
   **D.** Press on a new outer ring.

9. *Technician A* says still timing is the process of timing the valves.
   *Technician B* says crankshaft end play can be corrected by shims.
   Who is correct?
   **A.** A only **C.** Both A and B
   **B.** B only **D.** Neither A nor B

10. *Technician A* says when a fresh engine is started, it must be run between 1,400 and 2,000 rpm for at least 15 minutes.
    *Technician B* says the engine is cycled during the road test to seat the rings.
    Who is correct?
    **A.** A only **C.** Both A and B
    **B.** B only **D.** Neither A nor B

## ASE Practice Examination

A 50 question ASE practice exam, located in the appendix, is included to test students on the contents of the Shop Manual.

### Appendix A

#### ASE Practice Examination

1. During acceleration with the engine at normal operating temperature, the engine provides a noise similar to marbles rattling inside a metal can. The engine is throttle body injected with a distributor ignition (DI) system. The most likely cause of this noise could be:
   **A.** a rich air/fuel ratio.
   **B.** the engine thermostat stuck open.
   **C.** basic timing too far advanced.
   **D.** the EGR valve stuck open.

2. An engine has a clicking noise during acceleration.
   *Technician A* says the piston ring grooves may be worn.
   *Technician B* says the valve lifters may be sticking.
   Who is correct?
   **A.** A only **C.** Both A and B
   **B.** B only **D.** Neither A nor B

3. A port fuel-injected engine with a DI system backfires through the air intake only during acceleration, but idle operation is normal. The most likely cause of this problem is:
   **A.** A burned exhaust valve in one cylinder.
   **B.** A defective fuel injector.
   **C.** A defective spark plug.
   **D.** A cracked distributor cap.

4. An engine has a high-pitched squealing noise only at idle speed.
   *Technician A* says the alternator belt may be loose or dry.
   *Technician B* says the engine may have an intake manifold vacuum leak.
   Who is correct?
   **A.** A only **C.** Both A and B
   **B.** B only **D.** Neither A nor B

5. A vacuum gauge is connected to the intake manifold on a carbureted engine, and the vacuum gauge reading is 12 in. Hg with the engine idling. The most likely cause of this problem is:
   **A.** A sticking exhaust valve.
   **B.** Improper idle mixture screw adjustment.
   **C.** The PCV valve is stuck open.
   **D.** Weak valve springs.

6. During a cylinder balance test on a port fuel-injected engine, number five cylinder has a 20 rpm drop, and the other cylinders have a 60 rpm drop. The test is repeated five times at the same rpm, and during three of these tests, number five cylinder provides the same rpm drop as the other cylinders. The most likely cause of this problem is:
   **A.** A sticking exhaust valve.
   **B.** A burned exhaust valve.
   **C.** A defective spark plug.
   **D.** An intake manifold vacuum leak.

7. An engine has an accumulation of oil in the air cleaner. A large volume of crankcase vapors escapes from the PCV valve opening in the rocker arm cover with the engine idling. During a compression test, this engine has zero compression on number three cylinder.
   *Technician A* says the rings on number three cylinder are severely worn or broken.
   *Technician B* says the cylinder head gasket is blown out around number three cylinder.
   Who is correct?
   **A.** A only **C.** Both A and B
   **B.** B only **D.** Neither A nor B

8. All of these statements about a cylinder leakage test are true EXCEPT:
   **A.** The piston must be at TDC on the exhaust stroke in the cylinder being tested.
   **B.** If air escapes from the PCV valve opening during the test, the piston rings are worn.
   **C.** If air escapes from the throttle body during the test, an intake valve may be bent.
   **D.** Bubbles appearing in the radiator coolant during the test may indicate a leaking head gasket.

9. When discussing cylinder head installation:
   *Technician A* says when torquing head bolts the bolts at the outer ends of the head should be tightened first.
   *Technician B* says torque-to-yield head bolts may be reinstalled if they are tightened properly.
   Who is correct?
   **A.** A only **C.** Both A and B
   **B.** B only **D.** Neither A nor B

497

## Reviewers

We would like to extend a special thanks to those who saw things we overlooked and for their contributions to this text:

Christopher Bannister
New England Institute of Technology

Steve Michener
Mt. Hood Community College

Fred Raadsheer
British Columbia Institute of Technology

Russell E. Taylor
Northern Virginia Community College

John Thorp
Illinois Central College

## Contributing Companies

We would also like to thank these companies who provided technical art for this edition:

Actron Manufacturing Co.
American Council for an Energy Efficient Economy
American Honda Motor Co., Inc.
American Isuzu Motors, Inc.
Automotive Diagnostics, a division of SPX Corporation
Ballard Power Systems
Breton Publishers
Central Tools, Inc.
Champion Spark Plug Company
CRC Industries, Inc.
DaimlerChrysler Corporation
Detroit Diesel Allison
DuPont Automotive Finishes
Elgin Racing Cams
Federal-Mogul Corporation
Fel-Pro, Inc.
Goodson Shop Supplies
Hyundai Motor America
Jasper Engine and Transmission Exchange, Inc.
John Deere
JTG Associates
Kent-Moore Division, SPX Corp.
Kleer-Flo Company, Eden Prairie, MN
K-Line Industries
L.S. Starrett Co.
Lincoln Automotive
Mac Tools
Mercedes-Benz of N.A., Inc.
National Institute for Automotive Service Excellence (ASE)
Neway Manufacturing, Inc.
Nissan North America, Inc.
OTC Tool and Equipment, Division of SPX Corporation
Perfect Circle/Dana
Peterson Machine Tool, Inc.
Pro-Bal Industrial Balancers
SAE
Sioux Tools, Inc.
Snap-on Tools Company
Stant Manufacturing
Storm Vulcan Mattoni
Sun Electric Corporation
Sunnen Products Company
TRW, Incorporated
Valvoline Company
Winona-Van Norman

# Safety Practices

Upon completion and review of this chapter, you should be able to:

- ❑ Recognize shop hazards and take the necessary steps to avoid personal injury or property damage.
- ❑ Explain the purposes of the Occupational Safety and Health Act.
- ❑ Identify the necessary steps for personal safety in the automotive shop.
- ❑ Describe the reasons for prohibiting drug and alcohol use in the shop.
- ❑ Explain the steps required to provide electrical safety in the shop.
- ❑ Define the steps required to provide safe handling and storage of gasoline.
- ❑ Describe the necessary housekeeping safety steps.
- ❑ Explain the essential general shop safety practices.
- ❑ Define the steps required to provide fire safety in the shop.
- ❑ Describe typical fire extinguisher operating procedure.
- ❑ Explain four different types of fires, and the type of fire extinguisher required for each type of fire.
- ❑ Describe three other pieces of shop safety equipment other than fire extinguishers.
- ❑ Follow proper safety precautions while handling hazardous waste materials.
- ❑ Dispose of hazardous waste materials in accordance with state and federal regulations.

## Introduction

Safety is extremely important in the automotive shop! The knowledge and practice of safety precautions prevents serious personal injury and expensive property damage. Automotive students and technicians must be familiar with shop hazards and all types of safety, including personal, gasoline handling, housekeeping, general shop, fire, and hazardous material. The first step in providing a safe shop is learning about all types of safety precautions. However, the second, and most important, step in this process is applying our knowledge of safety precautions while working in the shop. In other words, we must actually develop safe working habits in the shop from our understanding of various safety precautions. When shop employees have a careless attitude toward safety, accidents are more likely to occur. All shop personnel must develop a serious attitude toward safety in the shop. The result of this serious attitude is that shop personnel will learn and adopt all shop safety rules.

Shop personnel must be familiar with their rights regarding hazardous waste disposal. These rights are explained in the right-to-know laws. Secondly, shop personnel must be familiar with hazardous materials in the automotive shop and the proper disposal methods for these materials according to state and federal regulations.

## Occupational Safety and Health Act

The **Occupational Safety and Health Act (OSHA)** was passed by the United States government in 1970. The purposes of this legislation are these:

1. To assist and encourage the citizens of the United States in their efforts to assure safe and healthful working conditions by providing research, information, education, and training in the field of occupational safety and health.
2. To assure safe and healthful working conditions for working men and women by authorizing enforcement of the standards developed under the Act.

The **Occupational Safety and Health Act (OSHA)** regulates working conditions in the United States.

Since approximately 25 percent of workers are exposed to health and safety hazards on the job, the OSHA is necessary to monitor, control, and educate workers regarding health and safety in the workplace. Employers and employees should be familiar with workplace hazardous materials information systems (WHMIS).

## Shop Hazards

Service technicians and students encounter many hazards in an automotive shop. When these hazards are known, basic shop safety rules and procedures must be followed to avoid personal injury. Some of the hazards in an automotive shop are these:

Shop hazards must be recognized and avoided to prevent personal injury.

1. Flammable liquids such as gasoline and paint must be handled and stored properly.
2. Flammable materials such as oily rags must be stored properly to avoid a fire hazard.
3. Batteries contain a corrosive sulfuric acid solution and produce explosive hydrogen gas while charging.
4. Loose sewer and drain covers may cause foot or toe injuries.
5. Caustic liquids such as those in hot cleaning tanks are harmful to skin and eyes.
6. High-pressure air in the shop compressed air system can be very dangerous if it penetrates the skin and enters the bloodstream.
7. Frayed cords on electric equipment and lights may result in severe electrical shock.
8. Hazardous waste material such as batteries and the caustic cleaning solution from a hot or cold cleaning tank must be handled properly to avoid harmful effects.
9. Carbon monoxide from vehicle exhaust is poisonous.
10. Loose clothing, jewelry, or long hair may become entangled in rotating parts on equipment or vehicles, resulting in serious injury.
11. Dust and vapors generated during some repair jobs are harmful. Asbestos dust generated during brake lining service and clutch service is a contributor to lung cancer.
12. High noise levels from shop equipment such as an air chisel may be harmful to the ears.
13. Oil, grease, water, or parts cleaning solutions on shop floors may cause someone to slip and fall, resulting in serious injury.

## Safety in the Automotive Shop

Each person in an automotive shop must follow certain basic shop safety rules to remove the danger from shop hazards. When all personnel in the automotive shop follow these basic shop safety rules, personal injury, vehicle damage, and property damage may be prevented.

Personal injury, vehicle damage, and property damage must be avoided by following safety rules regarding personal protection, substance abuse, electrical safety, gasoline safety, housekeeping safety, fire safety, and general shop safety.

### Personal Protection

1. Always wear safety glasses or a face shield in the shop (Figure 1-1).
2. Wear ear plugs or covers if high noise levels are encountered.
3. Always wear boots or shoes that provide adequate foot protection. Safety work boots or shoes with steel toe caps are best for working in the automotive shop. Most safety shoes also have slip-resistant soles. Footwear must protect against heavy falling objects, flying sparks, and corrosive liquids. Soles on footwear must protect against punctures by sharp objects. Sneakers and street shoes are not recommended in the shop.

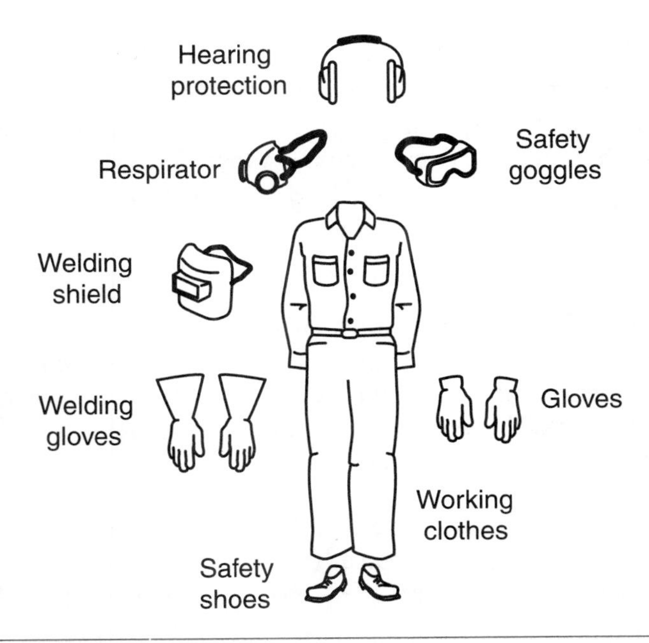


**Figure 1-1** Shop safety equipment, including safety goggles, respirator, welding shield, proper work clothes, ear protection, welding gloves, work gloves, and safety shoes.

4. Do not wear watches, jewelry, or rings when working on a vehicle. Severe burns occur when jewelry makes contact between an electric terminal and ground. Jewelry may catch on an object, resulting in painful injury.
5. Do not wear loose clothing, and keep long hair tied behind your head. Loose clothing or long hair is easily entangled in rotating parts.
6. Wear a respirator to protect your lungs when working in dusty conditions.

## Smoking, Alcohol, and Drugs in the Shop

Do not smoke when working in the shop. If the shop has designated smoking areas, smoke only in these areas. Do not smoke in customers' cars. Nonsmokers may not appreciate cigarette odor in their cars. A spark from a cigarette or lighter may ignite flammable materials in the workplace. The use of drugs or alcohol must be avoided while working in the shop. Even a small amount of drugs or alcohol affects reaction time. In an emergency situation, slow reaction time may cause personal injury. If a heavy object falls off the workbench, and your reaction time is slowed by drugs or alcohol, you may not get your foot out of the way in time, resulting in foot injury. When a fire starts in the workplace, and you are a few seconds slower getting a fire extinguisher into operation because of alcohol or drug use, it could make the difference between extinguishing a fire and having expensive fire damage.

The improper or excessive use of alcoholic beverages and/or drugs may be referred to as substance abuse.

## Electrical Safety

1. Frayed cords on electrical equipment must be replaced or repaired immediately.
2. All electric cords from lights and electric equipment must have a ground connection. The ground connector is the round terminal in a three-prong electrical plug. Do not use a two-prong adaptor to plug in a three-prong electrical cord. Three-prong electrical outlets should be mandatory in all shops.
3. Do not leave electrical equipment running and unattended.

## Gasoline Safety

Gasoline is a very explosive liquid! One exploding gallon of gasoline has a force equal to fourteen sticks of dynamite. It is the expanding vapors from gasoline that are extremely dangerous. These vapors are present even in cold temperatures. Vapors formed in gasoline tanks on cars are controlled, but vapors from a gasoline storage container may escape from the can, resulting in a hazardous situation. Therefore, gasoline storage containers must be placed in a well-ventilated space.

**AUTHOR'S NOTE:** I have experienced two gasoline fires that involved property damage and/or personal injury. In both cases, the fires were caused by individuals who did not understand the tremendous explosive power in gasoline, and these individuals did not follow basic gasoline and fire safety precautions. Safety precautions must be followed until they become automatic habits!

Approved gasoline storage cans have a flash-arresting screen at the outlet (Figure 1-2). These screens prevent external ignition sources from igniting the gasoline within the can while the gasoline is being poured. Follow these safety precautions regarding gasoline containers:

1. Always use approved gasoline containers that are painted red for proper identification.
2. Do not fill gasoline containers completely full. Always leave the level of gasoline at least one inch from the top of the container. This action allows expansion of the gasoline at higher temperatures. If gasoline containers are completely full, the gasoline will expand when the temperature increases. This expansion forces gasoline from the can and creates a dangerous spill.
3. If gasoline containers must be stored, place them in a well-ventilated area such as a storage shed. Do not store gasoline containers in your home or in the trunk of a vehicle.
4. When a gasoline container must be transported, be sure it is secured against upsets.
5. Do not store a partially filled gasoline container for long periods of time, because it may give off vapors and produce a potential danger.
6. Never leave gasoline containers open except while filling or pouring gasoline from the container.
7. Do not prime an engine with gasoline while cranking the engine.
8. Never use gasoline as a cleaning agent.

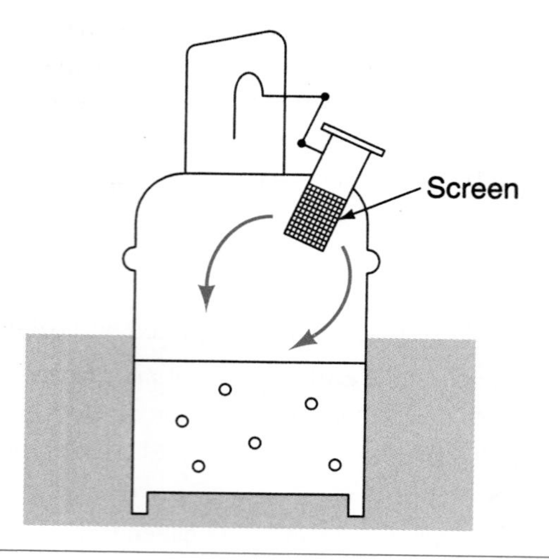


**Figure 1-2** Approved gasoline container.

## Housekeeping Safety

1. Keep shop floors clean! Always clean shop floors immediately after a spill.
2. Store paint and other flammable liquids in a closed steel cabinet (Figure 1-3).
3. Oily rags must be stored in approved, covered airtight containers (Figure 1-4). A slow generation of heat occurs from oxidation of oil on these rags. Heat may continue to be generated until the ignition temperature is reached. The oil and the rags then begin to burn, causing a fire. This action is called spontaneous combustion. However, if the oily rags are in an approved, airtight container, the fire cannot get enough oxygen to cause burning.
4. Keep the shop neat and clean. Always pick up tools and parts, and do not leave creepers lying on the floor.
5. Keep the workbenches clean. Do not leave heavy objects such as used parts on the bench after you are finished with them.

## General Shop Safety

1. All sewer covers must fit properly and be kept securely in place.
2. Always wear a face shield, protective gloves, and protective clothing when necessary. Gloves should be worn when working with solvents and caustic solutions, handling hot metal, or grinding metal. Various types of protective gloves are available. Shop coats and coveralls are the most common types of protective clothing.
3. Never direct high-pressure air from an air gun against human flesh. If this action is allowed, air may penetrate the skin and enter the bloodstream, causing serious health problems or death. Always keep air hoses in good condition. If an end blows off an air hose, the hose may whip around and result in personal injury. Use only Occupational Safety and Health Act (OSHA) approved air gun nozzles.
4. Handle all hazardous waste materials according to state and federal regulations. (These regulations are explained later in this chapter.)

**Figure 1-3** Paints and combustible material containers must be kept in an approved safety cabinet.

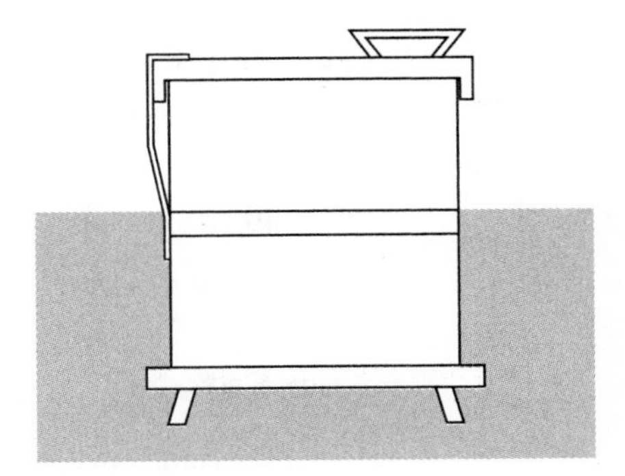

**Figure 1-4** Oily rags must be stored in approved airtight containers.

5. Always place a shop exhaust hose on the tailpipe of a vehicle if the engine is running in the shop, and be sure the shop exhaust fan is turned on.
6. Keep hands, long hair, jewelry, and tools away from rotating parts such as fan blades and belts on running engines. Remember that an electric-drive fan may start turning at any time.
7. When servicing brakes or clutches from manual transmissions, always clean asbestos dust from these components with an approved asbestos dust parts washer.
8. Always use the correct tool for the job. For example, never strike a hardened steel component, such as a piston pin, with a steel hammer. This type of component may shatter, and fragments may penetrate eyes or skin.
9. Follow the car manufacturer's recommended service procedures.
10. Be sure that the shop has adequate ventilation.
11. Make sure the work area has adequate lighting.
12. Use trouble lights with steel or plastic cages around the bulb. If an unprotected bulb breaks, it may ignite flammable materials in the area.
13. When servicing a vehicle, always apply the parking brake and place the transmission in park with an automatic transmission, or neutral with a manual transmission, if the engine is running. When the engine is stopped, place the transmission in park with an automatic transmission, or reverse with a manual transmission.
14. Avoid working on a vehicle parked on an incline.
15. *Never work under a vehicle unless the vehicle chassis is supported securely on jack stands.*

Jack stands may be called safety stands.

16. When one end of a vehicle is raised, place wheel chocks on both sides of the wheels remaining on the floor.
17. Be sure that you know the location of shop first-aid kits, eyewash fountains, and fire extinguishers.
18. Collect oil, fuel, brake fluid, and other liquids in the proper safety containers.
19. Use only approved cleaning fluids and equipment. Do not use gasoline to clean parts.
20. Obey all state and federal safety, fire, and hazardous material regulations.
21. Always operate equipment according to the equipment manufacturer's recommended procedure.
22. Do not operate equipment unless you are familiar with the correct operating procedure.
23. Do not leave running equipment unattended.
24. Be sure the safety shields are in place on rotating equipment.
25. All shop equipment must have regularly scheduled maintenance and adjustment.
26. Some shops have safety lines around equipment. Always work within these lines when operating equipment.
27. Be sure that shop heating equipment is well ventilated.
28. Do not run in the shop or engage in horseplay.
29. Post emergency phone numbers near the phone. These numbers should include a doctor, ambulance, fire department, hospital, and police.
30. Do not place hydraulic jack handles where someone can trip over them.
31. Keep aisles clear of debris.

## Fire Safety

1. Familiarize yourself with the location and operation of all shop fire extinguishers.
2. If a fire extinguisher is used, report it to management so the extinguisher can be recharged.
3. Do not use any type of open flame heater to heat the work area.
4. Do not turn on the ignition switch or crank the engine with a gasoline line disconnected.

5. Store all combustible materials such as gasoline, paint, and oily rags in approved safety containers.
6. Clean up gasoline, oil, or grease spills immediately.
7. Always wear clean shop clothes. Do not wear oil-soaked clothes.
8. Do not allow sparks and flames near batteries.
9. Welding tanks must be securely fastened in an upright position.
10. Do not block doors, stairways, or exits.
11. Do not smoke when working on vehicles.
12. Do not smoke or create sparks near flammable materials or liquids.
13. Store combustible shop supplies such as paint in a closed steel cabinet.
14. Gasoline must be kept in approved safety containers.
15. If a gasoline tank is removed from a vehicle, do not drag the tank on the shop floor.
16. Know the approved fire escape route from your classroom or shop to the outside of the building.
17. If a fire occurs, do not open doors or windows. This action creates extra draft, which makes the fire worse.
18. Do not put water on a gasoline fire, because the water will make the fire worse.
19. Call the fire department as soon as a fire begins, and then attempt to extinguish the fire.
20. If possible, stand six to ten feet from the fire and aim the fire extinguisher nozzle at the base of the fire with a sweeping action.
21. If a fire produces a lot of smoke in the room, remain close to the floor to obtain oxygen and avoid breathing smoke.
22. If the fire is too hot or the smoke makes breathing difficult, get out of the building.
23. Do not re-enter a burning building.
24. Keep solvent containers covered except when pouring from one container to another. When flammable liquids are transferred from bulk storage, the bulk container should be grounded to a permanent shop fixture such as a metal pipe. During this transfer process, the bulk container should be grounded to the portable container (Figure 1-5). These ground wires prevent the buildup of a static electric charge, which could result in a spark and disastrous explosion. Always discard or clean empty solvent containers, because fumes in these containers are a fire hazard.
25. Familiarize yourself with different types of fires and fire extinguishers, and know the type of extinguisher to use on each fire.

**Shop Manual**
Chapter 1, page 2

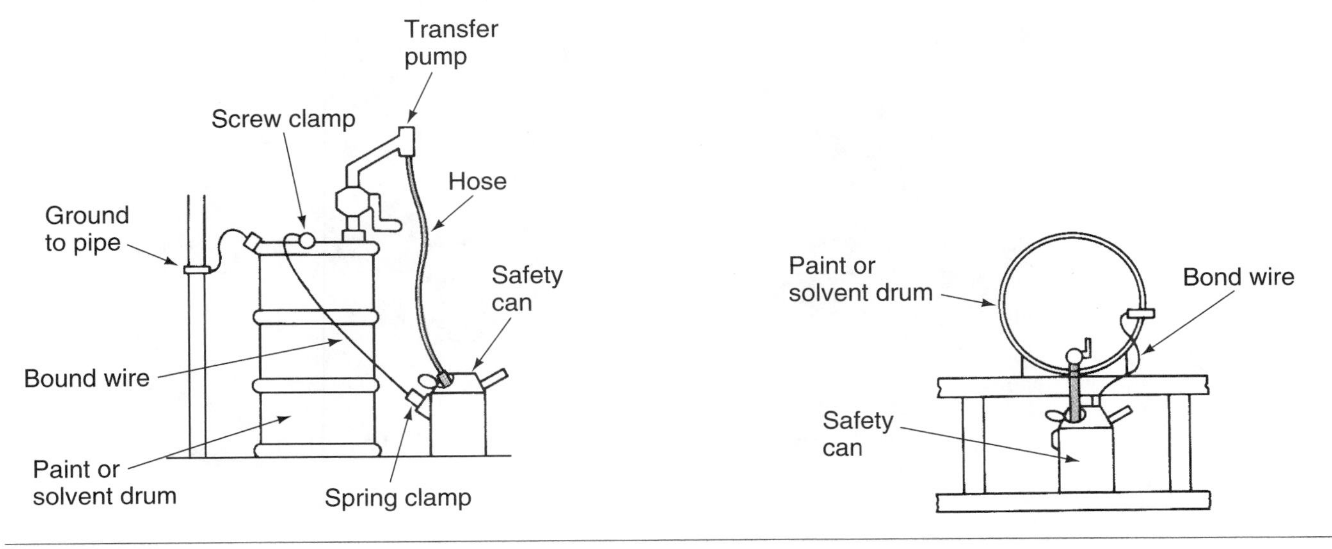


**Figure 1-5** Safe procedures for flammable liquid transfer. (Courtesy of DuPont Automotive Finishes)

# Shop Safety Equipment

Shop safety equipment must be easily accessible and in good working condition.

## Fire Extinguishers

Fire extinguishers are one of the most important pieces of safety equipment. All shop personnel must know the location of the fire extinguishers in the shop. If you have to waste time looking for an extinguisher after a fire starts, the fire could get out of control before you get the extinguisher into operation. Fire extinguishers should be located where they are easily accessible at all times. Everyone working in the shop must know how to operate the fire extinguishers. There are several different types of fire extinguishers, but their operation usually involves these steps:

1. Get as close as possible to the fire without jeopardizing your safety.
2. Grasp the extinguisher firmly and aim the extinguisher at the fire.
3. Pull a pin from the extinguisher handle.
4. Squeeze the handle to dispense the contents of the extinguisher.
5. Direct the fire extinguisher nozzle at the base of the fire, and dispense the contents of the extinguisher with a sweeping action back and forth across the fire. Most extinguishers discharge their contents in 8 to 25 seconds.
6. Always be sure the fire is extinguished.
7. Always keep an escape route open behind you so a quick exit is possible if the fire gets out of control.

**Multipurpose dry chemical fire extinguishers** may be used to extinguish several different types of fires.

**Halogen and halon fire extinguishers** contain bromine, chlorine, fluorine, or a mixture of these liquids. The use of these extinguishers produced toxic gases, and as a result, they are no longer allowed.

A decal on each fire extinguisher identifies the type of chemical in the extinguisher and provides operating information (Figure 1-6). Shop personnel should be familiar with the following types of fires and fire extinguishers:

1. Class A fires are those involving ordinary combustible materials such as paper, wood, clothing, and textiles. **Multipurpose dry chemical fire extinguishers** are used on these fires.
2. Class B fires involve the burning of flammable liquids such as gasoline, oil, paint, solvents, and greases. These fires may be extinguished with multipurpose dry chemical extinguishers. **Fire extinguishers** containing **halogen,** or **halon,** may be used to extinguish class B fires. The

**Figure 1-6** Types and sizes of fire extinguishers.

chemicals in this type of extinguisher attach to the hydrogen, hydroxide, and oxygen molecules to stop the combustion process almost instantly. However, the resultant gases from the use of halogen-type extinguishers are very toxic and harmful to the operator of the extinguisher.

3. Class C fires involve the burning of electrical equipment such as wires, motors, and switches. These fires may be extinguished with multipurpose dry chemical extinguishers.
4. Class D fires involve the combustion of metal chips, turnings, and shavings. Dry chemical extinguishers are the only type of extinguisher recommended for these fires.

Additional information regarding types of extinguishers for various types of fires is provided in Figure 1-7.

## Eyewash Fountains

Eye injuries may occur in various ways in an automotive shop. Some of the common eye accidents are these:

1. Thermal burns from excessive heat
2. Irradiation burns from excessive light, such as from an arc welder
3. Chemical burns from strong liquids, such as battery electrolyte
4. Foreign material in the eye
5. Penetration of the eye by a sharp object
6. A blow from a blunt object

Wearing safety glasses and observing shop safety rules will prevent most eye accidents. If a chemical gets in your eyes, it must be washed out immediately to prevent a chemical burn. An eyewash fountain is the most effective way to wash the eyes, and every shop should be equipped with some eyewash facility (Figure 1-8). Be sure you know the location of the eyewash fountain in the shop.

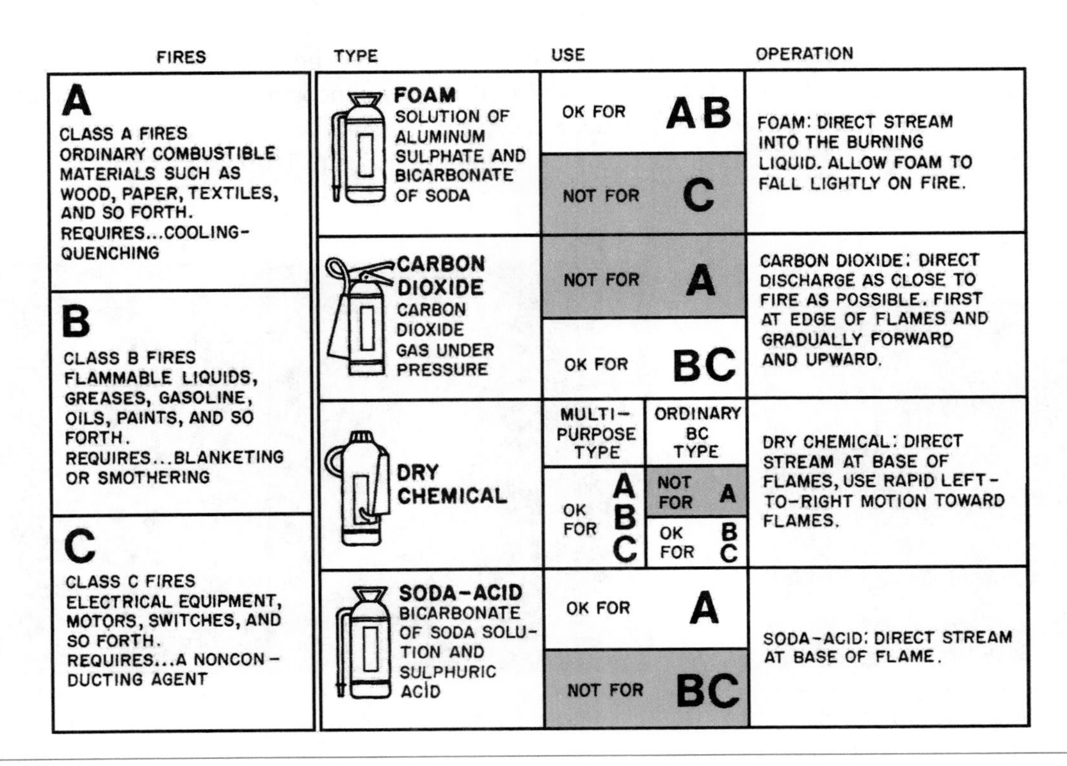


**Figure 1-7** Fire extinguisher selection.

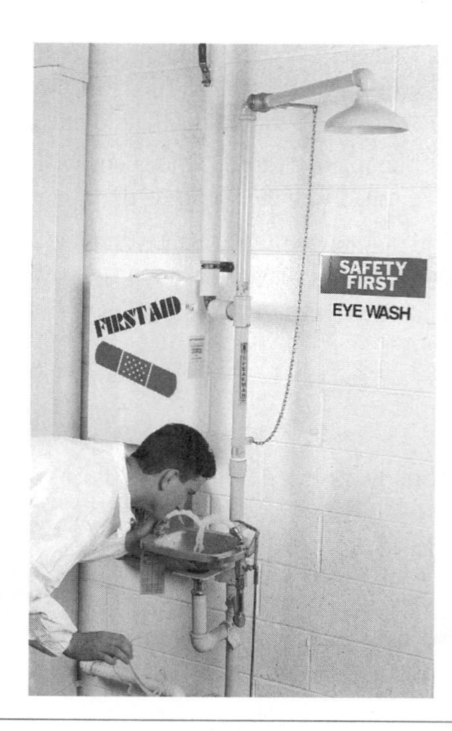


**Figure 1-8** Eyewash fountain. (Courtesy of Dupont Automotive Finishes)

## Safety Glasses and Face Shields

The mandatory use of eye protection with safety glasses or a face shield is one of the most important safety rules in a shop. Many shop insurance policies require the use of eye protection in the shop. Some automotive technicians have been blinded in one or both eyes because they did not bother to wear safety glasses. All safety glasses must be equipped with safety glass, and they should provide some type of side protection (Figure 1-9). When selecting a pair of safety glasses, they should feel comfortable on your face. If they are uncomfortable, you may tend to take them off, leaving the eyes unprotected. A face shield should be worn when handling hazardous chemicals or when using an electric grinder or buffer (Figure 1-10).

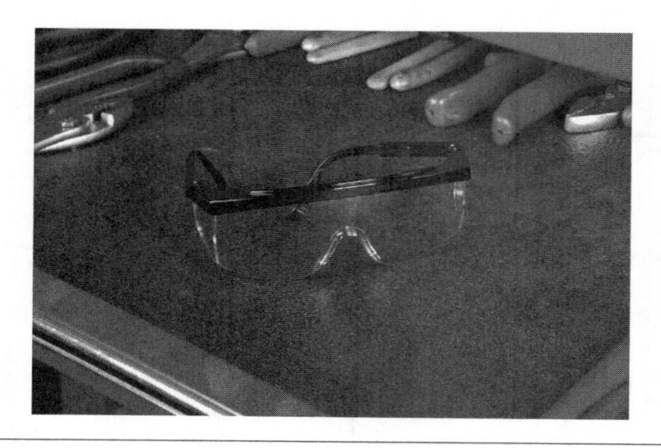

**Figure 1-9** Safety glasses with side protection must be worn in the automotive shop. (Courtesy of Goodson Shop Supplies)

**Figure 1-10** Face shield. (Courtesy of Goodson Shop Supplies)

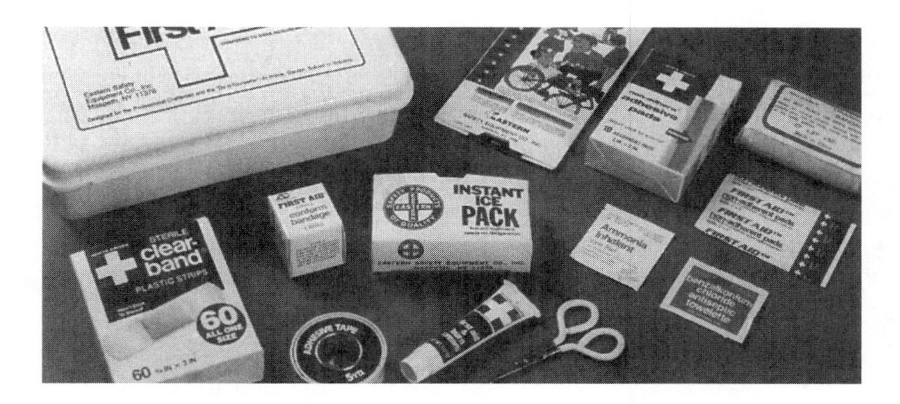


**Figure 1-11** First-aid kit.

### First-Aid Kits

First-aid kits should be clearly identified and conveniently located (Figure 1-11). These kits contain such items as bandages and ointment required for minor cuts. All shop personnel must be familiar with the location of first-aid kits. At least one of the shop personnel should have basic first-aid training, and this person should be in charge of administering first aid and keeping first-aid kits filled.

## Engine Rebuilding Specific Safety Concerns

**Shop Manual**
Chapter 1, page 10

The engine rebuilder is required to lift heavy objects in awkward positions. Always use the lifting procedures described in Chapter 1 of the Shop Manual when lifting intake manifolds, cylinder heads, exhaust manifolds, and so forth. If possible, use a hoist to lift these items. If a hoist is not available, recruit help.

When the engine requires removal from the vehicle, it is important to keep the work area clean. The process of removing the engine can be messy because coolant, oil, and gasoline may be spilled on the floor. Any spillage should be cleaned up as soon as it occurs.

The engine is removed from the vehicle by using an engine hoist. Engine hoists are available in several different designs. The technician must receive instruction on proper use prior to attempting to lift the engine. Before attaching the engine hoist, make sure it is in good operating condition. If the hoist is found to be defective, tag it, and report it to your supervisor. Also, check the hoist's capacity to assure it is capable of lifting the weight of the engine.

As the engine is lifted out of or into the vehicle, do not get under the engine. Do not place any part of your body under the engine. Guide the engine by standing off to the side and holding onto the top of the engine. If the engine comes loose from the hoist, do not attempt to catch it. As silly as this sounds, the natural instinct is to grab onto the engine to prevent damage. The engine is easier to replace than body parts. Let the engine fall.

Prior to starting an engine, make sure all fuel lines, electrical connections, belts, and hoses are properly connected. As the engine is being started, stand to the side in the event a hose, belt, or other component comes loose. Also, do not pour gasoline into the carburetor or throttle body. If necessary, use a squirt can to place a small amount of gasoline in the carburetor or throttle body. Do not put fuel into the carburetor or throttle body while the engine is being cranked. If the engine backfires, the fuel can ignite, which may result in severe burns. In addition, any sparks can ignite the fuel. Have a fire extinguisher next to the vehicle when the engine is being started. If a fire breaks out, it can be extinguished quickly without having to search for the extinguisher.

# Hazardous Waste Disposal

The EPA (Environmental Protection Agency) is a U.S. government department dedicated to reducing pollution and improving environmental quality.

Hazardous waste materials in automotive shops are chemicals or components that the shop no longer needs and that pose a danger to the environment and people if they are disposed of in ordinary garbage cans or sewers. However, it should be noted that no material is considered hazardous waste until the shop has finished using it and is ready to dispose of it. The **Environmental Protection Agency (EPA)** publishes a list of hazardous materials that is included in the Code of Federal Regulations. Waste is considered hazardous if it is included on the EPA list of hazardous materials, or if it has one or more of these characteristics:

1. ***Reactive.*** Any material that reacts violently with water or other chemicals is considered hazardous. When exposed to low-pH acid solutions, if a material releases cyanide gas, hydrogen sulphide gas, or similar gases, it is hazardous.
2. ***Corrosive.*** If a material burns the skin or dissolves metals and other materials, it is considered hazardous.
3. ***Toxic.*** Materials are hazardous if they leach one or more of eight heavy metals in concentrations greater than 100 times the primary drinking water standard.
4. ***Ignitable.*** A liquid is hazardous if it has a flash point below 140°F (60°C), and a solid is hazardous if it ignites spontaneously.

The automotive service industry is considered a generator of hazardous wastes. However, the vehicles it services are the real generators. New oil is not a hazardous waste, used oil is. Once you drain oil from an engine, you have generated the waste and now become responsible for the proper disposal of this hazardous waste. There are many other wastes that need to be handled properly after you have removed them, such as batteries, brake fluid, and transmission fluid.

Engine coolant should not be allowed to go down sewage drains. This is also true for all liquids drained from a car. Coolant should be captured and recycled or disposed of properly.

Filters for fluids (transmission, fuel, and oil filters) also need to be handled in designated ways. Used filters need to be drained and then crushed or disposed of in a special shipping barrel. Most regulations demand that oil filters be drained for at least 24 hours before they are disposed of or crushed.

The Resource Conservation and Recovery Act (RCRA) states that hazardous material users are responsible for hazardous materials from the time they become a waste until the proper waste disposal is completed.

Federal and state laws control the disposal of hazardous waste materials. Every shop employee must be familiar with these laws. Hazardous waste disposal laws include the **Resource Conservation and Recovery Act (RCRA).** This law basically states that hazardous material users are responsible for hazardous materials from the time they become a waste until the proper waste disposal is completed. Many automotive shops hire an independent hazardous waste hauler to dispose of hazardous waste material (Figure 1-12). The shop owner or manager should have a written contract with the hazardous waste hauler.

Rather than have hazardous waste material hauled to an approved hazardous waste disposal site, a shop may choose to recycle the material in the shop. Therefore, the user must store hazardous waste material properly and safely, and be responsible for the transportation of this material until it arrives at an approved hazardous waste disposal site and is processed according to the law.

The RCRA controls these types of automotive waste:

1. Paint and body repair products waste
2. Solvents for parts and equipment cleaning
3. Batteries and battery acid
4. Mild acids used for metal cleaning and preparation
5. Waste oil, engine coolants, or antifreeze
6. Air conditioning refrigerants
7. Engine oil filters

**Figure 1-12** Hazardous waste hauler. (Courtesy of DuPont Automotive Finishes)

Never, under any circumstances, use these methods to dispose of hazardous waste material:

1. Pour hazardous wastes on weeds to kill them.
2. Pour hazardous wastes on gravel streets to prevent dust.
3. Throw hazardous wastes in a dumpster.
4. Dispose of hazardous wastes anywhere but an approved disposal site.
5. Pour hazardous wastes down sewers, toilets, sinks, or floor drains.
6. Bury hazardous wastes in the ground.

**Right-to-know laws** inform workers regarding exposure to hazardous materials.

The **right-to-know laws** state that employees have a right to know when the materials they use at work are hazardous. The right-to-know laws started with the **Hazard Communication Standard** published by the Occupational Safety and Health Administration (OSHA) in 1983. This document was originally intended for chemical companies and manufacturers that required employees to handle hazardous materials in their work situation. At the present time, most states have established their own right-to-know laws. Meanwhile, the federal courts have decided to apply these laws to all companies, including automotive service shops. Under the right-to-know laws, the employer has three responsibilities regarding the handling of hazardous materials by its employees.

The **Hazard Communication Standard** is a code of federal regulations developed by OSHA regarding hazardous materials.

First, all employees must be trained about the types of hazardous materials they will encounter in the workplace. The employees must be informed about their rights under legislation regarding the handling of hazardous materials. All hazardous materials must be properly labeled, and information about each hazardous material must be posted on **material safety data sheets (MSDS)** available from the manufacturer (Figure 1-13). In Canada, MSDS sheets are called **workplace hazardous materials information systems (WHMIS).**

**Material safety data sheets (MSDS)** provide extensive information about hazardous materials.

The employer has a responsibility to place MSDS sheets where they are easily accessible by all employees. The MSDS sheets provide extensive information about the hazardous material such as:

1. Chemical name
2. Physical characteristics
3. Protective equipment required for handling
4. Explosion and fire hazards
5. Other incompatible materials
6. Health hazards such as signs and symptoms of exposure, medical conditions aggravated by exposure, and emergency and first-aid procedures

**Workplace hazardous materials information systems (WHMIS)** list information about hazardous materials.

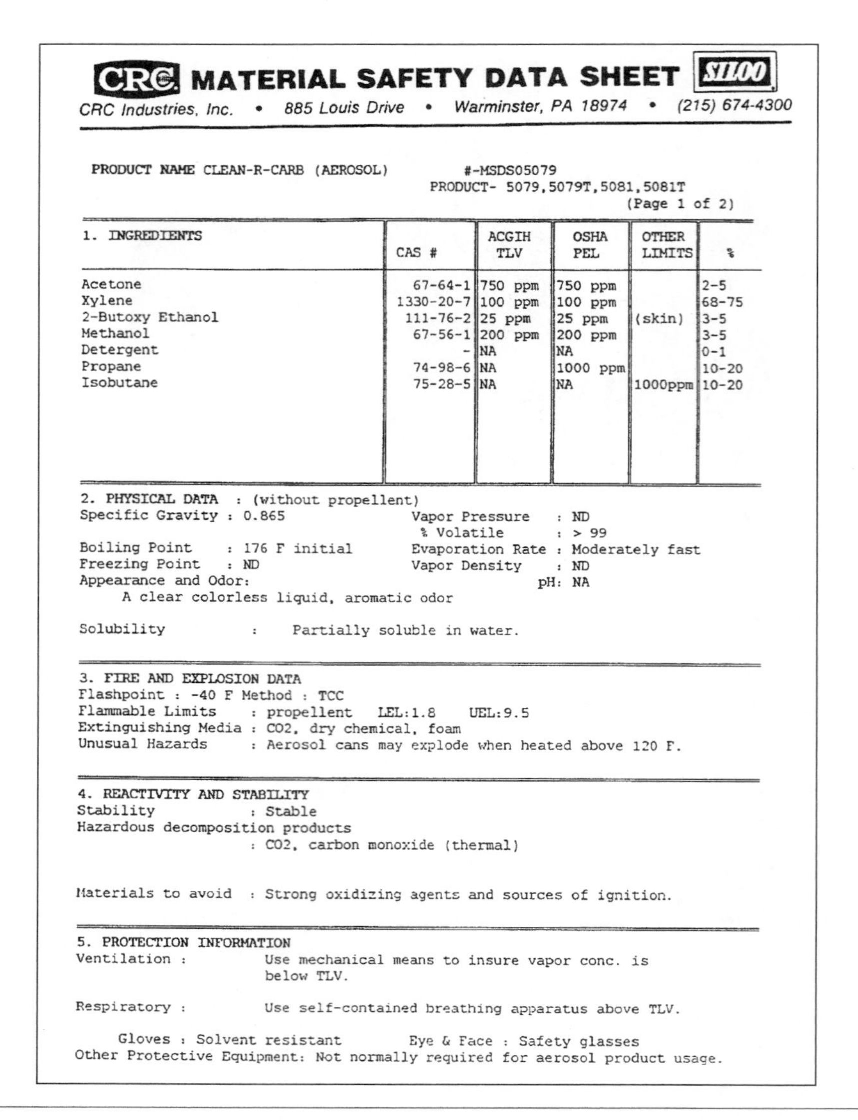

CRC MATERIAL SAFETY DATA SHEET SILOO

CRC Industries, Inc. • 885 Louis Drive • Warminster, PA 18974 • (215) 674-4300

PRODUCT NAME CLEAN-R-CARB (AEROSOL) #-MSDS05079
PRODUCT- 5079,5079T,5081,5081T
(Page 1 of 2)

| 1. INGREDIENTS | CAS # | ACGIH TLV | OSHA PEL | OTHER LIMITS | % |
|---|---|---|---|---|---|
| Acetone | 67-64-1 | 750 ppm | 750 ppm | | 2-5 |
| Xylene | 1330-20-7 | 100 ppm | 100 ppm | | 68-75 |
| 2-Butoxy Ethanol | 111-76-2 | 25 ppm | 25 ppm | (skin) | 3-5 |
| Methanol | 67-56-1 | 200 ppm | 200 ppm | | 3-5 |
| Detergent | - | NA | NA | | 0-1 |
| Propane | 74-98-6 | NA | 1000 ppm | | 10-20 |
| Isobutane | 75-28-5 | NA | NA | 1000ppm | 10-20 |

2. PHYSICAL DATA : (without propellent)
Specific Gravity : 0.865
Vapor Pressure : ND
% Volatile : > 99
Boiling Point : 176 F initial
Evaporation Rate : Moderately fast
Freezing Point : ND
Vapor Density : ND
Appearance and Odor:
pH: NA
A clear colorless liquid, aromatic odor

Solubility : Partially soluble in water.

3. FIRE AND EXPLOSION DATA
Flashpoint : -40 F Method : TCC
Flammable Limits : propellent LEL:1.8 UEL:9.5
Extinguishing Media : CO2, dry chemical, foam
Unusual Hazards : Aerosol cans may explode when heated above 120 F.

4. REACTIVITY AND STABILITY
Stability : Stable
Hazardous decomposition products
: CO2, carbon monoxide (thermal)

Materials to avoid : Strong oxidizing agents and sources of ignition.

5. PROTECTION INFORMATION
Ventilation : Use mechanical means to insure vapor conc. is below TLV.

Respiratory : Use self-contained breathing apparatus above TLV.

Gloves : Solvent resistant
Eye & Face : Safety glasses
Other Protective Equipment: Not normally required for aerosol product usage.

**Figure 1-13** Material safety data sheets (MSDS) inform employees about hazardous materials. (Courtesy of CRC Industries, Inc.)

7. Safe handling precautions
8. Spill and leak procedures

Second, the employer has a responsibility to make sure that all hazardous materials are properly labeled. The label information must include health, fire, and reactivity hazards posed by the material, and the protective equipment necessary to handle the material. The manufacturer must supply all warning and precautionary information about hazardous materials, and this information must be read and understood by the employee before handling the material.

Third, employers are responsible for maintaining permanent files regarding hazardous materials. These files must include information on hazardous materials in the shop, proof of employee training programs, and information about accidents such as spills or leaks of hazardous materials. The employer's files must also include proof that employees' requests for hazardous material information such as MSDS sheets have been met. A general right-to-know compliance procedure manual must be maintained by the employer.

**Shop Manual**
Chapter 1, page 15

## A BIT OF HISTORY

Prior to 1970, workers did not have adequate protection from unsafe and unhealthy working conditions. Protection from unsafe and/or unhealthy working conditions was mainly left up to individual business owners. As a result, some workers were unnecessarily exposed to unsafe and unhealthy conditions. In 1970, the Occupational Safety and Health Act (OSHA) provided workers with protection from unsafe and unhealthy working conditions, and this act authorized enforcement of safety standards.

# Summary

- ❑ The United States Occupational Safety and Health Act of 1970 assured safe and healthful working conditions, and authorized enforcement of safety standards.
- ❑ Many hazardous materials and conditions can exist in an automotive shop, including flammable liquids and materials, corrosive acid solutions, loose sewer covers, caustic liquids, high-pressure air, frayed electric cords, hazardous waste materials, carbon monoxide, improper clothing, harmful vapors, high noise levels, and spills on shop floors.
- ❑ The danger regarding the labeling and handling of hazardous conditions and materials may be avoided by applying the necessary safety precautions. These precautions include all areas of safety such as personal safety, gasoline safety, housekeeping safety, general shop safety, fire safety, and hazardous waste handling safety.
- ❑ The automotive shop must supply the necessary shop safety equipment, and all shop personnel must be familiar with the location and operation of this equipment. Shop safety equipment includes gasoline safety cans, steel storage cabinets, combustible material containers, fire extinguishers, eyewash fountains, safety glasses and face shields, first-aid kits, and hazardous waste disposal containers.
- ❑ Workplace hazardous material information systems (WHMIS) provide information regarding the labeling and handling of hazardous materials.

### Terms to Know

Environmental Protection Agency (EPA)
Halogen and halon fire extinguishers
Hazard Communication Standard
Material Safety Data Sheets (MSDS)
Multipurpose dry chemical fire extinguisher
Occupational Safety and Health Act (OSHA)
Resource Conservation and Recovery Act (RCRA)
Right-to-know laws
Workplace Hazardous Materials Information Systems (WHMIS)

# Review Questions

## Short Answer Essays

1. Explain the purposes of the Occupational Safety and Health Act.
2. Define twelve shop hazards, and explain why each hazard is dangerous.
3. Describe five steps that are necessary for personal protection in the automotive shop.
4. Explain why smoking is dangerous in the shop.
5. Describe the danger in drug or alcohol use in the shop.
6. Explain three safety precautions related to electrical safety in the shop.
7. Define six essential safety precautions regarding gasoline handling.
8. Describe five steps required to provide housekeeping safety in the shop.
9. Describe how used oil filters need to be disposed of.
10. Describe typical fire extinguisher operation.

## Fill-in-the-Blanks

1. The poisonous gas in vehicle exhaust is __________ __________.

2. Safety boots with __________ toe caps are best for working in the shop.

3. One gallon of gasoline has a force equal to __________ sticks of dynamite.

4. Breathing asbestos dust may cause __________ __________.

5. Class C fires involve the burning of __________ equipment.

6. Irradiation eye burns may be caused by excessive light from a(n) __________ .

7. Hazardous wastes in an automotive shop include:
   A. __________ B. __________ C. __________
   D. __________ E. __________ F. __________

8. The right-to-know laws state that employees have a right to know when the __________ they handle at work are__________.

9. Material safety data sheets (MSDS) supply specific information regarding __________ __________.

10. Hazardous materials must never be dumped in __________, __________, __________, or __________.

## Multiple Choice

1. While discussing shop hazards:
   *Technician A* says high-pressure air from an air gun may penetrate the skin.
   *Technician B* says air in the bloodstream may be fatal.
   Who is correct?
   **A.** A only **C.** Both A and B
   **B.** B only **D.** Neither A nor B

2. While discussing personal protection in the shop:
   *Technician A* says jewelry, such as rings or watches, may cause serious burns if they make contact between an electric terminal and ground on the vehicle.
   *Technician B* says sneakers are suitable footwear in the automotive shop.
   Who is correct?
   **A.** A only **C.** Both A and B
   **B.** B only **D.** Neither A nor B

3. While discussing fire fighting:
   *Technician A* says halogen-type fire extinguishers produce no harmful gases when they are used to extinguish a fire.
   *Technician B* says that multipurpose dry chemical fire extinguishers may only be used on type D fires.
   Who is correct?
   **A.** A only **C.** Both A and B
   **B.** B only **D.** Neither A nor B

4. While discussing fire fighting:
   *Technician A* says water should be sprayed on a gasoline fire.
   *Technician B* says if a fire occurs inside a building, the doors and windows should be opened.
   Who is correct?
   **A.** A only **C.** Both A and B
   **B.** B only **D.** Neither A nor B

5. While discussing hazardous waste disposal:
*Technician A* says the right-to-know laws require employers to train employees regarding hazardous waste materials.
*Technician B* says the right-to-know laws do not require employers to keep permanent records regarding hazardous waste materials.
Who is correct?
**A.** A only  **C.** Both A and B
**B.** B only  **D.** Neither A nor B

6. While discussing material safety data sheets (MSDS):
*Technician A* says these sheets explain employers' and employees' responsibilities regarding hazardous material handling and disposal.
*Technician B* says these sheets contain specific information about hazardous materials.
Who is correct?
**A.** A only  **C.** Both A and B
**B.** B only  **D.** Neither A nor B

7. While discussing hazardous materials:
*Technician A* says a solid that ignites spontaneously is considered a hazardous material.
*Technician B* says a liquid with a flash point below 140°F (60°C) is considered a hazardous material.
Who is correct?
**A.** A only  **C.** Both A and B
**B.** B only  **D.** Neither A nor B

8. While discussing hazardous waste disposal:
*Technician A* says certain types of hazardous waste may be poured down a floor drain.
*Technician B* says hazardous waste users are responsible for hazardous waste materials from the time they become waste until the proper waste disposal is completed.
Who is correct?
**A.** A only  **C.** Both A and B
**B.** B only  **D.** Neither A nor B

9. While discussing engine safety precautions:
*Technician A* says the engine oil pan may be removed while the engine is suspended on an engine hoist.
*Technician B* says gasoline may be poured into the throttle body while starting an engine.
Who is correct?
**A.** A only  **C.** Both A and B
**B.** B only  **D.** Neither A nor B

10. While discussing hazardous waste disposal:
*Technician A* says air-conditioning refrigerants are considered hazardous waste materials.
*Technician B* says information about spill and leak procedures for hazardous waste materials is contained in material safety and data sheets (MSDS).
Who is correct?
**A.** A only  **C.** Both A and B
**B.** B only  **D.** Neither A nor B

# Theory of Engine Operation

Upon completion and review of this chapter, you should be able to:

- ❒ Describe the basic laws of physics involved with engine operation.
- ❒ Classify engines according to the number of cycles, the number of cylinders, cylinder arrangement, and valve train type.
- ❒ Define the four-stroke cycle theory.
- ❒ Describe the different cylinder arrangements and the advantages of each.
- ❒ Describe the different valve trains used in modern engines.
- ❒ Define engine measurement terms such as bore and stroke, displacement, compression ratio, engine efficiency, horsepower and torque, horsepower losses, mechanical efficiency, and thermal efficiency.
- ❒ Describe the relationship between compression ratio and engine power output.
- ❒ Define mechanical, volumetric, and thermal efficiencies, and describe factors that affect each.
- ❒ Describe the basic operation of alternative engine designs, including diesel, Miller-cycle, and Stratified Charge.
- ❒ Describe basic fuel cell operation.

## Introduction

Today's engines are designed to meet the demands of the automobile buying public and many government-mandated emissions and fuel economy regulations. High performance, fuel economy, reduced emissions, and reliability are demanded. To meet these demands, manufacturers are producing engines using lightweight blocks and cylinder heads, nontraditional materials such as powdered metals and composites, and computerized engine designs. Today's technician is called upon to diagnose and service these advanced engines properly.

The **internal combustion engine** used in automotive applications utilizes several laws of physics and chemistry to operate. Although engine sizes, designs, and construction vary greatly, they all operate on the same basic principles. This chapter discusses these basic principles and engine designs.

**Internal combustion engines** burn their fuels within the engine.

## Major Engine Components

The automobile engine is a metal block containing a series of chambers or holes. The volume capacity of these holes varies in relation to the position of a rotating shaft. There is a port for each chamber that allows the admission of the combustible mixture and another port for the expulsion of burned gases. The volume of the combustion chambers must be variable in order for the engine to make use of the expansion of the burning gases. This ability enables the chamber to compress the gases before combustion and to purge itself of the burned material and refill itself with a fresh combustible charge after combustion has taken place.

**Shop Manual**
Chapter 2, page 27

The **cylinder block** is the structure that the rest of the engine is built upon (Figure 2-1). The block contains the cylinder bores, coolant passages, and oil passages. The bottom of the block is also machined with bores to house the crankshaft, and is known as the crankcase. The hollow jackets of the upper block add rigidity to the entire structure and contain the liquid coolant that carries the heat away from the cylinders and the other parts of the block. Usually, the water and oil pumps are mounted directly to the block.

The **cylinder block** is the main structure of the engine. Most of the other engine components are attached to the block

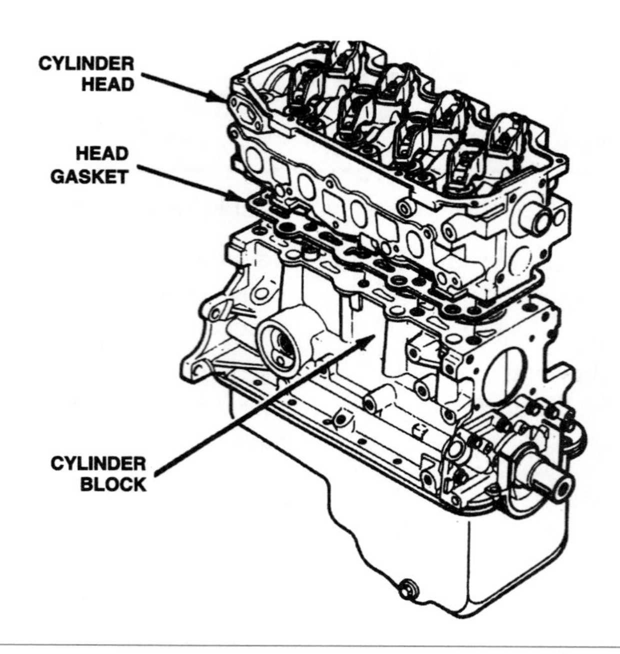


Figure 2-1 Cylinder block and cylinder head. (Courtesy of Federal-Mogul Corporation)

The **crankshaft** is a mechanical device that converts the reciprocating motions of the pistons into rotary motion.

**Shop Manual**
Chapter 2, page 29

A **piston** is an engine component in the form of a hollow cylinder that is enclosed at the top and open at the bottom. Combustion forces are applied to the top of the piston to force it down.

A **connecting rod** is the link between the piston and crankshaft.

The **crankshaft** (Figure 2-2) is a long iron or steel fabrication made up of bearing points called journals. These journals turn on their own axes, allowing for the rotating motion that is necessary for engine operation. The crankshaft is also equipped with counterweighted crank throws, or crankpins, which are located several inches away from the center of the shaft, and which, of course, turn in a circle as the crankshaft turns. The crank throws are centered under the cylinders that are machined into the upper block.

**Pistons** equipped with sealing rings are located in the cylinders and are linked to the crankpins with **connecting rods** (Figure 2-3). The rods are connected to the pistons with piston pins and bushings, and are connected to the crank throws at the lower end with insert bearings. The force of combustion works on the top of the piston and is transferred to the crankshaft through the connecting rod. As the piston moves up and down in the cylinder, it rotates the crankshaft. This process converts the

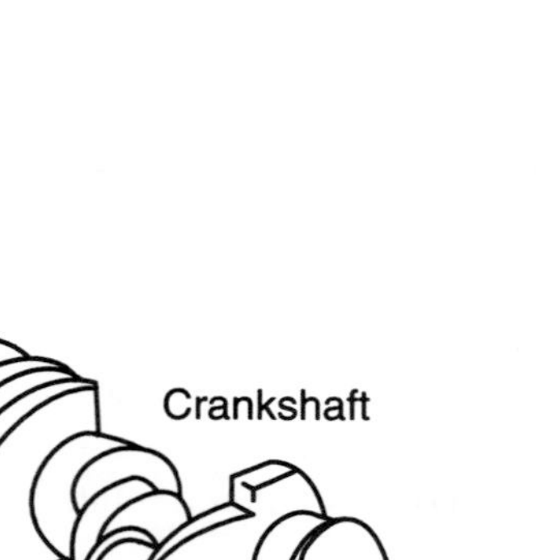


Figure 2-2 Crankshaft and related components.

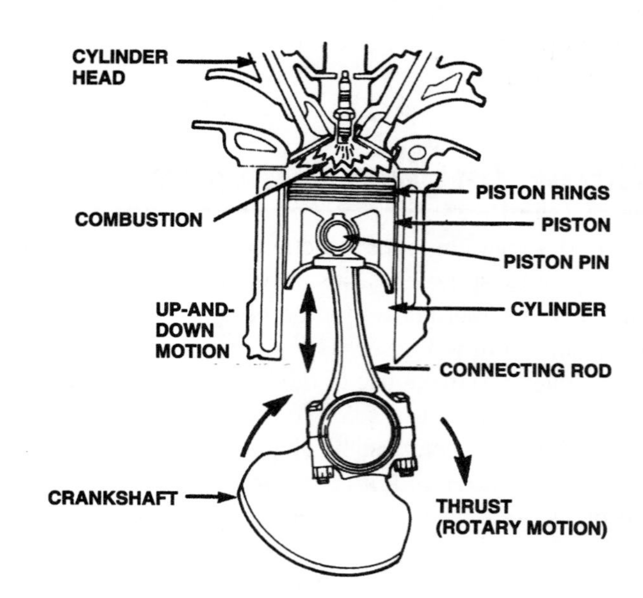


Figure 2-3 Piston and connecting rod. (Courtesy of Federal-Mogul Corporation)

reciprocating piston motion into rotary motion, which is used to turn the wheels. A flywheel at the rear of the crankshaft provides a large, stable mass for smoothing out the rotation.

The **cylinder heads** form tight covers for the tops of the cylinders and contain machined chambers into which the air/fuel mixture is forced (Figure 2-4). In OHC engines, the cylinder head also supports the camshaft. The area of the cylinder head located above the cylinder bore is called the **combustion chamber**. The valves in each cylinder are opened and closed by the action of the **camshaft** and related valve train. The camshaft is connected to the crankshaft by a gear and chain or belt, and is driven at one-half crankshaft speed. In other words, for every complete circle the crankshaft makes, the camshaft makes a half-circle. The camshaft may be mounted in the engine block itself, or on or in the cylinder head, depending on design.

The camshaft has lobes that are used to open and close the valves (Figure 2-5). A camshaft mounted in the block operates the valves remotely through pushrods and rocker arms. In some OHC (overhead cam shaft) engines, the camshaft lobes contact the rocker arms. The cylinder heads also have threaded holes for the spark plugs that screw right into the heads so that their tips protrude into the combustion chambers.

Lubricating oil for the engine is normally stored in a pan at the bottom of the engine and is force-fed to almost all parts of the engine by a gear pump.

On most engines, the **cylinder head** contains the valves, valve seats, valve guides, valve springs, and the upper portion of the combustion chamber.

The **combustion chamber** is the volume of the cylinder above the piston with the piston at TDC.

The **camshaft** is the shaft containing lobes to operate the engine valves.

**Shop Manual**
Chapter 2, page 30

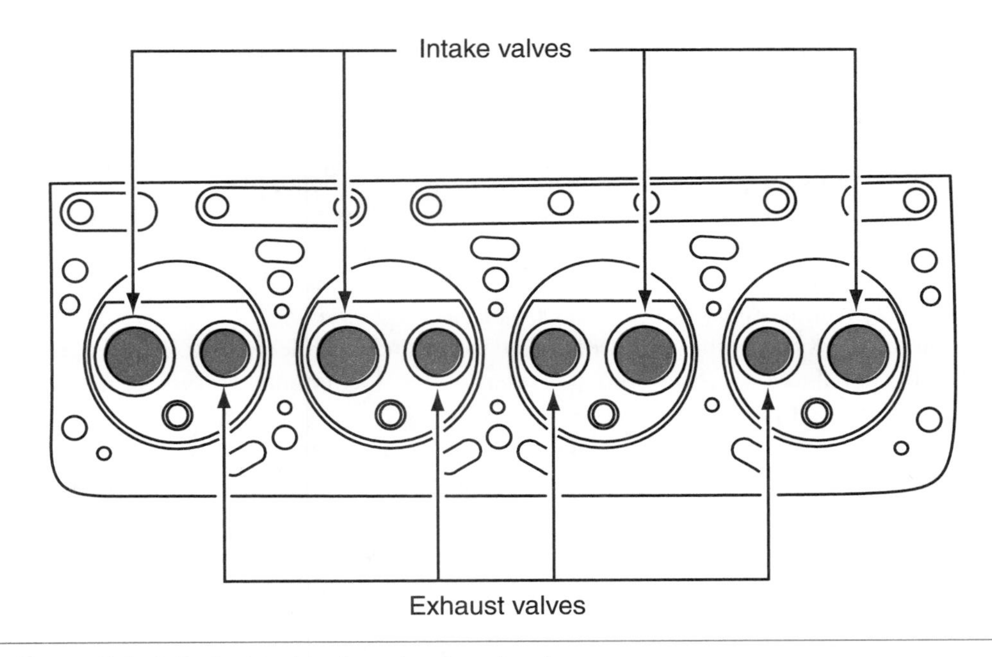


**Figure 2-4** Cylinder head and combustion chambers.

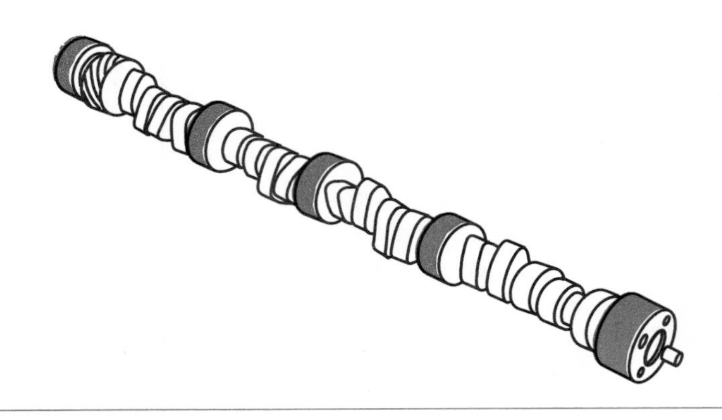

**Figure 2-5** Camshaft.

# Engine Operation

**Thermodynamics** is the study of the relationship between heat energy and mechanical energy.

One of the many laws of physics utilized within the automotive engine is **thermodynamics.** The driving force of the engine is the expansion of gases. Gasoline (a liquid fuel) will change state to a gas if it is heated or burned. Gasoline must be mixed with oxygen before it can burn. In addition, the air/fuel mixture must be burned in a confined area in order to produce power. Gasoline that is burned in an open container produces very little power, but if the same amount of fuel is burned in an enclosed container, it will expand with force. When the air/fuel ratio changes states, it also expands as the molecules of the gas collide with each other and bounce apart. Increasing the temperature of the gasoline molecules increases their speed of travel, causing more collisions and expansion.

Heat is generated by compressing the air/fuel mixture within the combustion chamber. Igniting the compressed mixture causes the heat, pressure, and expansion to multiply. This process releases the energy of the gasoline so it can produce work. The igniting of the mixture is a controlled burn, not an explosion. The controlled combustion releases the fuel energy at a controlled rate in the form of heat energy. The heat, and consequential expansion of molecules, increases the pressure inside the combustion chamber. Typically, the pressure works on top of a piston that is connected to a crankshaft. As the piston is driven, it causes the crankshaft to rotate.

**Torque** is a rotating force around a pivot point.

The engine produces **torque,** which is applied to the drive wheels. As the engine drives the wheels to move the vehicle, a certain amount of work is done. The rate of work being performed is measured in horsepower.

## Energy and Work

In engineering terms, in order to have work, there must be motion. Using this definition, work can be measured by combining distance and weight and expressed as foot-pound (ft.-lb.) or Newton-meter (Nm). These terms describe how much weight can be moved a certain distance. A foot-pound is the amount of energy required to lift one pound of weight one foot in distance. The amount of work required to move a 500-pound weight five feet is 2,500 foot-pounds (3,390 Nm). In the metric system, the unit used to measure force is called Newton-meters (Nm). One foot pound is equal to 1.355 Nm. Torque is measured as the amount of force in newtons multiplied by the distance that the force acts in meters.

Work is a unit of energy in that energy is the cause and work is the effect. Basically, energy is anything that is capable of resulting in motion. Common forms of energy include electrical, chemical, heat, radiant, and mechanical.

**Potential energy** is energy that is not being used at a given time, but which can be used.

Working energy is called **kinetic energy.**

There are two types of energy: potential and kinetic. **Potential energy** is available to be used for a purpose, but is not being used at this point in time. **Kinetic energy** is energy that is in motion.

Energy cannot be created or destroyed; however, it can be stored, controlled, and changed to other forms of energy. For example, the vehicle's battery stores chemical energy that is changed to electrical energy when a load is applied. An automotive engine transforms heat energy into mechanical work.

# Engine Classifications

Engine classification is usually based on the number of cycles, the number of cylinders, cylinder arrangement, and valve train type. In addition, engine displacement is commonly used to identify the engine.

Sometimes, several different methods can be used to identify the same engine. For example, a Chrysler 2.0 liter DOHC is also identified as a 420A and D4FE. In this case, the engine has a displacement of 2.0 liters and uses dual overhead camshafts (DOHC). The 420A is derived from four valves per cylinder, 2.0 liter displacement, and American built. The D4FE means the engine uses dual overhead camshafts with four valves per cylinder and has front exhaust (the exhaust manifold is mounted facing the front of the vehicle).

## Cycles

This section will define many of the terms used by automotive manufacturers to classify their engines. Most automotive and truck engines are four-stroke cycle engines. A **stroke** is the movement of the piston from one end of its travel to the other; for example, if the piston is at the top of its travel and then is moved to the bottom of its travel, one stroke has occurred. Another stroke occurs when the piston is moved from the bottom of its travel to the top again. A **cycle** is a sequence that is repeated. In the four-stroke engine, four strokes are required to complete one cycle.

The **stroke** is the amount of piston travel from TDC to BDC measured in inches or millimeters.

A **cycle** is a complete sequence of events.

The internal combustion engine must draw in an air/fuel mixture, compress the mixture, ignite the mixture, then expel the exhaust. This is accomplished in four piston strokes (Figure 2-6). The

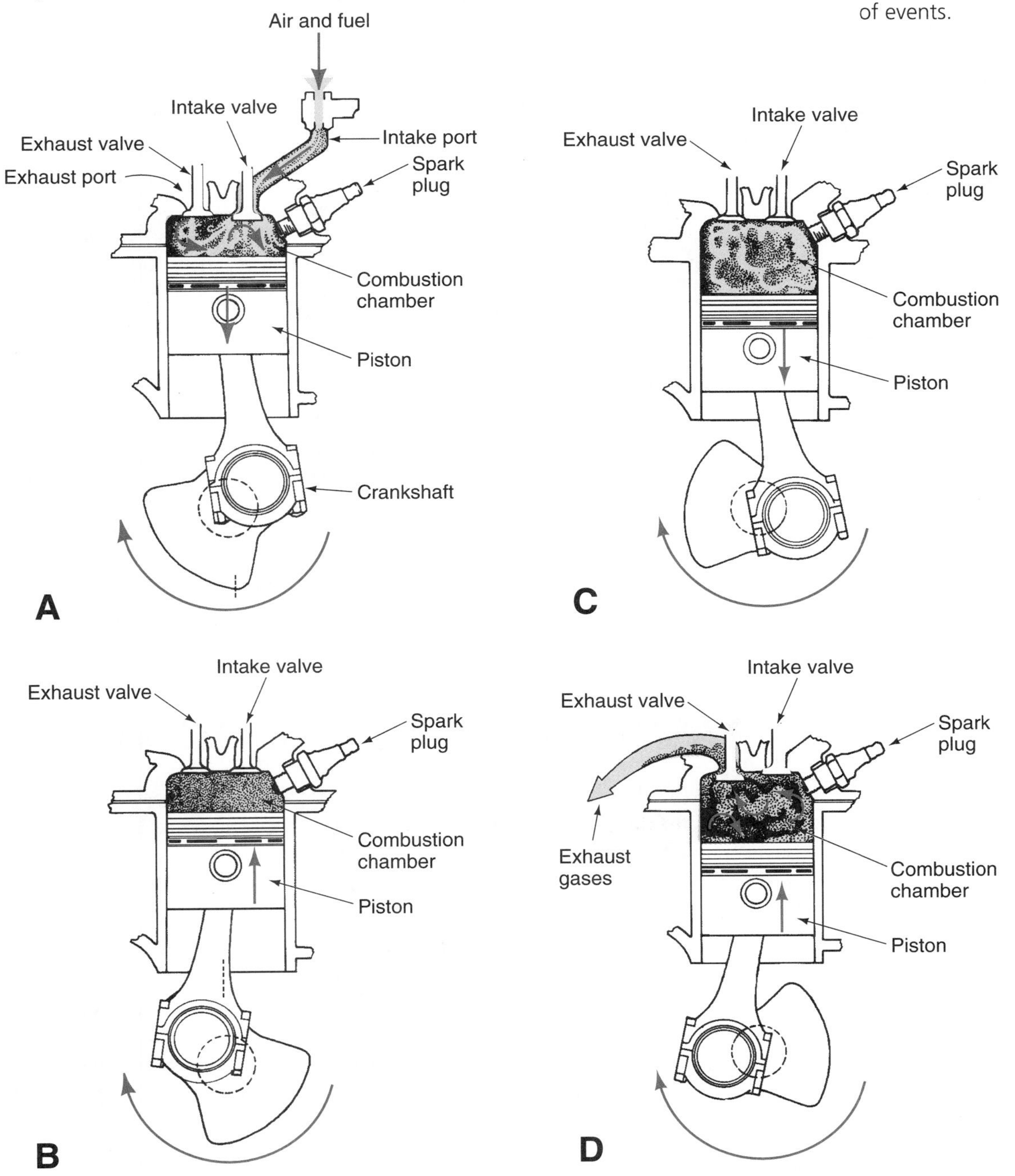


**Figure 2-6** The four strokes of an automotive engine: (A) intake stroke, (B) compression stroke, (C) power stroke, and (D) exhaust stroke. (Courtesy of Breton Publishers)

**Shop Manual**
Chapter 2, page 34

**Top dead center (TDC)** is a term used to indicate that the piston is at the very top of its stroke.

**Bottom dead center (BDC)** is a term used to indicate that the piston is at the very

**Valve overlap** is the length of time, measured in degrees of crankshaft revolution, that the intake and exhaust valves of the same combustion chamber are open simultaneously.

**Reciprocating** is an up-and-down or back-and-forth motion.

process of drawing in the air/fuel mixture is actually accomplished by atmospheric pressure pushing it into a low-pressure area created by the downward movement of the piston.

The first stroke of the cycle is the intake stroke (see Figure 2-6A). As the piston moves down from **top dead center (TDC),** the intake valve is opened so the vaporized air/fuel mixture can be pushed into the cylinder by atmospheric pressure. As the piston moves downward in its stroke, a vacuum is created (low pressure). Since high pressure moves toward low pressure, the air/fuel mixture is pushed past the open intake valve and into the cylinder. After the piston reaches **bottom dead center (BDC),** the intake valve is closed, and the stroke is completed. Closing the intake valve after BDC allows an additional amount of air/fuel mixture to enter the cylinder, increasing the volumetric efficiency of the engine. Even though the piston is at the end of its stroke and no more vacuum is created, the additional mixture enters the cylinder because it weighs more than air alone.

The compression stroke begins as the piston starts its travel back to TDC (see Figure 2-6B). The intake and exhaust valves are both closed, trapping the air/fuel mixture in the combustion chamber. The movement of the piston toward TDC compresses the mixture. As the molecules of the mixture are pressed tightly together, they begin to heat. When the piston reaches TDC, the mixture is fully compressed, and a spark is induced in the cylinder by the ignition system. Compressing the mixture provides for better burning and for intense combustion.

When the spark occurs in the compressed mixture, the rapid burning causes the molecules to expand, beginning the power stroke (see Figure 2-6C). The expanding molecules create a pressure above the piston and push it downward. The downward movement of the piston in this stroke is the only time the engine is productive concerning power output. During the power stroke, the intake and exhaust valves remain closed.

The exhaust stroke of the cycle begins when the piston reaches BDC of the power stroke (see Figure 2-6D). Just prior to the piston reaching BDC, the exhaust valve is opened. The upward movement of the piston back toward TDC pushes out the exhaust gases from the cylinder past the exhaust valve and into the vehicle's exhaust system. As the piston approaches TDC, the intake valve opens and the exhaust valve is closed a few degrees after TDC. The degrees of crankshaft rotation when both the intake and exhaust valves are open is called **valve overlap.** The cycle is then repeated again as the piston begins the intake stroke. Most engines in use today are referred to as **reciprocating.** Power is produced by the up and down movement of the piston in the cylinder. This linear motion is then converted to rotary motion by a crankshaft.

## A BIT OF HISTORY

The first workable internal combustion engine was created by Jean Joseph Lenoir in 1860. Dr. Nikolaus Otto designed the first successful four-stroke engine in 1866. All previous internal combustion engines did not compress the air/fuel mixture. They attempted to draw the mixture in during a downward movement of the piston, then ignite it. The expansion of the gases would force the piston down the remainder of its travel. This design was used in an attempt to make the pistons double acting (a power stroke each way). Dr. Otto's engine used the downward stroke to ingest the air/fuel mixture, then an upward stroke to compress it. Many laughed at his idea, since only one stroke was used to produce power. However, comparing Otto's engine to other engines of that era, his was lighter and able to run almost twice as fast, and required only 7 percent of the cylinder displacement to produce the same amount of horsepower with the same amount of fuel. Both Lenoir's and Otto's engines were powered by cooking gas. In 1876, Dr. Otto successfully converted his engine to burn gasoline.

## Number of Cylinders

One cylinder would not be able to produce sufficient power to meet the demands of today's vehicles. Most automotive and truck engines use three, four, five, six, eight, ten, or twelve cylinders (Figure 2-7). The number of cylinders used by the manufacturer is determined by the amount of work required from the engine.

Vehicle manufacturers attempt to achieve a balance among power, economy, weight, and operating characteristics. An engine having more cylinders generally runs smoother than those having three or four cylinders, because there is less crankshaft rotation between power strokes; however, adding more cylinders increases the weight of the vehicle and the cost of production.

## Cylinder Arrangement

Engines are also classified by the arrangement of the cylinders. The cylinder arrangement used is determined by vehicle design and purpose. The most common engine designs are in-line and V-type.

The in-line engine places all of its cylinders in a single row (Figure 2-8). Advantages of this engine design include ease of manufacturing and serviceability. The disadvantage of this engine

**Figure 2-7** The automotive engine can be designed with several different configurations. (Courtesy of DaimlerChrysler Corporation)

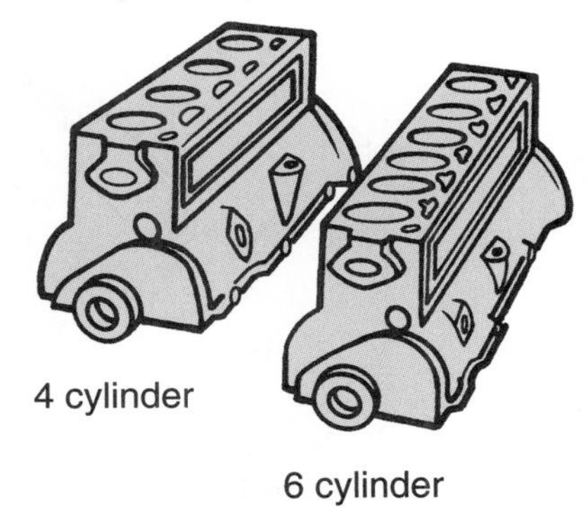


**Figure 2-8** In-line engines are designed for ease of construction and service.

A **transverse-mounted engine** faces from side to side instead of front to back.

The opposed cylinder engine may be called a pancake engine.

An **overhead valve (OHV)** engine has the valves mounted in the cylinder head and the camshaft mounted in the block.

design is the block height. Since the engine is tall, aerodynamic design of the vehicle is harder to achieve. Most manufacturers overcome this disadvantage by installing a **transverse-mounted engine** in the engine compartment of front-wheel-drive vehicles.

The V-type engine has two rows of cylinders set 60 to 90 degrees from each other (Figure 2-9). The V-type design allows for a shorter block height, thus easier vehicle aerodynamics. The length of the block is also shorter than in-line engines with the same number of cylinders.

A common engine design used for rear engine vehicles, as well as the front-wheel-drive Subaru, is the opposed cylinder engine (Figure 2-10). This engine design has two rows of cylinders directly across from each other. The main advantage of this engine design is the very small vertical height.

Another engine design is the slant cylinder (Figure 2-11). This design is similar to in-line engines, except the cylinders are placed at a slant. This design reduces the height of the engine and allows for more aerodynamic vehicle designs.

## Valve Train Type

Engines can also be classified by their valve train types. The three most commonly used valve trains are the:

1. **Overhead valve (OHV).** The intake and exhaust valves are located in the cylinder head, while the camshaft and lifters are located in the engine block (Figure 2-12). Valve train components of this design include lifters, pushrods, and rocker arms.

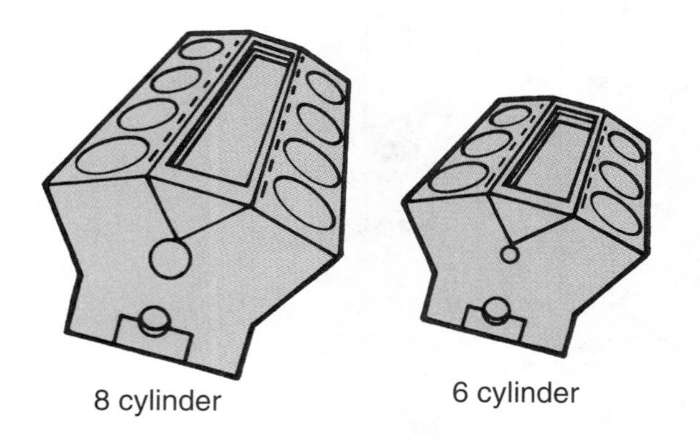


Figure 2-9 The V-type engine design allows for lower hood lines.

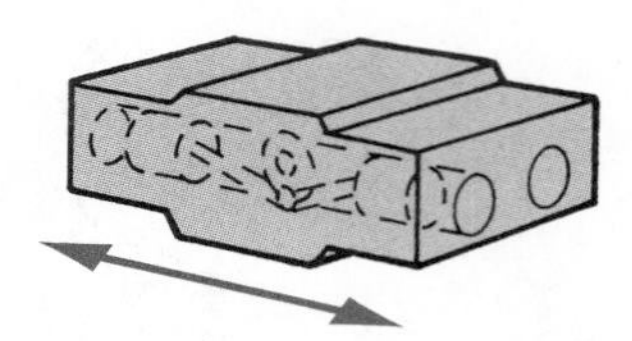

Figure 2-10 Opposed cylinder engine design.

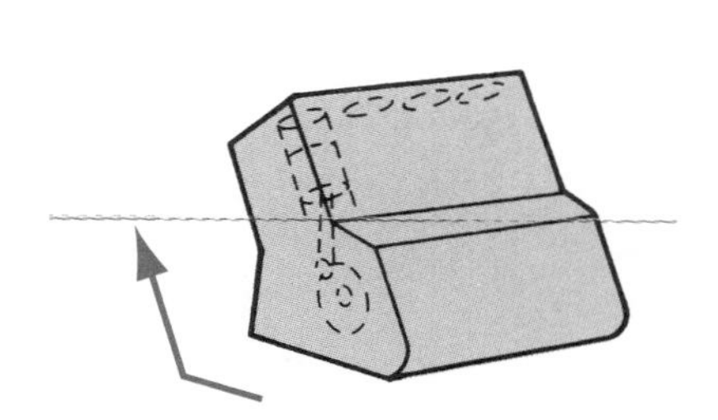

Figure 2-11 Slant cylinder design lowers the overall height of the engine.

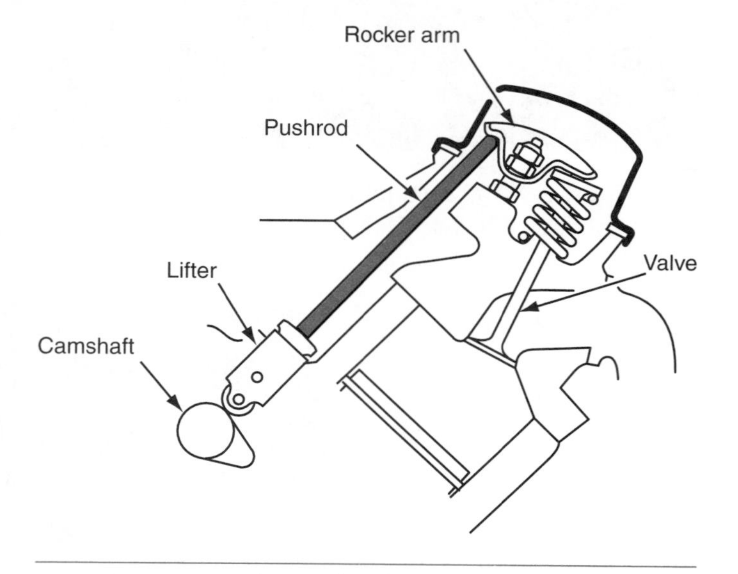


Figure 2-12 The overhead valve engine has been one of the most popular valve designs.

2. **Overhead cam (OHC).** The intake and exhaust valves are located in the cylinder head along with the camshaft (Figure 2-13). The valves are operated directly by the camshaft and a follower, eliminating many of the moving parts required in the OHV engine. If a single camshaft is used in each cylinder head, the engine is classified as a single overhead cam (SOHC) engine. This designation is used even if the engine has two cylinder heads with one camshaft each.
3. **Dual overhead cam (DOHC).** The DOHC uses separate camshafts for the intake and exhaust valves. A DOHC V8 engine is equipped with a total of four camshafts (Figure 2-14).

An **overhead cam (OHC)** engine has a single camshaft mounted above each cylinder head.

A **dual overhead cam (DOHC)** engine has two camshafts mounted above each

## Ignition Types

Most automotive engines use a mixture of gasoline and air, which is compressed and then ignited with a spark plug. This type of engine is referred to as a **spark-ignition (SI) engine.** Diesel engines, on the other hand, do not use spark plugs. The fuel and air mixture is ignited by the heat created during the compression stroke. These engines are referred to as **compression-ignition (CI) engines.**

A **spark-ignition (SI) engine** ignites the air/fuel ratio in the combustion chamber by a spark across the spark plug electrodes.

A **compression-ignition (CI) engine** ignites the air/fuel mixture in the cylinders from the heat of compression.

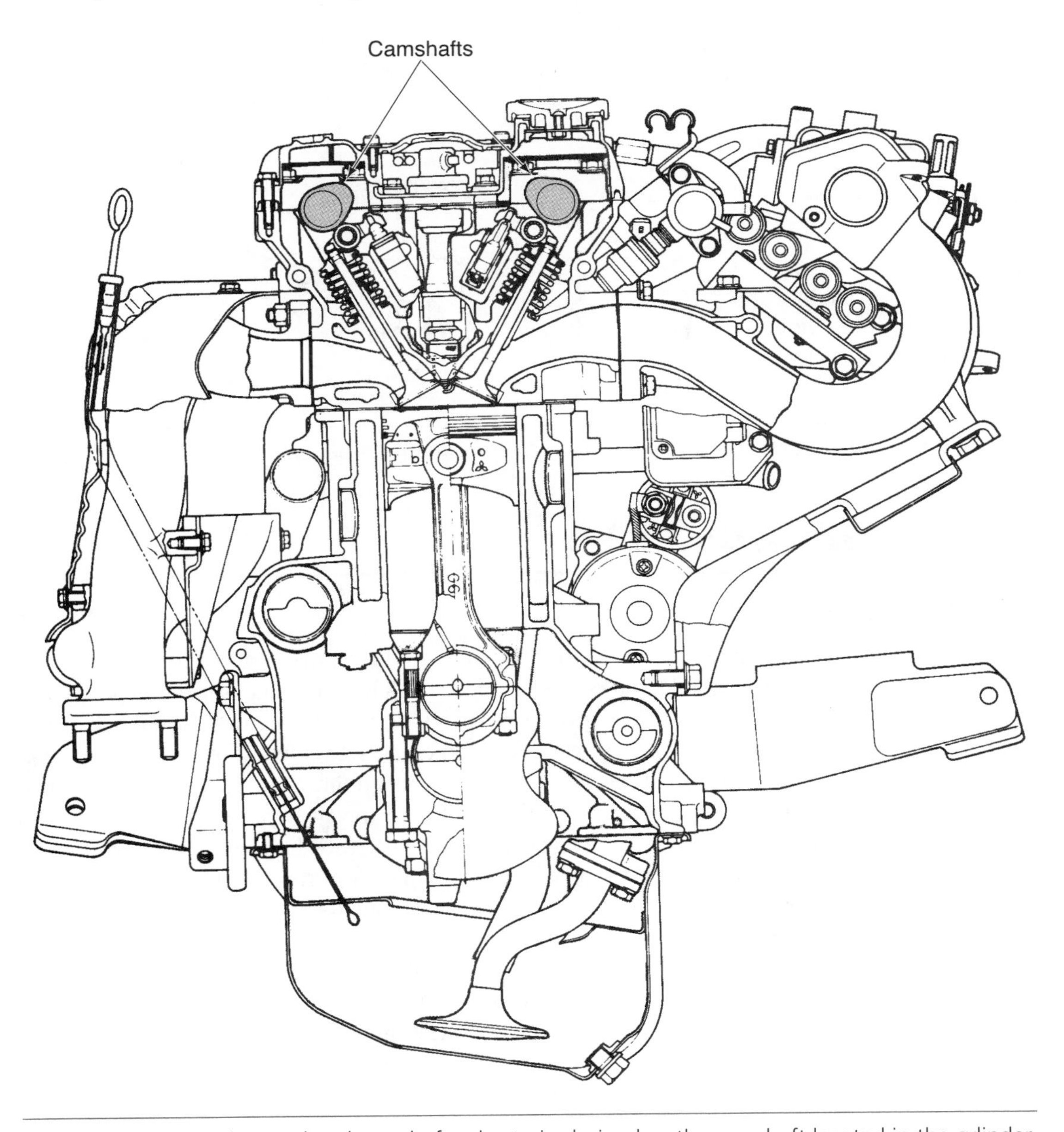


Figure 2-13 The overhead camshaft valve train design has the camshaft located in the cylinder head. (Courtesy of Hyundai Motor America)

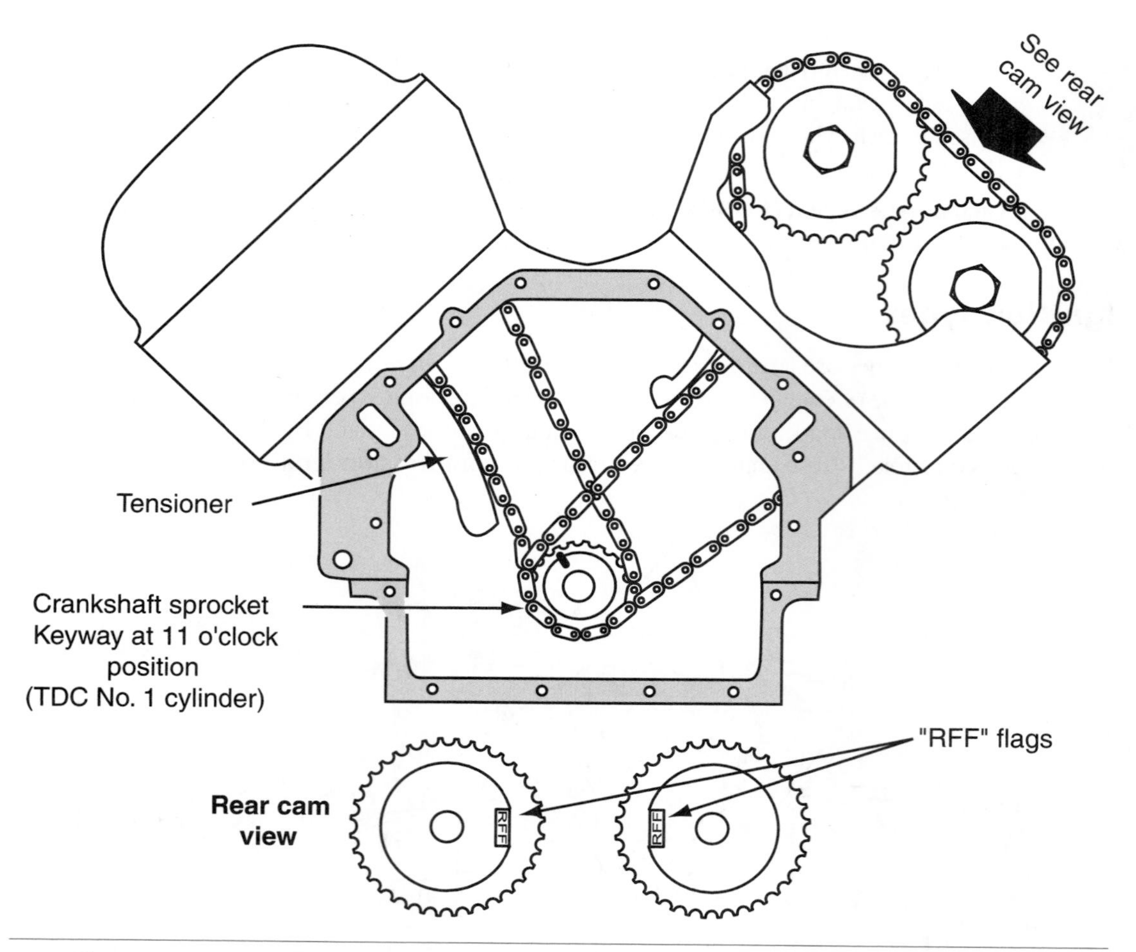


Figure 2-14 Dual overhead cam engines have two camshafts per cylinder head.

# Engine Vibration

## Balance Shafts

Many engine designs have inherent vibrations. The in-line four-cylinder engine is an example of this problem. Some engine manufacturers use balance shafts to counteract this tendency. Four-cylinder engine vibration occurs at two positions of piston travel. The first is when a piston reaches TDC. The inertia of the piston moving upward creates an upward force in the engine. The second vibration occurs when all pistons are level. At this point, the downward moving pistons have accelerated and have traveled more than half their travel distance. The engine forces are downward because the pistons have been accelerating since they have to travel a greater distance than the upward moving pistons to reach the point that all pistons are level.

The following formula (with values plugged in) demonstrates the principle that piston travel is further at the top 90 degrees verses the bottom 90 degrees of crankshaft rotation. First, determine the total length. This is rod length, from the center of the big end to the center of the small end, and the crankshaft throw (stroke divided by 2). Next, determine the amount the piston drops in the first 90 degrees of crankshaft rotation (TDC to 90 degrees). Once this is known, the final amount of piston travel can be determined. The following formula is for a 2.5 liter four-cylinder engine (the values are in millimeters):

1. Total length = rod + throw (stroke/2)

   209 = 157 + 52

2. Travel the first 90 degrees of crankshaft rotation:

   $209 - 157^2 + 52^2 = 209 - 21945 = 148$

3. Total – difference = piston travel to midpoint

   209 – 148 = 61 (travel for first 90 degrees of crankshaft rotation)

4. Bottom half of travel = distance left – rod length + throw

   148 – 157 + 52 = 43 (travel for last 90 degrees of crankshaft rotation)

The balance shafts rotate at twice the speed of the crankshaft. This creates a force that counteracts crankshaft vibrations. At TDC, the forces of the piston will exert upward; to offset the vibration, the crankshaft and the balance shaft(s) lobes have a downward force (Figure 2-15). At 90 degrees of crankshaft rotation, all pistons are level. Remember, the pistons traveling downward have traveled more than half of their travel distance to reach this point. The engine forces are in a downward direction since these pistons have been accelerated to reach the midpoint at the same time as the upward moving pistons, which need to travel a lesser distance. Since the balance shaft(s) rotate at twice the crankshaft speed, the lobes of the balance shaft(s) are facing up, exerting an upward force to counteract the downward forces of the pistons (Figure 2-16). At 180 degrees of crankshaft rotation, the upward moving pistons reach TDC, and again exert an upward force. The balance shaft(s) have their lobes facing downward to exert a downward force to counteract the upward force (Figure 2-17). Finally, at 270 degrees of crankshaft rotation, the pistons are level again. At this point, the lobes of the balance shaft(s) are exerting an upward force to counteract the downward force of the pistons (Figure 2-18).

Some V6 engines introduced in the 1960s have a 90-degree block and three-throw crankshaft with four main bearings. A 90-degree block has a 90-degree angle between the center of the cylinder bores in each bank. This engine has two connecting rods: one from each cylinder bank attached to each crankshaft throw, with each throw 120 degrees apart. The two connecting rods on each crankshaft throw are on the same horizontal plane. In this type of engine design,

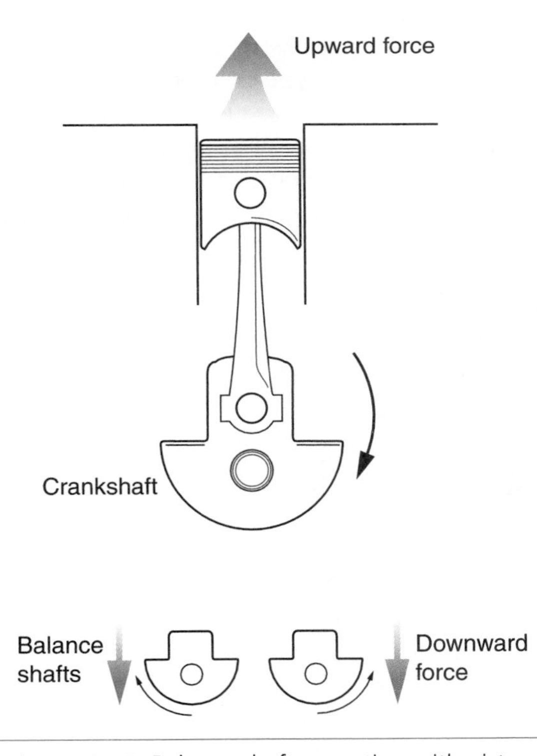


Figure 2-15 Balance shaft operation with piston moving upward.

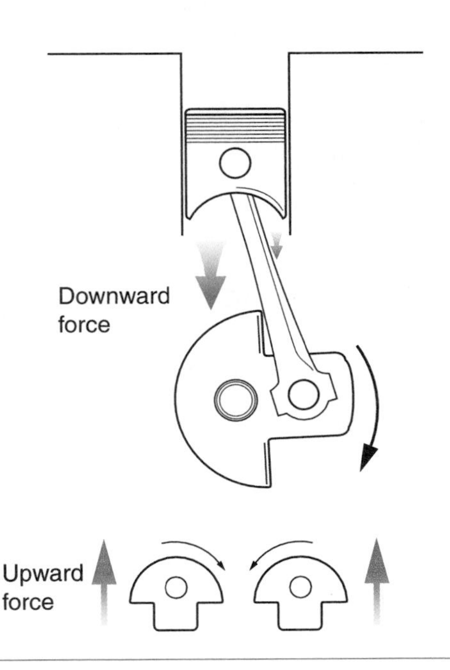


Figure 2-16 Balance shaft operation with piston moving downward.

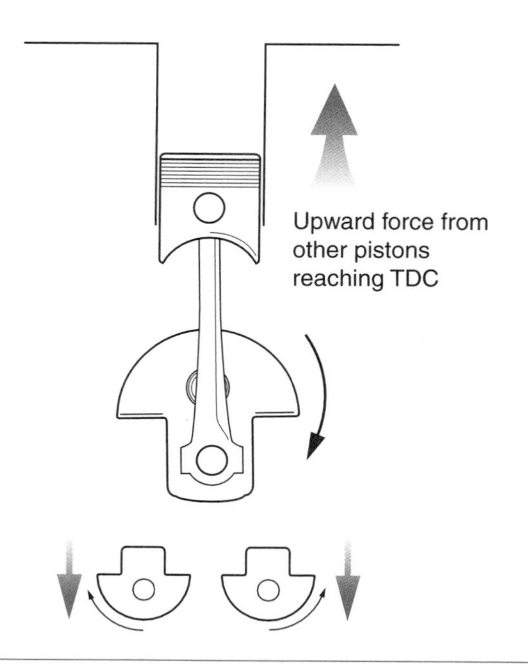


Figure 2-17 Balance shaft operation in relation to other pistons.

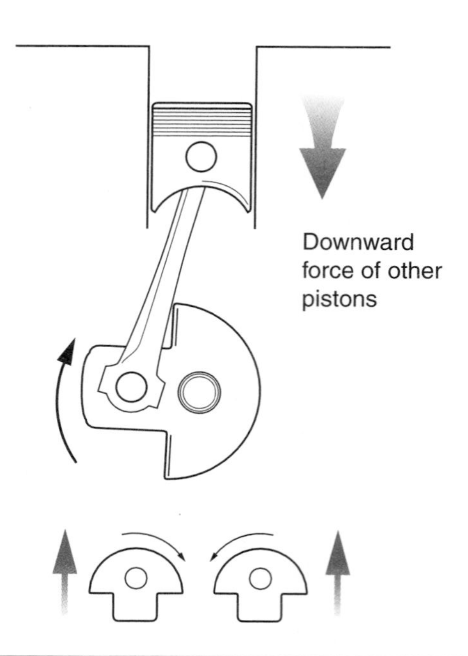


Figure 2-18 Balance shaft operation with other pistons moving downward.

The **splay angle** is the angle between two offset crankshaft throws mounted beside each other on some V6 engines.

A **splayed crankshaft** is a crankshaft in which each pair of side-by-side journals is

the pistons do not reach TDC at even intervals. This engine has firing intervals of 150 degrees – 90 degrees, 150 degrees – 90 degrees, 150 degrees – 90 degrees. The uneven firing intervals created severe engine vibrations. The manufacturer installed very flexible engine mounts to try and keep the engine vibrations from reaching the chassis and passenger compartment.

In 1977, this engine was modified by the manufacturer so the crankshaft has offset pins for each adjacent pair of connecting rods (Figure 2-19). The crankshaft pins for each pair of adjacent connecting rods are offset 30 degrees. The offset angle between the adjacent crankshaft pins may be referred to as the **splay angle,** and this type of crankshaft may be called a **splayed crankshaft.** This crankshaft design provided even piston firing intervals of 120 degrees and greatly reduced vibrations.

Some V6 engines have a 60-degree block and an 18-degree splay angle between the adjacent crankshaft pins. This design is narrower than a 90-degree block, and may be more suitable for transverse mounting in front-wheel-drive vehicles in which space is a major consideration.

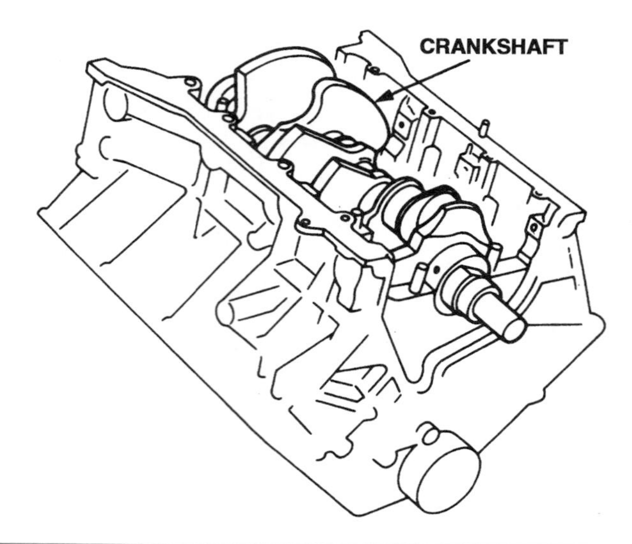


Figure 2-19 Crankshaft with offset crankpins. (Courtesy of American Honda Motor Co., Inc.)

The 18-degree splay angle between the adjacent crankshaft pins provided an uneven firing pattern of 132 degrees – 108 degrees, 132 degrees – 108 degrees, 132 – 108 degrees. This uneven firing pattern produced some engine vibrations. A 60-degree V6 engine must have a splay angle of 60 degrees between the adjacent crankshaft pins to be even firing.

Even firing or uneven firing V6 engines may have a balance shaft to reduce engine vibrations. Some balance shafts on overhead valve (OHV) engines are mounted in the block V above the camshaft (Figure 2-20). Bearing journals on the balance shaft are mounted on bushings in the block. This type of engine has two gears mounted on the front end of the camshaft. The inner camshaft gear is meshed with the gear on the balance shaft, and the timing chain surrounds the outer camshaft gear and the crankshaft gear to drive the camshaft (Figure 2-21). On some dual overhead cam (DOHC) engines, the balance shaft is mounted in the block above the crankshaft (Figure 2-22). The balance shaft journals are supported on bushings in the block. A primary

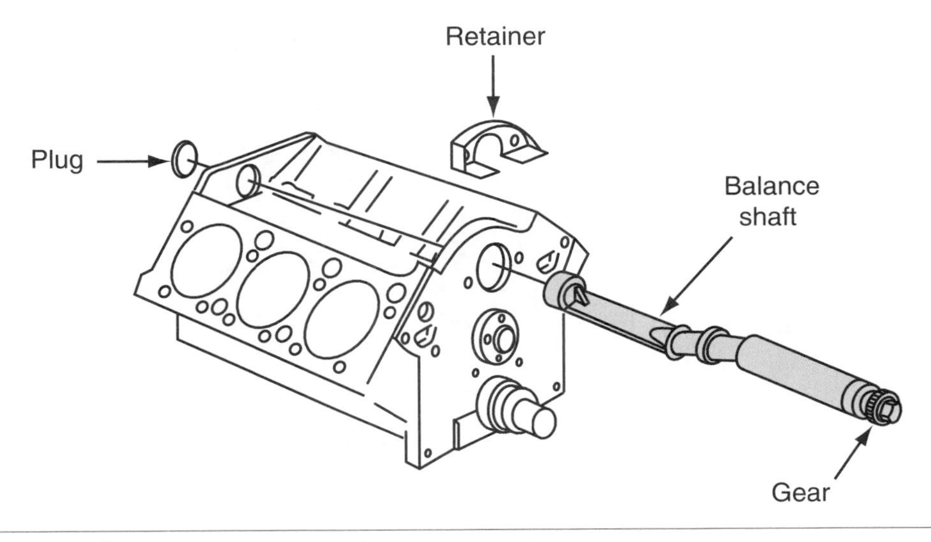


Figure 2-20 Balance shaft in overhead valve (OHV) V6 engine.

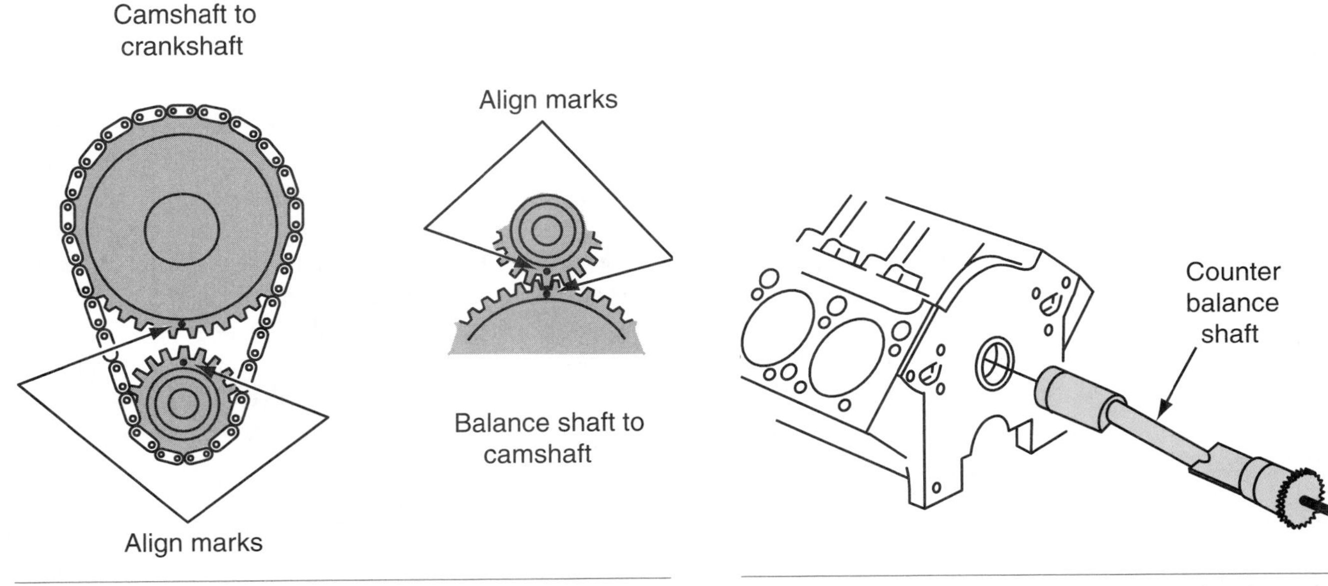


Figure 2-21 Balance shaft, camshaft, and crankshaft sprockets.

Figure 2-22 Balance shaft dual overhead cam (DOHC) V6 engine.

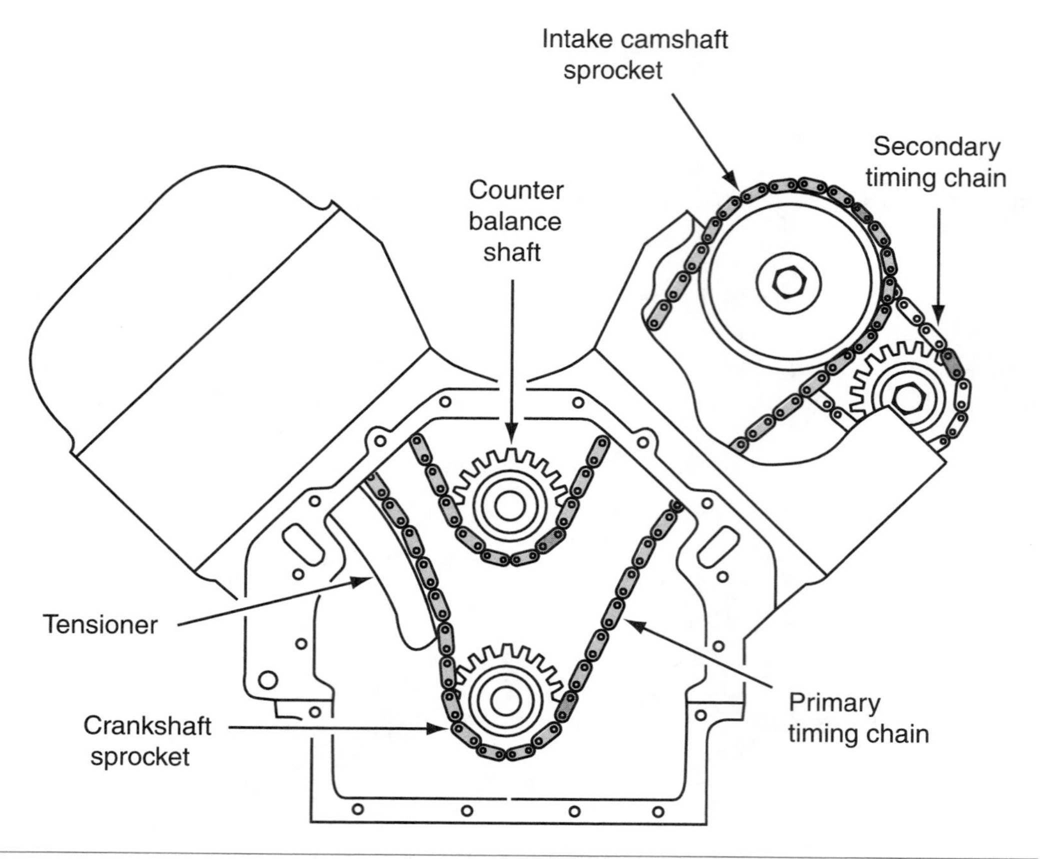


**Figure 2-23** Balance shaft drive chain dual overhead cam (DOHC) V6 engine.

**Displacement** is the measure of engine volume. The larger the displacement, the greater the power output.

A **bore** is the diameter of a hole.

If the bore is larger than its stroke, the engine is referred to as oversquare. If the bore is smaller than the stroke, the engine is referred to as undersquare. A square engine has the same bore size as it does stroke

**Stroke** is the distance that the piston travels from top dead center (TDC) to bottom dead center (BDC) or vice versa.

timing chain surrounds the crankshaft sprocket, intake camshaft sprockets, and the balance shaft sprocket (Figure 2-23). Secondary timing chains are connected from the intake to the exhaust camshaft sprockets. All balance shafts must be properly timed in relation to the crankshaft, or severe engine vibrations will result.

## Engine Displacement

**Displacement** is the volume of the cylinder between TDC and BDC measured in cubic inches, cubic centimeters, or liters (Figure 2-24). Most engines today are identified by their displacement in liters. Engine displacement is an indicator of its size and power output. The more fuel that can be burned, the greater amount of energy that can be produced. To burn more fuel, more air must also enter the cylinders. Engine displacement measurements are an indicator of the amount or mass of air that can enter the engine. The mass of air allowed to enter the cylinders is controlled by a throttle plate. If the throttle plate is closed, a smaller mass of air is allowed to enter the cylinder, and the power output is reduced. As the throttle plate is opened, the mass of air allowed to enter the cylinders is increased, resulting in greater power output. The more air that is allowed, the more fuel that can be added to it. As engine speed increases, it will eventually reach a point at which no more air can enter. Any increase in speed beyond this point causes a loss in power output. The major factor in determining the maximum mass of air that can enter the cylinders is engine displacement. The amount of displacement is determined by the number of cylinders, cylinder bore, and length of the piston stroke (Figure 2-25).

**Bore** and **Stroke.** The **bore** of the cylinder is its diameter measured in inches (in.) or millimeters (mm). The stroke is the amount of piston travel from TDC to BDC measured in inches or millimeters. An oversquare engine is used when high rpm (revolutions per minute) output is

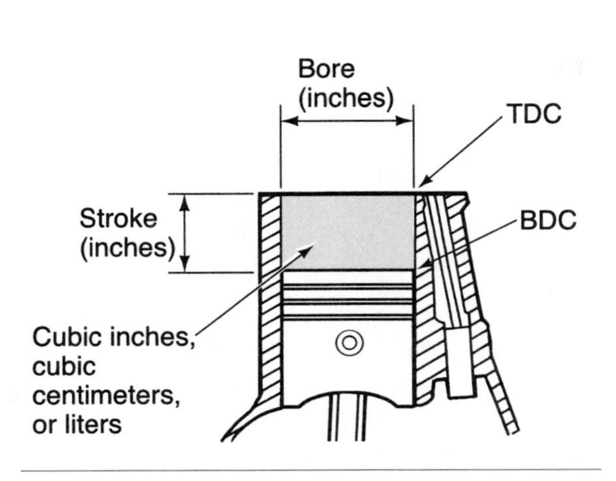


Figure 2-24 Displacement is the volume of the cylinder between TDC and BDC.

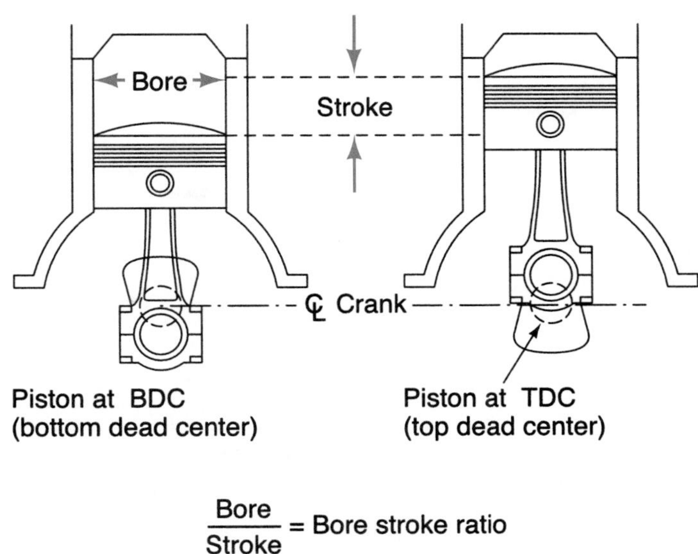


Figure 2-25 The bore and stroke determine the amount of displacement of the cylinder.

required. This meets the requirements of most automotive engines. Most truck engines are designed as undersquare since they deliver more low-end torque.

## Calculating Engine Displacement

Mathematical formulas can be used to determine the displacement of an engine. As with many mathematical formulas, there is more than one method that can be used. The first is as follows:

Engine displacement = 0.785 ´ B2 ´ S ´ N

where 0.785 is a constant

B = bore

S = stroke

N = number of cylinders

The second formula is:

Engine displacement = π ´ R2 ´ S ´ N

where π = 3.1416

R = bore radius (diameter/2)

S = stroke length

N = number of cylinders

Pi times the radius squared is equal to the area of the cylinder cross section. This area is then multiplied by the stroke, and this product is multiplied by the number of cylinders.

To calculate total cubic inch displacement (CID) of a V6 engine with a bore size of 3.800 inches (9.65 cm) and stroke length of 3.400 inches (8.64 cm) would be as follows:

Engine displacement = 0.785 × 3.8002 × 3.400 × 6 = 231.24 CID

or

Engine displacement = 3.1416 × 1.92 × 3.400 × 6 = 231.35 CID

The difference in decimal values is due to rounding pi to four digits. Round the product to the nearest whole number. In this case, the CID would be expressed as 231.

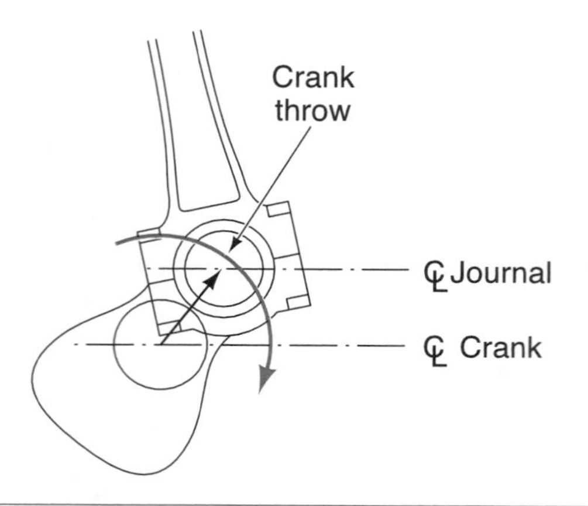


**Figure 2-26** An alternate method of determining the piston stroke. Double the crankshaft throw to determine the stroke.

Most engine manufacturers designate engine size by metric displacement. To calculate engine displacement in cubic centimeters or liters, simply use the metric measurements in the displacement formula. In the preceding example, the result would be 3,789.56 cc, which would be rounded off to 3.8 liters. Another way to convert to metric measurements is to use the CID and multiply it by 16.4 (1 cubic inch is equal to 16.44 cc). Use a service manual to determine the bore and stroke of the engine.

An alternate method of determining the stroke is to measure the amount of crankshaft throw (Figure 2-26). The stroke of the engine is twice the crankshaft throw. Use the bore and stroke to calculate the total displacement of the engine assigned to you. After figuring the total displacement of the engine, review your results with your instructor.

## Direction of Crankshaft Rotation

The Society of Automotive Engineers (SAE) standard for engine rotation is counterclockwise when viewed from the rear of the engine (flywheel side). All automotive engines rotate in this direction. When viewing the engine from the front, the rotation is clockwise.

## Engine Measurements

As discussed previously, engine displacement is a measurement used to determine the output of the engine and can be used to identify the engine. Other engine measurements include compression ratio, horsepower, torque, and engine efficiency. Some of these are also used to identify the engine; for example, an engine with 3.8 L displacement may be available in different horsepower and torque ratings.

The **compression ratio** is a comparison between the volume above the piston at BDC and the volume above the piston at TDC. Compression ratio is the measure used to indicate the amount the piston compresses each intake charge.

### Compression Ratio

The **compression ratio** expresses a comparison between the volume the air/fuel mixture is compressed into at TDC and the total volume of the cylinder (piston at BDC). The ratio is the result of dividing the total volume of the cylinder by the volume with the piston at TDC (Figure 2-27).

Compression ratios affect engine efficiency. Higher compression ratios result in increased cylinder pressures, which compress the air/fuel molecules tighter. This is desirable because it causes the flame to travel faster. The drawback to high compression ratios is the need for higher grades of gasoline and increased emissions.

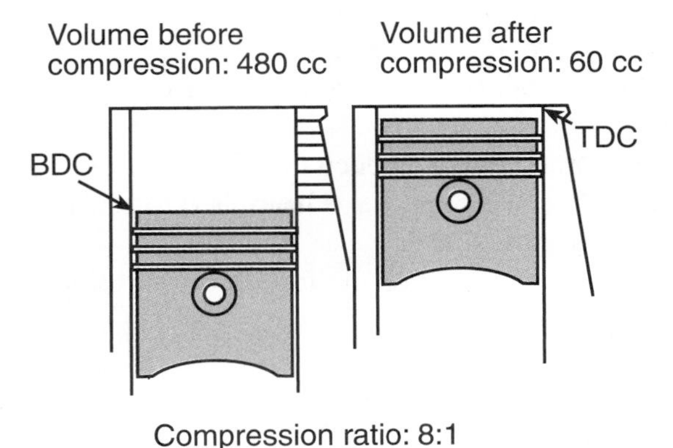


**Figure 2-27** The compression ratio is a measurement of the amount the air/fuel mixture will be compressed.

Theoretically, more power can be produced from a higher compression ratio. An increase in the compression ratio results in a higher degree of compression. Combustion occurs at a faster rate because the molecules of the air/fuel mixture are packed tighter. It is possible to change the compression ratio of an engine by performing some machining operations or changing piston designs. Increasing the compression ratio may increase the power output of the engine, but the higher compression ratio increases the compression temperature. This can cause **preignition** and detonation. To counteract this tendency, higher octane gasoline must be used.

**Shop Manual**
Chapter 2, page 45

**Preignition** is an explosion in the combustion chamber resulting from the air/fuel mixture igniting prior to the spark being delivered from the ignition system.

Most automotive engines have a compression ratio between 8:1 (expressed as eight to one) and 10:1. The usable compression ratio of an engine is determined by the following factors:

- The temperature at which the fuel will ignite.
- The temperature of the air charge entering the engine.
- The density of the air charge.
- Combustion chamber design.

## Calculating Compression Ratio

When calculating the total volume above the piston, the combustion chamber in the cylinder head must also be considered. The formula for calculating compression ratio is:

CR = (VBDC + VCH)/(VTDC + VCH)

where CR = compression ratio

VBDC = volume above the piston at BDC

VCH = combustion chamber volume in the cylinder head

VTDC = volume above the piston at TDC

Example: Volume above the piston at BDC = 56 cubic inches

Combustion chamber volume = 4.5 cubic inches

Volume above the piston at TDC = 1.5 cubic inches

The compression ratio would be:

$$\frac{56 + 4.5}{1.5 + 4.5} = 10:1$$

## Horsepower and Torque

Torque is a mathematical expression for rotating or twisting force around a pivot point. As the pistons are forced downward, this pressure is applied to a crankshaft that rotates. The crankshaft

**Horsepower** is the measure of the rate of work.

**Brake horsepower** is the usable power produced by an engine. It is measured on a dynamometer using a brake to load the engine.

**Indicated horsepower** is the amount of horsepower the engine can theoretically produce.

**Gross horsepower** is the maximum engine output as measured on a dynamometer.

**Net horsepower** is the maximum engine horsepower as measured on a dynamometer.

transmits this torque to the drivetrain and ultimately to the drive wheels. **Horsepower** is the rate at which an engine produces torque.

To convert terms of force applied in a straight line to force applied rotationally, the formula is: torque = force × radius

If a 10-pound force is applied to a wrench one foot in length, 10 foot-pounds (ft.-lb.) of torque is produced. If the same 10-pound force is applied to a wrench two feet in length, the torque produced is 20 ft.-lb. (Figure 2-28).

Horsepower can be determined once the torque output of an engine at a given rpm is known. Use the following formula:

Horsepower = torque – rpm/5,252

Torque and horsepower will peak at some point in the rpm range (Figure 2-29). This graph is a mathematical representation of the relationship of torque and horsepower in one engine. The graph indicates this engine torque peaks at about 1,700 rpm, while brake horsepower peaks at about 3,500 rpm. The third line in the graph represents the amount of horsepower required to overcome internal resistances or friction in the engine. The amount of **brake horsepower** is less than **indicated horsepower** due to this and other factors.

The Society of Automotive Engineers (SAE) standards for measuring horsepower include ratings for gross and net horsepower. **Gross horsepower** expresses the maximum power developed by the engine without additional accessories operating. These accessories include the air cleaner and filter, cooling fan, charging system, and exhaust system. **Net horsepower** expresses the amount of horsepower the engine develops as installed in the vehicle. Net horsepower is about 20 percent lower than gross horsepower. Most manufacturers express horsepower as SAE net horsepower.

For example, if the torque of the engine is measured at the following values, a graph can be plotted to show the relationship between torque and horsepower (Figure 2-30):

Torque output @ 1,500 rpm = 105 lb.-ft.

Torque output @ 2,800 rpm = 110 lb.-ft.

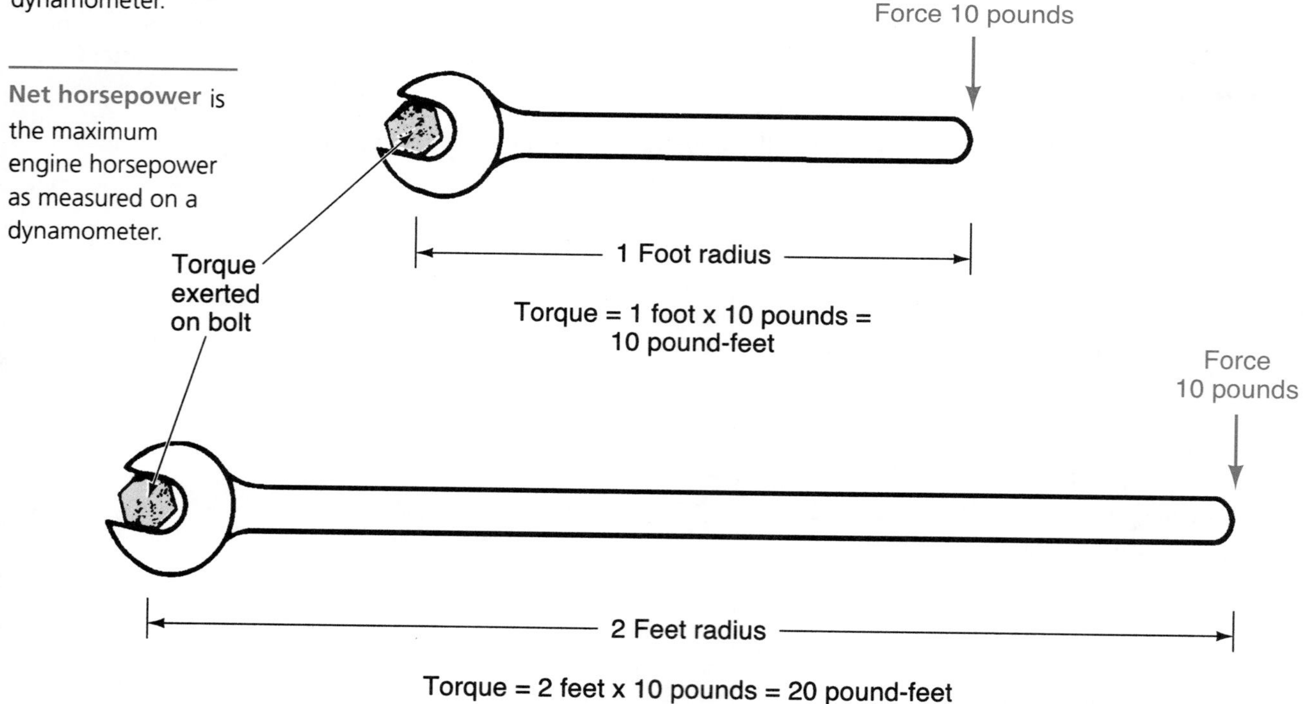


**Figure 2-28** Example of how torque can be increased.

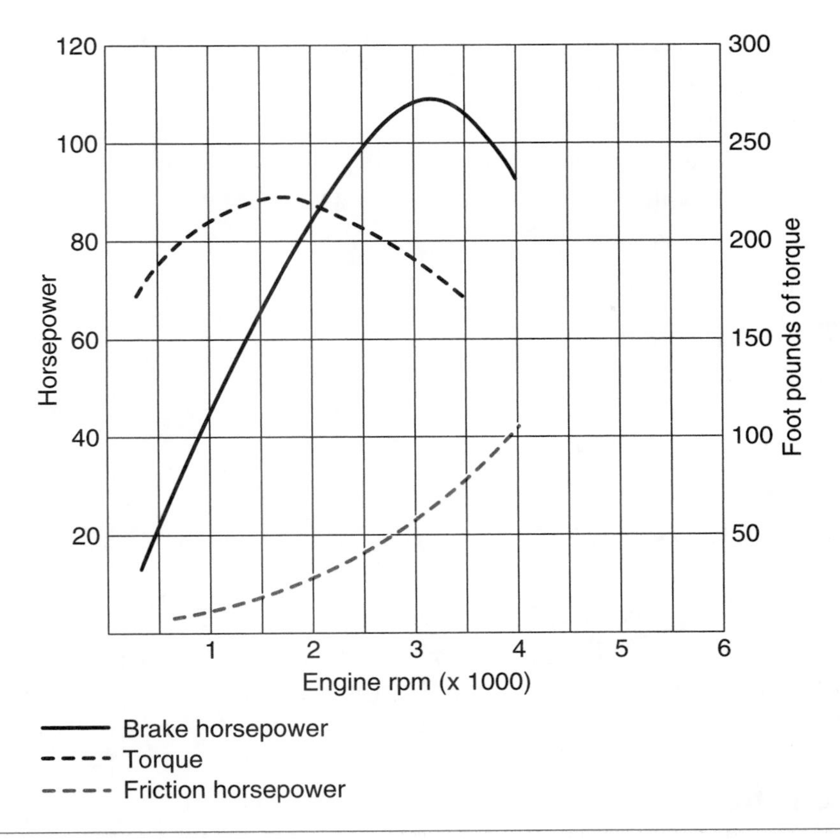


**Figure 2-29** Example of torque and horsepower curves.

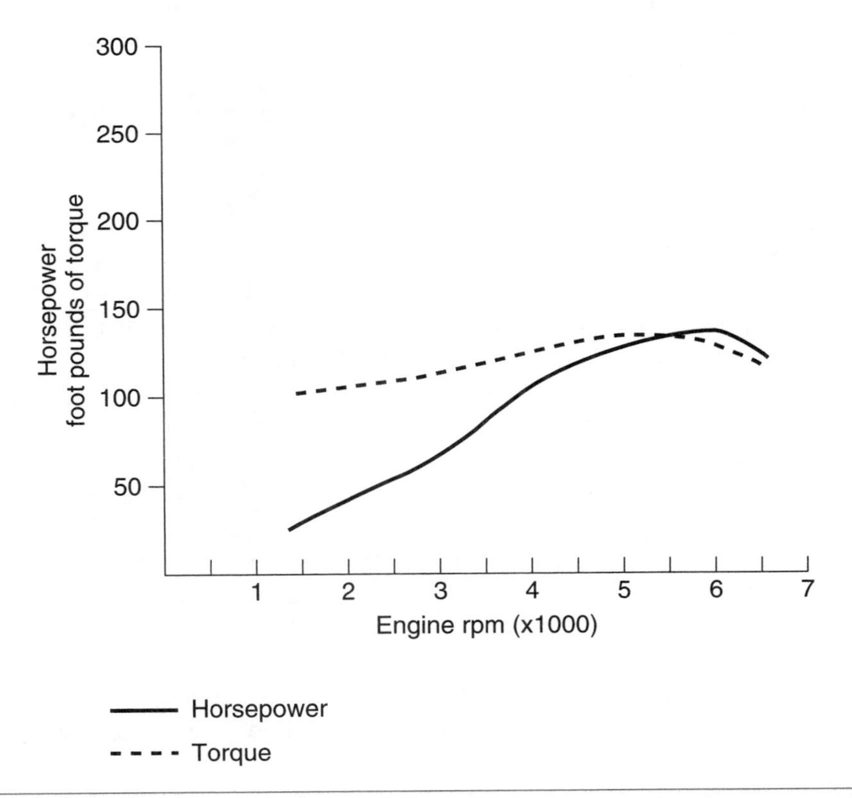


**Figure 2-30** It is possible to graph the relationship between torque and horsepower. Highest torque output is usually at lower engine speeds, while peak horsepower is at higher engine speeds.

Torque output @ 3,500 rpm = 115 lb.-ft.

Torque output @ 5,000 rpm = 130 lb.-ft.

Torque output @ 6,000 rpm = 125 lb.-ft.

Torque output @ 6,500 rpm = 100 lb.-ft.

In this instance, the horsepower is as follows:

Horsepower @ 1,500 rpm = 30

Horsepower @ 2,800 rpm = 59

Horsepower @ 3,500 rpm = 77

Horsepower @ 5,000 rpm = 124

Horsepower @ 6,000 rpm = 143

Horsepower @ 6,500 rpm = 124

The formula for determining volumetric efficiency is:

Volumetric efficiency = aov/aip

where aov = actual air output volume

aip = maximum possible air input volume

An example for calculating the volumetric efficiency is an engine with 350 CID displacement that actually has an air/fuel volume of 341 inches. The actual amount digested is less than the available space, and this engine is 97 percent efficient.

## Engine Efficiency

**Efficiency** is a measure of a device's ability to convert energy into work.

**Mechanical efficiency** is a comparison of the power actually delivered by the crankshaft to the power developed within the cylinders at the same rpm.

**Volumetric efficiency** is the measurement of the amount of air/fuel mixture that actually enters the combustion chamber compared to the

There are several different measurements used to describe the **efficiency** of an engine. Efficiency is mathematically expressed as output divided by input. The terms used to define the input and output must be the same. Some of the most common efficiencies of concern are mechanical, volumetric, and thermal efficiencies.

**Mechanical efficiency** is a comparison between brake horsepower and indicated horsepower; in other words, it is a ratio of the power actually delivered by the crankshaft to the power developed within the cylinders at the same rpm. The formula used to calculate mechanical efficiency is:

Mechanical efficiency = brake horsepower/indicated horsepower

Ideally, mechanical efficiency would be 100 percent. However, the power delivered will always be less due to power lost in overcoming friction between moving parts in the engine. The mechanical efficiency of an internal combustion four-stroke engine is approximately 90 percent. If the engine's brake horsepower is 120 hp and the indicated horsepower is 140 hp, the mechanical efficiency of the engine is 85.7 percent.

**Volumetric efficiency** is a measurement of the amount of air/fuel mixture that actually enters the combustion chamber compared to the amount that could be ingested at that speed. As the volumetric efficiency increases, the power developed by the engine increases proportionately. Ideally, 100 percent volumetric efficiency is desired, but actual efficiency is reduced due to pumping losses. It must be realized that a given mass of air/fuel mixture occupies different volumes under different conditions. Atmospheric pressures and temperatures will affect the volume of the mixture.

Pumping losses are the result of restrictions to the flow of air/fuel mixture into the engine. These restrictions are the result of intake manifold passages, throttle plate opening, valves, and cylinder head passages. There are some machining techniques used by builders of performance engines that will reduce these restrictions. Pumping losses are influenced by engine speed; for example, an engine may have a volumetric efficiency of 75 percent at 1,000 rpm and increase to an efficiency of 85 percent at 2,000 rpm, then drop to 60 percent at 3,000 rpm. As engine speed increases, volumetric efficiency may drop to 50 percent.

If the airflow is drawn in at a low engine speed, the cylinders can be filled close to capacity. A required amount of time is needed for the airflow to pass through the intake manifold and pass the intake valve to fill the cylinder. As engine speed increases, the amount of time the intake valve is open is not long enough to allow the cylinder to be filled. Because of the effect of engine speed on volumetric efficiency, engine manufacturers use intake manifold design, valve port design, camshaft timing, and exhaust tuning to improve the engine's breathing. The use of turbochargers and superchargers, which increase the pressure in the engine's intake system, will bring volumetric efficiency to over 100 percent.

**Thermal efficiency** is a measurement comparing the amount of energy present in a fuel and the actual energy output of the engine. Thermal efficiency is measured in British thermal units (Btu). The formula used to determine a gasoline engine's thermal efficiency has a couple of constants. First, 1 hp equals 42.4 Btus per minute. Second, gasoline has approximately 110,000 Btus per gallon. Knowing this, the formula is as follows:

**Thermal efficiency** is the difference between potential and actual energy developed in a fuel measured in Btus per pound or gallon.

Thermal efficiency = (bhp – 42.4 Bpmin)/(110,000 Bpg – gpmin)

where bhp = brake horsepower

Bpmin = Btu per minute

Bpg = Btu per gallon

gpmin = gallons used per minute

As a whole, gasoline engines waste about two-thirds of the heat energy available in gasoline. Approximately one-third of the heat energy is carried away by the engine's cooling system. This is required to prevent the engine from overheating. Another third is lost in hot exhaust gases. The remaining third of heat energy is reduced by about 5 percent due to friction inside of the engine. Another 10 percent is lost due to friction in drivetrain components. Due to all of the losses of heat energy, only about 19 percent is actually applied to the driving wheels.

Of the 19 percent applied to the driving wheels, an additional amount is lost due to the rolling resistance of the tires against the road. Resistance to the vehicle moving through the air requires more of this energy. By the time the vehicle is actually moving, the overall vehicle efficiency is about 15 percent.

# Other Engine Designs

## Two-Stroke Engines

Throughout the history of the automobile, the gasoline four-stroke internal combustion engine has been the standard. As technology has improved, other engine designs have been used. This section of the chapter provides an overview of these other engine designs.

A **reed valve** is a one-way check valve. The reed opens to allow the air/fuel mixture to enter from one direction, while closing to prevent movement in the other direction.

Two-stroke engines are capable of producing a power stroke every revolution. Valves are not used in most two-stroke engines; instead, the piston movement covers and uncovers intake and exhaust ports. The air/fuel mixture enters the crankcase through a **reed valve** or **rotary valve** (Figure 2-31). Upward movement of the piston causes pressure to increase in the cylinder above the piston, and creates a slight vacuum below the piston. Atmospheric pressure pushes the air/fuel mixture into the low-pressure area below the piston. At the same time, a previously ingested air/fuel mixture is being compressed above the piston. This action combines the compression stroke and intake stroke.

When the compressed air/fuel mixture is ignited, the piston moves down the cylinder. This slightly compresses the air/fuel mixture in the crankcase. The air/fuel mixture in the crankcase is directed to the top of the piston through the transfer port to the intake port. The reed valve or rotary valve closes to prevent the mixture from escaping from the crankcase. The downward piston movement opens the exhaust port, and the pressure of the expanding gases pushes out the spent emissions. Continued downward movement of the piston uncovers the intake port and allows the air/fuel mixture to enter above the piston. This action combines the power and exhaust strokes.

A **rotary valve** is a valve that rotates to cover and uncover the intake port. A rotary valve is usually designed as a flat disc that is driven

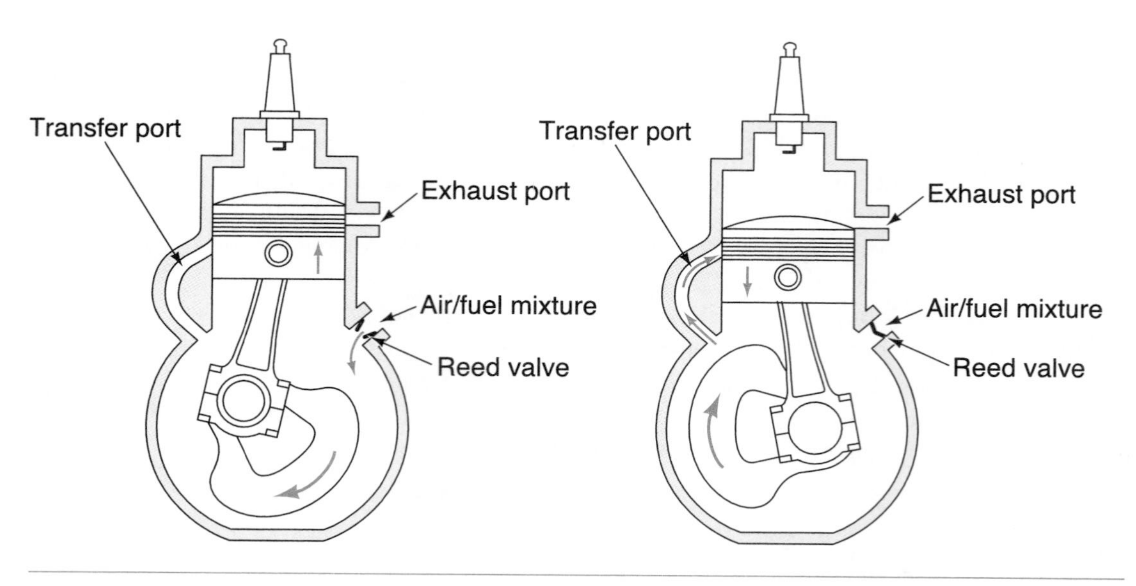


Figure 2-31 Two-stroke cycle engine design.

Most gasoline two-stroke engines do not use the crankcase as a sump for oil since it is used for air/fuel intake. To provide lubrication to the engine, the oil must be mixed with the fuel or injected into the crankcase. This results in excessive exhaust emissions. Diesel-design two-stroke engines may have a sump. Many automotive engine manufacturers are trying to find ways to reduce the emission levels of the two-stroke engine to allow for its use in automobiles. The advantage of the two-stroke engine is that it produces more power per cubic inch of displacement than the four-stroke engine.

## Diesel Engines

Diesel and gasoline engines have many similar components; however, the diesel engine does not use an ignition system consisting of spark plugs and coils. Instead of using a spark delivered by the ignition system, the diesel engine uses the heat produced by compressing air in the combustion chamber to ignite the fuel. Fuel injectors are used to supply fuel into the combustion chamber. The fuel is sprayed, under pressure, from the injector as the piston is completing its compression stroke. The temperature increase generated by compressing the air (approximately 1,000°F) is sufficient to ignite the fuel as it is injected into the cylinder. This begins the power stroke (Figure 2-32).

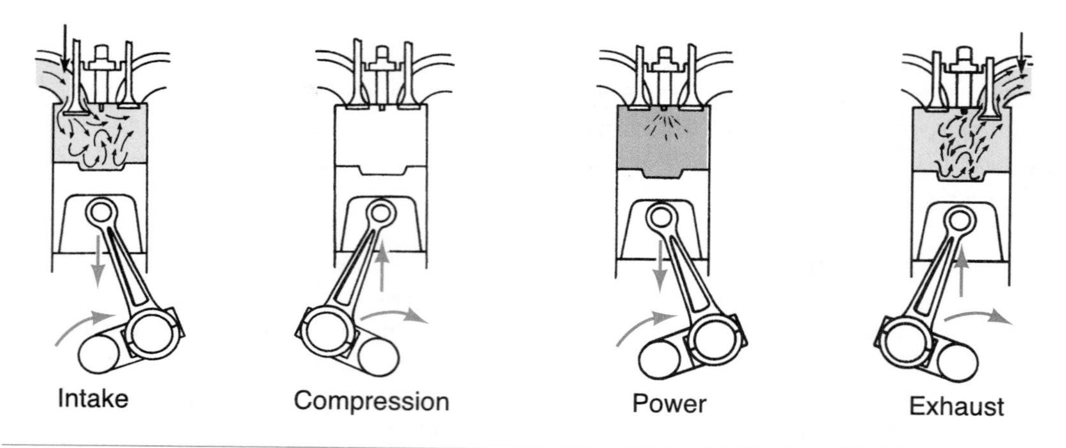


Figure 2-32 The four strokes of a diesel engine.

Since starting the diesel engine is dependent on heating the intake air to a high enough level to ignite the fuel, a method of preheating the intake air is required to start a cold engine. Some manufacturers use **glow plugs** to accomplish this. Another method includes using a heater grid in the air intake system.

**Glow plugs** are threaded into the combustion chamber and use electrical current to heat the intake air on diesel engines

In addition to ignition methods, there are other differences between gasoline and diesel engines (Figure 2-33). The combustion chambers of the diesel engine are designed to accommodate the different burning characteristics of diesel fuel. There are three common combustion chamber designs:

An open-type combustion chamber is located directly inside of the piston. The fuel is injected directly into the center of the chamber. Turbulence is produced by the shape of the chamber.

A precombustion chamber is a smaller, second chamber connected to the main combustion chamber. Fuel is injected into the precombustion chamber, where the combustion process is started. Combustion then spreads to the main chamber. Since the fuel is not injected directly on top of the piston, this type of diesel injection may be called indirect injection.

A turbulence combustion chamber is a chamber designed to create turbulence as the piston compresses the air. When the fuel is injected into the turbulent air, a more efficient burn is achieved.

Diesel engines can be either four-stroke or two-stroke designs. Two-stroke engines complete all cycles in two strokes of the piston, much like a gasoline two-stroke engine (Figure 2-34).

## Miller Cycle Engine

The Miller-cycle engine is a modification of the four-stroke cycle engine. Some late-model vehicles such as the Mazda Millennia are powered by a Miller-cycle engine. In this type of engine, a supercharger supplies highly compressed air through an intercooler into the combustion cylinders. The intake valves remain open longer compared to a conventional four-stroke cycle engine. This action prevents the upward piston movement from compressing the air/fuel mixture in the cylinder until the piston has moved one-fifth of the way upward on the compression stroke. The later valve closing

| | Gasoline | Diesel |
|---|---|---|
| Intake | Air/fuel | Air |
| Compression | 8–10 to 1<br>130 psi<br>545°F | 13–25 to 1<br>400–600 psi<br>1,000°F |
| Air/fuel mixing point | Carburetor or before intake valve with fuel injection | Near TDC by injection |
| Combustion | Spark ignition | Compression ignition |
| Power | 464 psi | 1,200 psi |
| Exhaust | 1,300°–1,800°F<br>CO = 3% | 700°–900°F<br>CO = 0.5% |
| Efficiency | 22–28% | 32–38% |

Figure 2-33 Comparisons between gasoline and diesel engines.

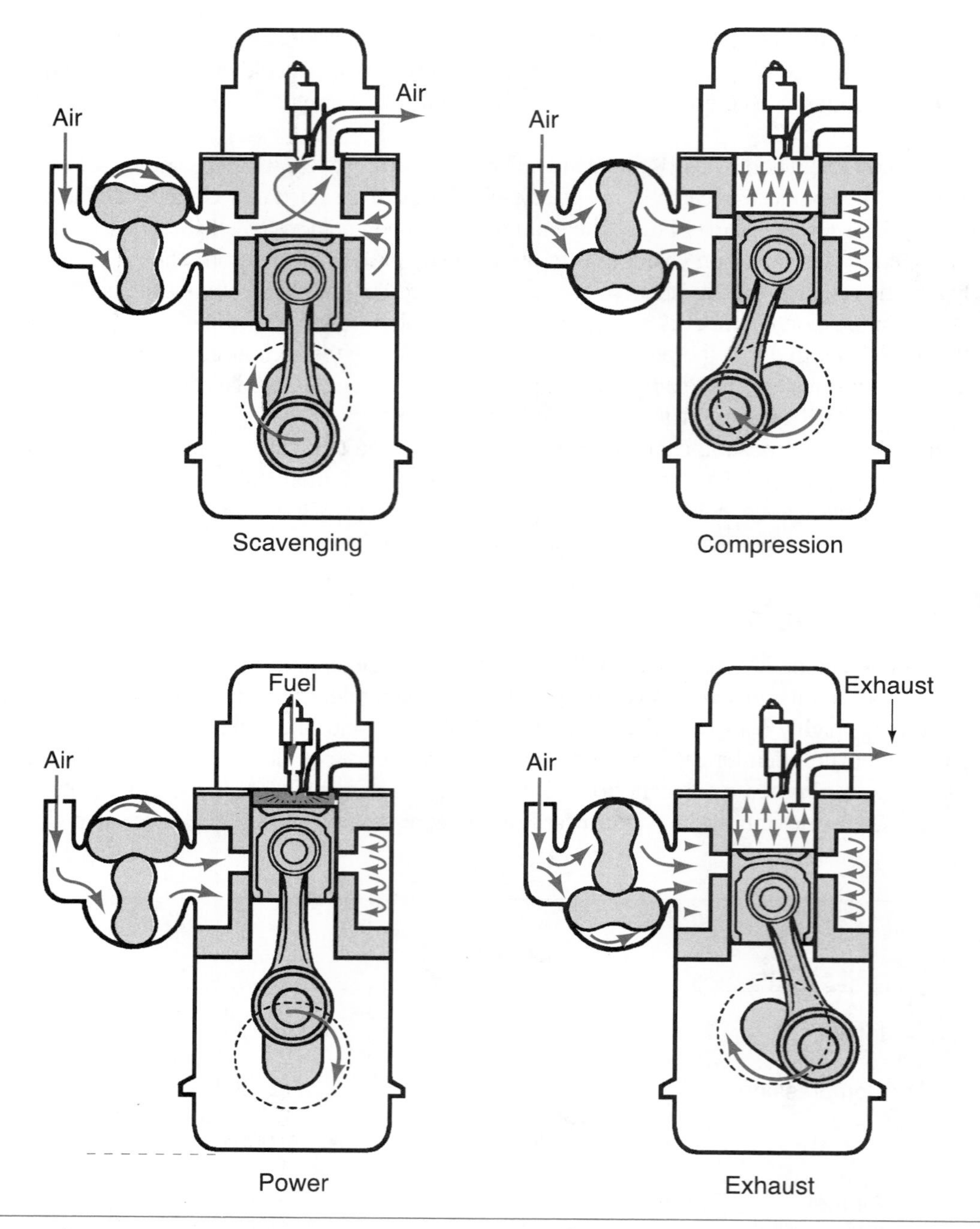


**Figure 2-34** Some diesel engines are two-stroke engines.

allows the supercharger to force more air into the cylinder. The shorter compression stroke lowers cylinder temperatures. The greater volume of air in the cylinder creates a longer combustion time, which maintains downward pressure on the piston for a longer time on the power stroke. This action increases engine torque and efficiency. The Miller-cycle principle allows an engine to produce a high horsepower and torque for the cubic inch displacement (CID) of the engine.

## Stratified Charge Engine

The stratified charge engine uses a precombustion chamber to ignite the main combustion chamber (Figure 2-35). The advantage of this system is increased fuel economy and reduced emission levels.

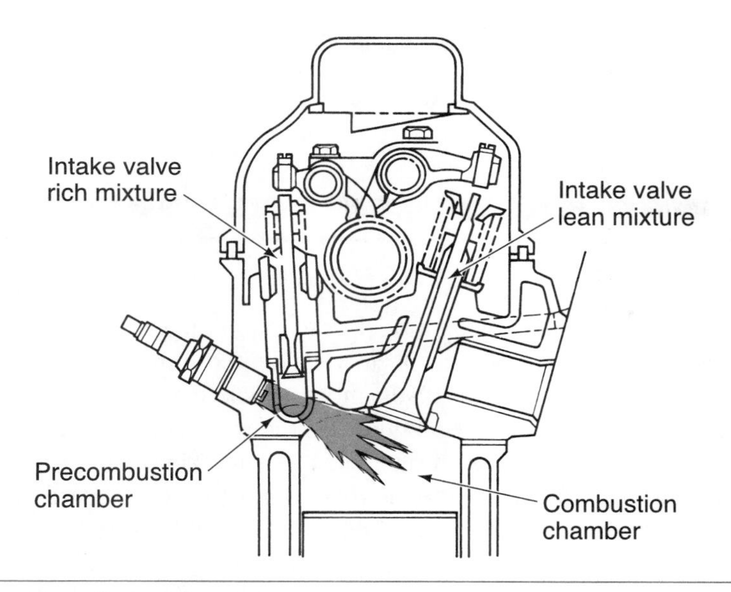


**Figure 2-35** Stratified engine design.

In this engine type, the air/fuel mixture is stratified to produce a small rich mixture at the spark plug while providing a lean mixture to the main chamber. During the intake stroke, a very lean air/fuel mixture enters the main combustion chamber. At the same time, a very rich mixture is pushed past the auxiliary intake valve and into the precombustion chamber (Figure 2-36). At

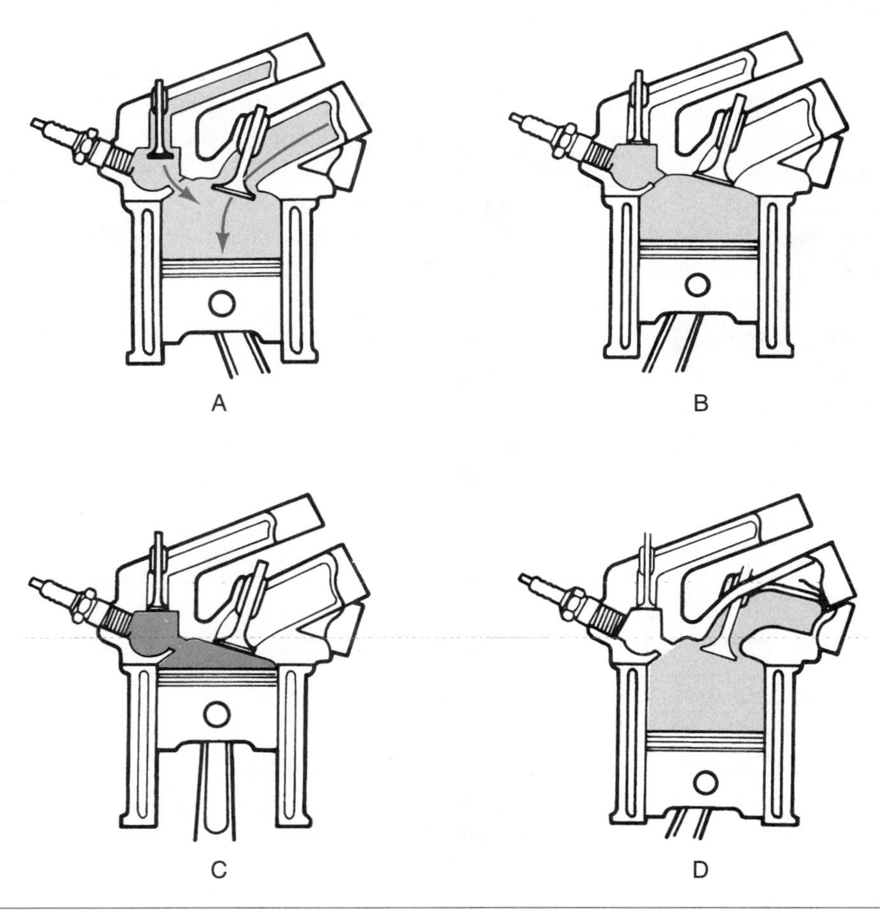


**Figure 2-36** The stratified engine uses the rich mixture in the precombustion chamber to ignite the lean mixture in the main chamber.

the completion of the compression stroke, the spark plug fires to ignite the rich mixture in the precombustion chamber. The burning rich mixture then ignites the lean mixture in the main combustion chamber.

# Vehicles with Alternate Power Sources

## Electric Vehicles

In this section, our objective is to provide a brief description of several alternate vehicle power sources. These power sources are being sold in very limited numbers, or they are still in the research and development stage. In the 1990s, most vehicle manufacturers began developing electric vehicles, and these developments continue in the twenty-first century. The main reason for this action was more stringent emission standards. As emission standards become more stringent, it becomes increasingly difficult and expensive to meet these standards with gasoline and diesel engines. The California Air Resources Board (CARB) established a low-emission vehicles/clean fuel program to further reduce mobile source emissions in California during the late 1990s. In this program, emission standards are established for five vehicle types: conventional vehicle (CV), transitional low emission vehicle (TLEV), low emission vehicle (LEV), ultra low emission vehicle (ULEV), and zero emission vehicle (ZEV) (Figure 2-37). The CARB has developed a sales-weighting and emissions credit system for introducing the TLEV, LEV, ULEV, and ZEV vehicles into the California market. With present technology, ZEV standards cannot be achieved with a gasoline-powered vehicle. The electric vehicle does meet ZEV standards, and so there is a need for this type of vehicle.

General Motors introduced the EV1 electric car to the market in 1996. The original battery pack in this car contained twenty-six 12V batteries that delivered electrical energy to a three-phase 102 kilowatt (kW) AC electric motor. This motor drives the front wheels. The driving range before the batteries become discharged is 70 miles in city driving or 90 miles of highway driving. A 1.2kW charger in the vehicle trunk will recharge the batteries in 15 hours. If the batteries are charged with an external 6.6 kW charger operating on a 220V/30 ampere charging source, the batteries can be recharged in 3 hours. The EVI accelerates from 0 to 60 mph (97 kmh) in 9 seconds, and a top speed of 80 mph (130 kmh) may be obtained. Equipment on the EVI includes dual air bags, antilock brakes, CD player, cruise control, power steering, power windows, and an air-conditioning system using a heat pump principle. This car is equipped with a Galileo electronic brake system that employs a computer and sensors at each wheel to direct power assist braking, regenerative/friction brake blending, four-wheel antilock braking, traction assist, tire pressure monitoring, and system diagnostics. In 1998, nickel/metal/hydride batteries were installed in EV1 vehicles, which extended the driving range between battery charges to 160 miles (257 km).

| | CV | TLEV | LEV | ULEV | ZEV |
|---|---|---|---|---|---|
| NMOG | 0.25[c] | 0.125 | 0.075 | 0.040 | 0.0 |
| CO | 3.4 | 3.4 | 3.4 | 1.7 | 0.0 |
| $NO_x$ | 0.4 | 0.4 | 0.2 | 0.2 | 0.0 |

(a) Higher (less stringent) standards were established at 100,000 miles.
(b) NMOG (non-methane organic gases) are NMHC + ketones + aldehydes + alcohols.
(c) Emission standards of NMHC.

Figure 2-37 California tailpipe emission standards in g/mile for five-passenger car vehicle types at 50,000 miles. (Courtesy of the American Council for an Energy-Efficient Economy)

## Hybrid Vehicles

In 1993 President Clinton signed the Partnership for a New Generation of Vehicles (PNGV). This signing initiated a cooperative program between the automotive manufacturers and government agencies to develop a new generation of cars by the year 2004 with the following criteria:

1. Fuel economy of 80 mpg of gasoline or gasoline equivalent
2. Acceleration from 0 to 60 mph in 12 seconds
3. Seating capacity of five to six passengers
4. Luggage capacity of 475 liters
5. Exhaust emissions in grams per mile (gpm), hydrocarbon (HC) 0.125, carbon monoxide (CO) 1.7, nitrous oxides ($NO_x$) 0.2
6. Comfort, gradability, ride, handling, noise level, and vibration equivalent to a 1994 sedan

The PNGV requirements present a challenge for automotive manufacturers, especially the fuel economy requirement. Producing hybrid vehicles is one way of meeting the PNGV fuel economy requirements.

A hybrid vehicle has two different power sources. In most hybrid vehicles, the power sources are a small displacement gasoline or diesel engine and an electric motor. Toyota is developing a hybrid vehicle with a 1.5L dual overhead cam (DOHC), gasoline direct-injection four-cylinder engine (Figure 2-38). In a direct-injection engine, the injectors deliver the gasoline directly into the combustion chamber. This type of injection system must operate at higher pressure so the fuel can be discharged out of the injectors into the high compression pressure in the combustion chambers. The engine is mounted transversely under the hood. The nickel/metal/hydride battery pack is installed behind the rear seat, leaving adequate storage room in the trunk (Figure 2-39). The generator/starter, power split device, propulsion motor/regenerator, and sprocket chain to the final drive are mounted in an aluminum case that is bolted to the back of the engine. This case is about the same length as a conventional four-speed automatic transaxle (Figure 2-40). An automatic controller operates the system and determines whether power to the drive wheels is supplied from the electric motor, the gasoline engine, or both. Charging the batteries with an external charger is not necessary because the batteries are charged while the engine is running. The system operation may be summarized as follows:

1. During startup, low speed, and low speed deceleration, only the electric propulsion motor supplies power to the drive wheels.
2. During the normal speed range, power from the engine is divided by the power split device so that some of the engine power is supplied to the drive wheels, and engine power is also supplied to the generator. The electric power supplied from the generator drives the electric propulsion motor, and power from this motor is also supplied to the drive wheels.

Figure 2-38 Hybrid power system. (Reprinted with permission)

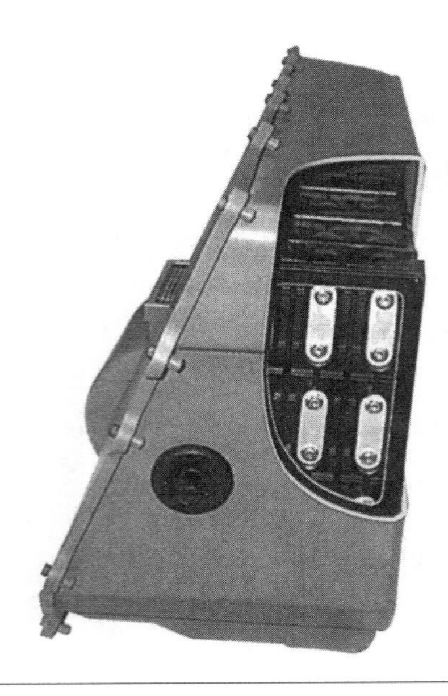

**Figure 2-39** Nickel/metal/hydride battery pack. (Reprinted with permission)

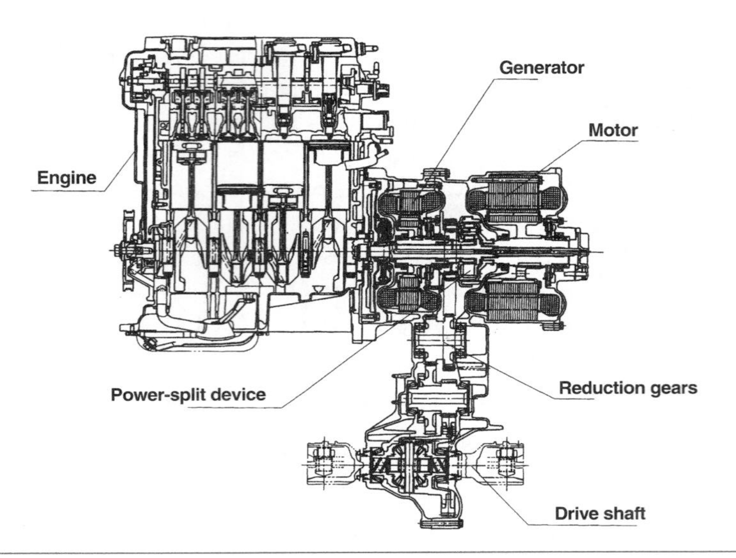


**Figure 2-40** Hybrid power system with 1.5L gasoline engine and electric propulsion motor. (Reprinted with permission)

3. Full-throttle operation is the same as normal speed driving operation except the battery also supplies power to the propulsion motor to maximize motor output to the drive wheels.
4. During deceleration, the wheels drive the propulsion motor, and this motor supplies power to recharge the batteries.
5. The battery state of charge is continually monitored by the controller, and power is supplied from the generator to recharge the battery whenever this action is required.
6. When the vehicle is brought to a stop, the engine is shut off.

A specially designed compact sedan is being developed for the hybrid drive system. The hybrid system is designed to meet ULEV emission standards, and fuel economy should meet PNGV requirements. The Honda Insight (a hybrid car) has recently been introduced in Honda dealerships.

## Fuel Cell–Powered Vehicles

A fuel cell–powered vehicle is an electric vehicle with some important differences. An electric motor supplies torque to the drive wheels, but the fuel cell produces and supplies electric power to the electric motor, whereas in an electric vehicle, this power is supplied by the batteries. Most of the vehicle manufacturers, in cooperation with some independent laboratories, are involved in fuel cell research and development programs. A number of prototype fuel cell vehicles have been produced. Several vehicle manufacturers are committed to having a fuel cell vehicle available in the dealership showrooms by the year 2004.

Fuel cells electrochemically combine oxygen from the air with hydrogen from a hydrocarbon fuel to produce electricity. The oxygen and hydrogen are fed to the fuel cell as "fuel" for the electrochemical reaction. There are different types of fuel cells, but the most common type is the proton exchange membrane (PEM) fuel cell. Each individual fuel cell contains two electrodes separated by a membrane. Electrical power output from one individual fuel cell is very low, and so many fuel cells are contained in a fuel cell stack to supply enough electric energy to the electric drive motor (Figure 2-41). Hydrogen is supplied to one of the electrodes in each fuel cell. This electrode is coated with a catalyst that separates the electrons and protons in the hydrogen. The movement of electrons supplies electrical power to the drive motor. The protons move through the proton exchange membrane to the other electrode. Oxygen from the air is supplied to this electrode, and oxygen from the air combines with the hydrogen protons to form water vapor, which is emitted from the vehicle (Figure 2-42). The only emission from a fuel cell is water vapor ($H_2O$).

One of the major areas of research and development is the source of hydrogen fuel in fuel cell vehicles. One solution is to store hydrogen on-board the vehicle. This concept has several complications. Using hydrogen fuel would require a whole new refueling system in all areas of the country, which is very expensive. Gaseous hydrogen can be stored in large cylinders containing a hydride material something like steel wool. These cylinders would require a large storage space on the vehicle. Liquid hydrogen must be refrigerated to approximately −400°F to keep it in liquid form. Storing liquid hydrogen on a vehicle involves the use of a special double-walled insulated fuel tank. Storing liquid hydrogen on a vehicle also involves some safety concerns, because as the fuel tank warms up, the pressure increases, and this may activate the pressure relieve valve. This action discharges flammable hydrogen into the atmosphere, creating a source of danger and pollution.

An on-board reformer may be used to extract hydrogen from liquid fuels such as gasoline or methanol. The main disadvantage to this system is the space required by the reformer. Using methanol would also require a new refueling system across the country. Recent developments include a multi-reformer that operates on different types of fuels. Hydrogen can be obtained by electrolysis from water; however, this process requires a lot of electrical energy. With present

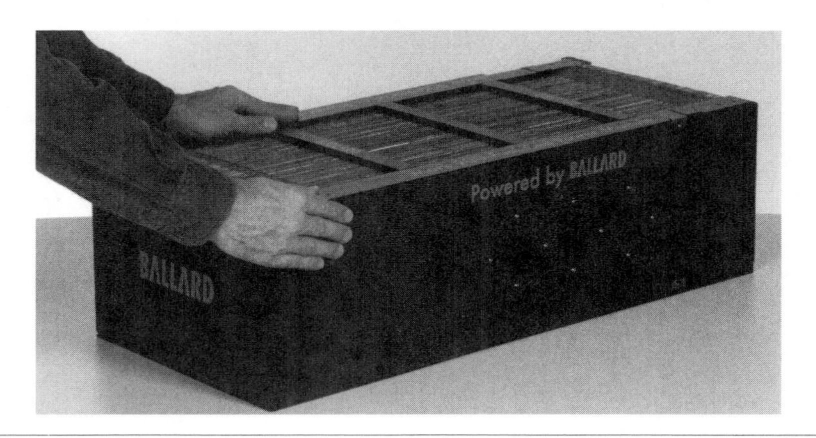


Figure 2-41 Fuel cell stack. (Courtesy of Ballard Power Systems)

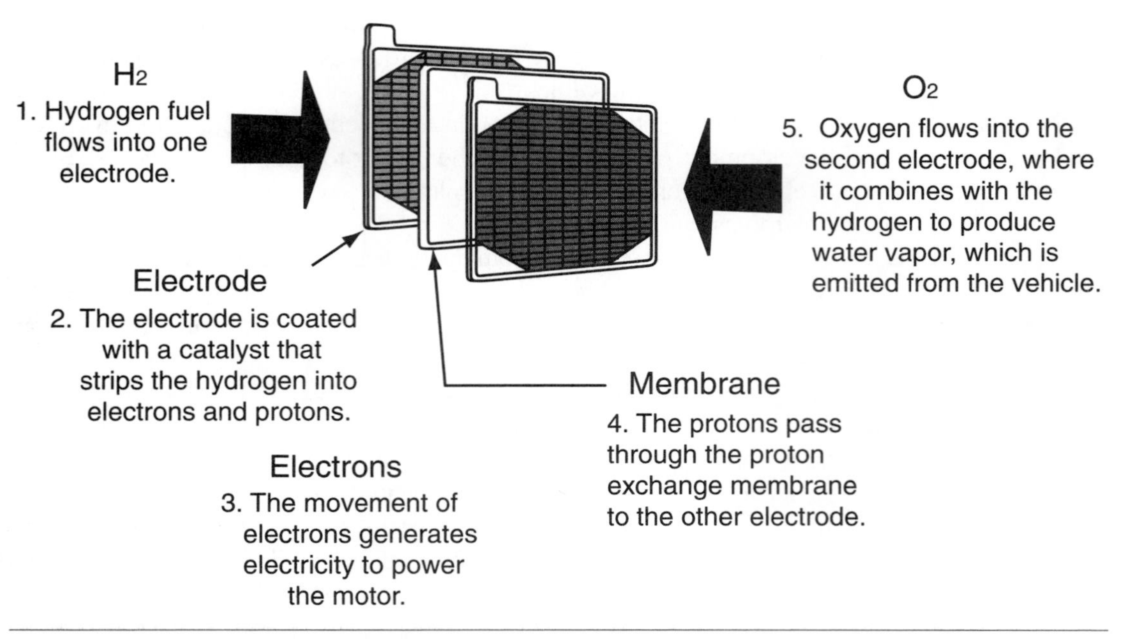


**Figure 2-42** Fuel cell operation.

technology, this method of obtaining hydrogen is not an option. It appears that the first fuel cell vehicles offered to customers will have on-board reformers that extract hydrogen from some well-known fuel. Nissan's fuel cell vehicle has an on-board reformer that extracts hydrogen from methanol (Figure 2-43). The two major obstacles to be overcome in the development of fuel cell vehicles are cost and the on-board space required by the system components. Research and development is taking place at a rapid pace, and components are quickly being downsized and improved. As production of fuel cell vehicle components increases, the price will be reduced.

**AUTHOR'S NOTE:** In the last 30 years, I have seen many significant changes in automotive technology. Engines have gone from no electronic control to complete electronic control of ignition, fuel, and emission systems. We have just experienced the

**Figure 2-43** Fuel cell car with on-board methanol reformer. (Courtesy of Nissan North America, Inc.)

introduction of hybrid cars in some dealership showrooms, and these vehicles will increase in numbers as more hybrid cars, sport utility vehicles (SUVs), and light-duty trucks are manufactured in the next two to five years. In the next five to ten years, we may see the internal combustion engine gradually replaced by electric-drive vehicles powered by fuel cells. The major question with the introduction of fuel cell vehicles is what fuel will be used to supply hydrogen for the fuel cell. Will an on-board reformer extract hydrogen from gasoline or methanol, or will hydrogen be stored on board to supply the fuel cell? Engineers generally agree that a hydrogen refueling infrastructure is still 10 to 20 years in the future.

## Engine Identification

Proper engine identification is required for the technician and/or machinist to obtain correct specifications, service procedures, and parts. The vehicle identification number (VIN) is one source for information concerning the engine. The VIN is located on a metal tag that is visible through the vehicle's windshield (Figure 2-44). The VIN provides the technician with a variety of information concerning the vehicle (Figure 2-45). The eighth digit represents the engine code. The service manual provides information concerning interpretation of the engine code.

A vehicle code plate is also located on the vehicle body, usually inside the driver's side door or under the hood. This plate provides information concerning paint codes, transmission codes, and so forth (Figure 2-46). The engine sales code is also stamped on the vehicle code plate.

The engine itself may have different forms of identification tags or stamped numbers (Figure 2-47). In addition, color coding of engine labels may be used to indicate transitional low emission vehicles (TLEV) and flex fuel vehicles (FFV). These numbers may provide build date codes required to order proper parts. Use the service manual to determine the location of any engine identification tags or stamps and the methods to interpret them; for example: a Jeep engine with the build date code of *401HX23* identifies a 2.5L engine with a multipoint fuel-injection system, 9:1 compression ratio, built on January 23, 1994. This is determined by the first digit representing the year of manufacture (4 = 1994). The second and third digits represent the month of manufacture (01 = January). The fourth and fifth digits represent the engine type, fuel system, and compression ratio (HX = 2.5L engine with a multipoint fuel-injection system, 9:1 compression ratio). The sixth and seventh digits represent the day of manufacture (23).

**Figure 2-44** The VIN number can be see through the windshield.

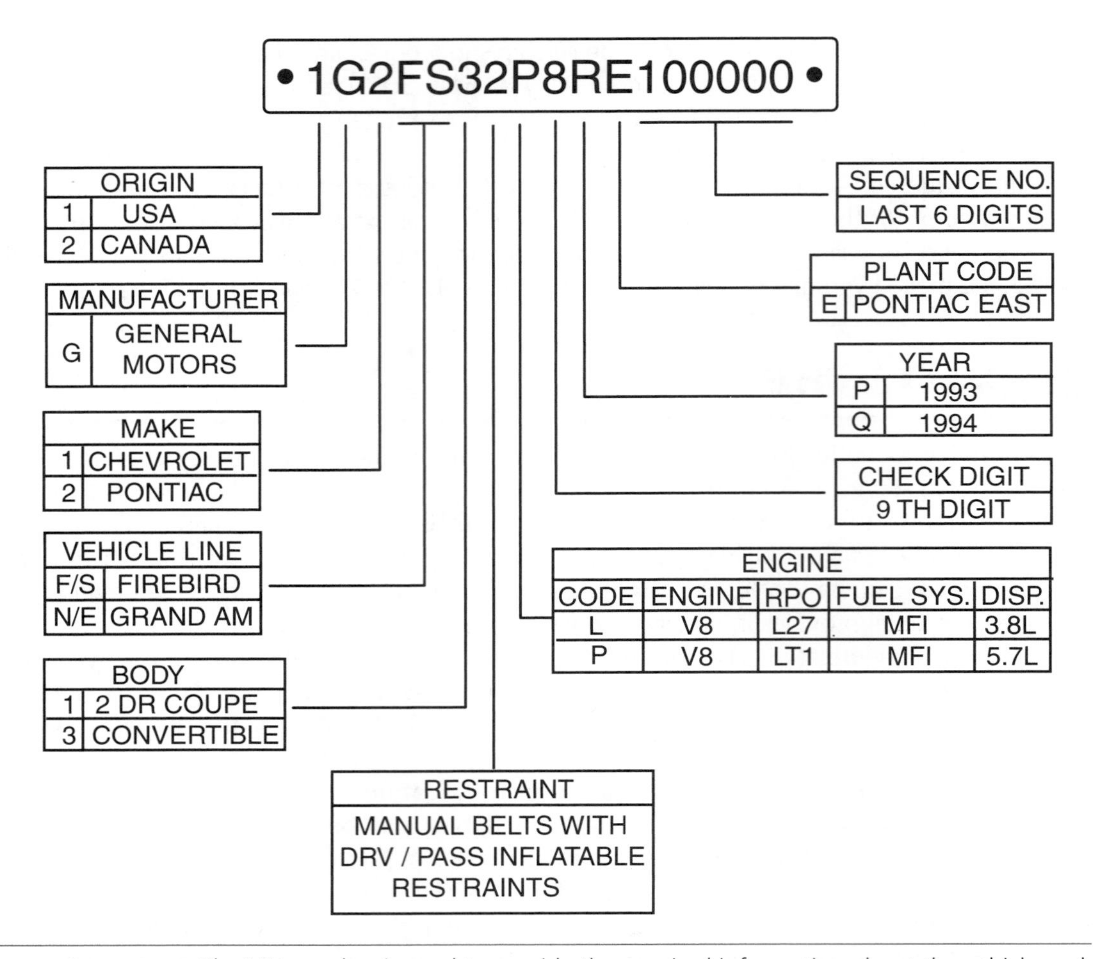


**Figure 2-45** The VIN number is used to provide the required information about the vehicle and engine.

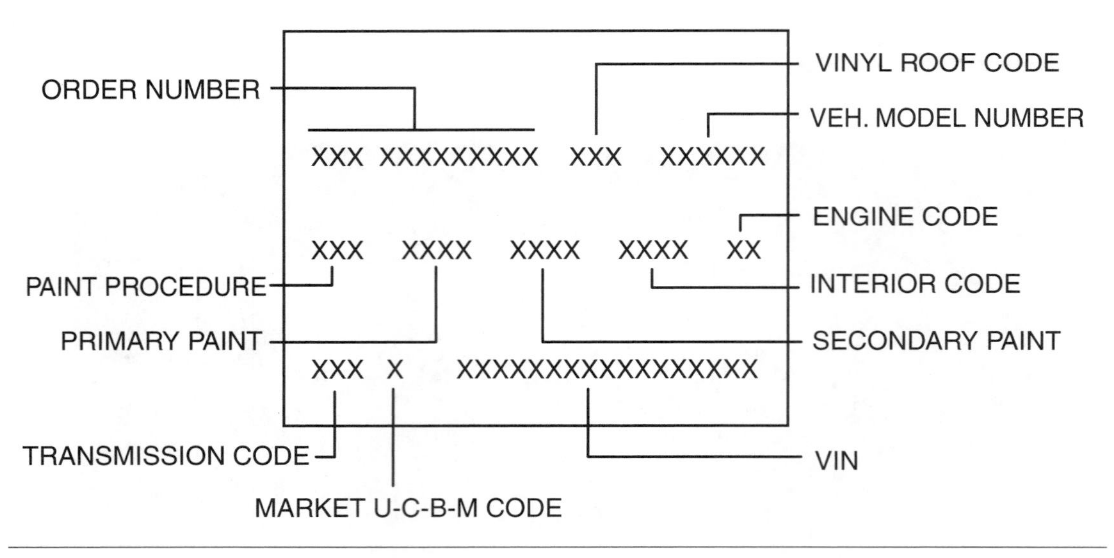


**Figure 2-46** Vehicle code plates provide sales code information about the vehicle.

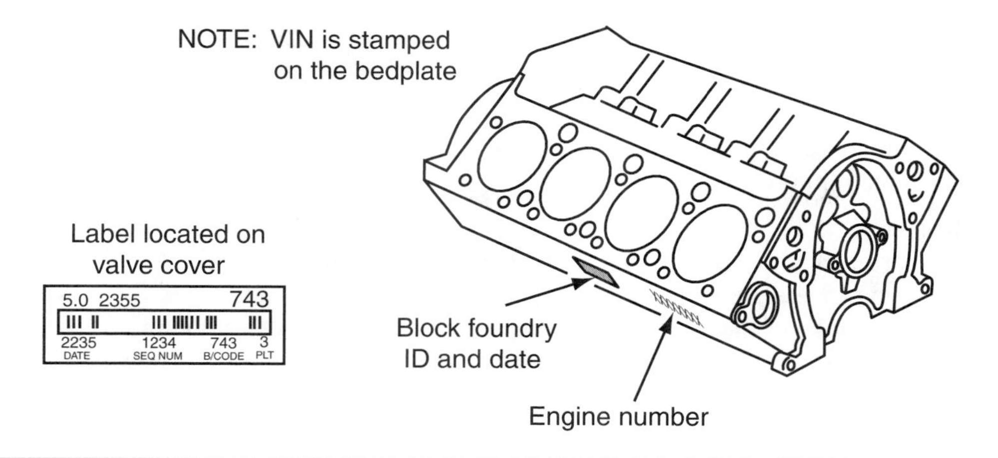


**Figure 2-47** The engine may have identification numbers stamped on it.

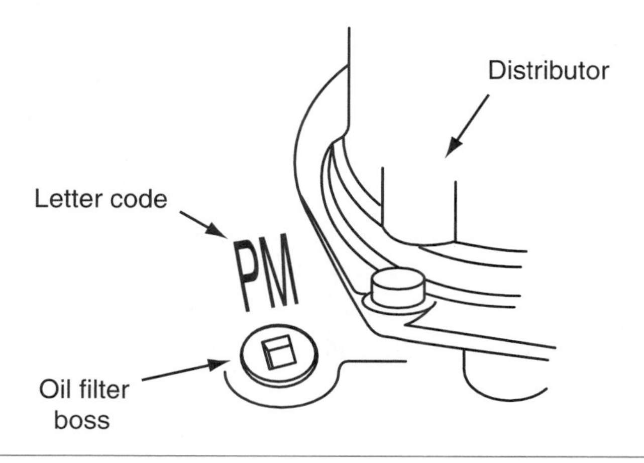


**Figure 2-48** Additional markings indicate if components are undersize or oversize.

Some engines are manufactured with oversized or undersized components. This occurs due to mass production of the engines. Components that may be different sizes than nominal include:

- Cylinder bores
- Camshaft bearing bores
- Crankshaft main bearing journals
- Connecting rod journals

Engines with oversize or undersize components (from the factory) are usually identified by a letter code stamped on the engine block (Figure 2-48). This information is helpful when ordering parts and measuring for wear.

## Summary

❒ One of the many laws of physics utilized within the automotive engine is thermodynamics. The driving force of the engine is the expansion of gases. Heat is generated by compressing the air/fuel mixture within the engine. Igniting the compressed mixture causes the heat, pressure, and expansion to multiply.

Terms to Know

Bore

Bottom dead center (BDC)

Brake horsepower

Camshaft

Combustion chamber

Compression-ignition (CI) engines

Compression ratio

Connecting rods

Crankshaft

Cycle

Cylinder block

Cylinder heads

Displacement

Dual overhead cam (DOHC)

Efficiency

Glow plugs

Gross horsepower

Horsepower

Indicated horsepower

Internal combustion engines

Kinetic energy

Mechanical efficiency

Net horsepower

Overhead cam (OHC)

Overhead valve (OHV)

Pistons

Preignition

Potential energy

Reed valve

Reciprocating

Rotary valve

Spark-ignition (SI) engine

- ❒ Engine classification is usually based on the number of cycles, the number of cylinders, cylinder arrangement, and valve train type. In addition, engine displacement is used to identify the engine.
- ❒ Most automotive and truck engines are four-stroke cycle engines. A stroke is the movement of the piston from one end to the other.
- ❒ The first stroke of the cycle is the intake stroke. The compression stroke begins as the piston starts its travel back to TDC. When the spark is introduced to the compressed mixture, the rapid burning causes the molecules to expand, and this is the beginning of the power stroke. The exhaust stroke of the cycle begins with the upward movement of the piston back toward TDC and pushes out the exhaust gases from the cylinder past the exhaust valve and into the vehicle's exhaust system.
- ❒ The three most commonly used valve trains are the overhead valve (OHV), overhead cam (OHC), and dual overhead cam (DOHC).
- ❒ Other engine designs include diesel, Miller-cycle, and stratified charge.
- ❒ Alternate vehicle power systems include electric-drive, hybrid, and fuel cell.
- ❒ There are three common combustion chamber designs used in the diesel engine: open combustion chamber—combustion chamber is located directly inside of the piston; precombustion chamber—a smaller, second chamber connected to the main combustion chamber; and the turbulence combustion chamber—a chamber designed to create turbulence as the piston compresses the air.
- ❒ Displacement is the volume of the cylinder between TDC and BDC measured in cubic inches, cubic centimeters, or liters. Engine displacement is an indicator of its power output.
- ❒ Torque is a turning or twisting force. As the pistons are forced downward, this pressure is applied to a crankshaft that rotates. The crankshaft transmits this torque to the drivetrain, and ultimately to the drive wheels.
- ❒ Horsepower is the rate at which an engine produces torque.

# Review Questions

## Short Answer Essays

1. Describe how the basic laws of thermodynamics are used to operate the automotive engine.
2. List and describe the four strokes of the four-stroke cycle engine.
3. Describe the different cylinder arrangements and the advantages of each.
4. Describe the different valve trains used in modern engines.
5. Describe the basic operation of the stratified charge engine.
6. Define the term cycle.
7. Define the term stroke.
8. Explain how the air/fuel mixture is ingested into the cylinder.
9. Explain the basic operation of the two-stroke engine.
10. Describe the basic operation of the Miller-cycle engine.

## Fill-in-the-Blanks

1. Most automotive and truck engines are ________________ cycle engines.
2. During the intake stroke, the intake valve is __________, while the exhaust valve is __________ .
3. The __________ engine places all of the cylinders in a single row.
4. An __________ __________ engine has the valve mechanisms in the cylinder head, while the camshaft is located in the engine block.
5. A __________ is a sequence that is repeated.
6. In a four-stroke engine, four strokes are required to complete one __________ .
7. Instead of using a spark produced by the ignition system, the diesel engine uses the __________ produced by compressing air in the combustion chamber to ignite the fuel.
8. A ______ __________ engine has a power stroke every revolution.
9. In a two-cycle engine the air/fuel mixture enters the crankcase through a __________ __________ or __________ __________ .
10. The __________ __________ engine uses combustion from the precombustion chamber to ignite the air/fuel mixture in the main combustion chamber.

Terms to Know

Splay angle

Splayed crankshaft

Stroke

Thermal efficiency

Thermodynamics

Top dead center (TDC)

Torque

Transverse-mounted engine

Valve overlap

Volumetric efficiency

## Multiple Choice

1. Increasing the compression ratio of an engine is being discussed.
   *Technician A* says increased compression ratios result in a reduction of power produced.
   *Technician B* says higher compression ratio engines require higher octane gasoline.
   Who is correct?
   **A.** A only **C.** Both A and B
   **B.** B only **D.** Neither A nor B
2. Engine measurement is being discussed.
   *Technician A* says the bore is the distance the piston moves within the cylinder.
   *Technician B* says the stroke is the diameter of the cylinder.
   Who is correct?
   **A.** A only **C.** Both A and B
   **B.** B only **D.** Neither A nor B
3. *Technician A* says to determine horsepower, torque must be known first.
   *Technician B* says horsepower will usually peak at a lower engine speed than torque.
   Who is correct?
   **A.** A only **C.** Both A and B
   **B.** B only **D.** Neither A nor B
4. *Technician A* says mechanical efficiency is a comparison between brake horsepower and indicated horsepower.
   *Technician B* says volumetric efficiency is a measurement of the amount of air/fuel mixture that actually enters the combustion chamber compared to the amount that could be drawn in.
   Who is correct?
   **A.** A only **C.** Both A and B
   **B.** B only **D.** Neither A nor B
5. During the compression stroke in a four-cycle engine:
   **A.** The intake valve is open and the exhaust valve is closed.
   **B.** The exhaust valve is open and the intake valve is closed.
   **C.** Both valves are open.
   **D.** Both valves are closed.

6. All of these statements about diesel engines are true EXCEPT:
   **A.** The air/fuel mixture is ignited by the heat of compression.
   **B.** Glow plugs may be used to heat the air/fuel mixture during cold starting.
   **C.** Spark plugs are located in each combustion chamber.
   **D.** Diesel engines may be two-cycle or four-cycle.

7. While discussing volumetric efficiency:
   *Technician A* says volumetric efficiency is not affected by intake manifold design.
   *Technician B* says volumetric efficiency decreases at high engine speeds.
   Who is correct?
   **A.** A only
   **B.** B only
   **C.** Both A and B
   **D.** Neither A nor B

8. In a two-stroke cycle engine:
   **A.** A power stroke is produced every crankshaft revolution.
   **B.** Intake and exhaust valves are located in the cylinder head.
   **C.** The lubricating oil is usually contained in the crankcase.
   **D.** The crankcase is not pressurized.

9. In an engine with a balance shaft:
   **A.** The balance shaft does not have to be timed to the crankshaft.
   **B.** The balance shaft turns at twice the crankshaft speed.
   **C.** The balance shaft may be mounted on the cylinder head.
   **D.** The balance shaft increases engine torque.

10. In a fuel cell car:
   **A.** The fuel cell supplies power directly to the drive wheels.
   **B.** An on-board reformer may be required to separate hydrogen from methanol.
   **C.** A single fuel cell supplies adequate power for a small compact car.
   **D.** The fuel cell uses hydrogen and oxygen to produce electricity.

# Engine Operating Systems

Upon completion and review of this chapter, you should be able to:

- ❏ Explain the purpose of the starting system.
- ❏ List and identify the components of the starting system.
- ❏ Explain the purpose of the battery.
- ❏ List the purposes of the engine's lubrication system.
- ❏ Describe the function of the lubrication system.
- ❏ List and describe the operation of the major components of the lubrication system.
- ❏ Describe the basic types and purposes of additives formulated into engine oil.
- ❏ Explain the purpose of the Society of Automotive Engineers' (SAE) classifications of oil.
- ❏ Explain the purpose of the American Petroleum Institute's (API) classifications of oil.
- ❏ Describe the two basic types of oil pumps: rotary and gear.
- ❏ Describe the purpose of the cooling system.
- ❏ Explain the operation of the thermostat.
- ❏ Describe the function of the radiator.
- ❏ Explain the function of the water pump.
- ❏ Explain the direction of coolant flow in a reverse flow cooling system.
- ❏ Explain the advantages of a reverse flow cooling system.
- ❏ Describe the purpose of antifreeze and explain its characteristics.
- ❏ Explain the purpose of the emission control systems.

## Introduction

There are three basic requirements for the engine to run: the correct air/fuel mixture; compression (good mechanical condition); and spark delivered at the correct time. Delivery of the correct air/fuel mixture is the function of the fuel system, and delivery of the spark at the correct time is the responsibility of the ignition system. For the engine to start, the engine must be cranked at a rotational speed high enough to start the combustion process so the engine will run on its own. This is the purpose of the starting system. Providing maximum engine service life is a responsibility of the lubrication and cooling systems. In addition, today's engines must meet stringent emission requirements, which is the function of the emission control systems. All of these systems work together to keep the engine operating properly.

Many of the service procedures involving the engine's subsystems are performed as routine maintenance. Today's technician will be required to service these systems when performing such tasks as oil changes, filter changes, cooling system flushes, and so forth. The technician may also need to replace worn oil or water pumps to correct poor performance of these systems. Even though these tasks are covered in the context of engine rebuilding, keep in mind that most of these services can be performed with the engine in the vehicle.

## The Starting System

The internal combustion engine must be rotated before it will run under its own power. The starting system is a combination of mechanical and electrical parts working together to start the engine. The starting system is designed to change the electrical energy that is being stored in the battery into mechanical energy. For this conversion to be accomplished, a starter or cranking motor is used. The starting system includes the following components (Figure 3-1):

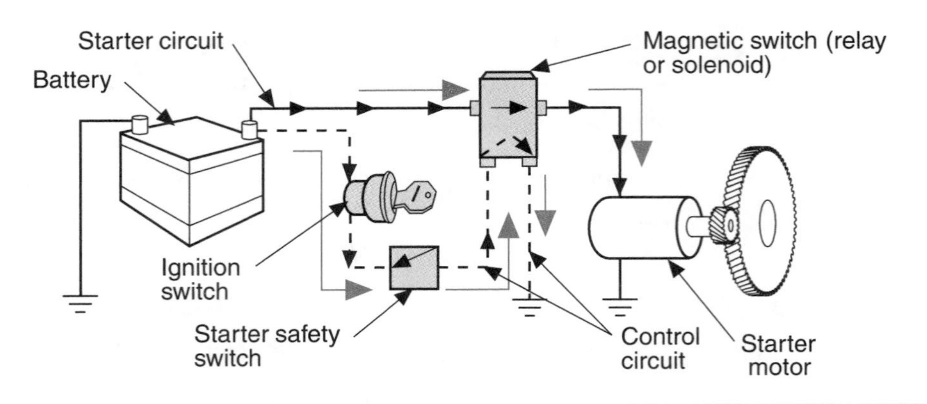


Figure 3-1 Major components of the starting system. The solid line represents the starting circuit. The dashed line indicates the starter control circuit.

1. Battery
2. Cable and wires
3. Ignition switch
4. Starter solenoid or relay
5. Starter motor
6. Starter drive and flywheel ring gear
7. Starting safety switch

## A BIT OF HISTORY

In the early days of the automobile, the vehicle did not have a starter motor. The operator had to use a starting crank to turn the engine by hand. Charles F. Kettering invented the first electric "self-starter," which was developed and built by Delco Electric. The self-starter first appeared on the 1912 Cadillac and was actually a combination starter and generator.

**Shop Manual**
Chapter 3, page 100

## The Battery

An automotive battery is an electrochemical device that provides for and stores potential electrical energy. When the battery is connected to an external load such as a starter motor, an energy conversion occurs, causing an electrical current to flow through the circuit. Electrical energy is produced in the battery by the chemical reaction that occurs between two dissimilar plates that are immersed in an electrolyte solution.

When discharging the battery (current flowing from the battery), the battery changes chemical energy into electrical energy. It is through this change that the battery releases stored energy. During charging (current flowing through the battery from the charging system), electrical energy is converted into chemical energy. Because of this, the battery can store energy until it is needed.

The largest demand is placed on the battery when it must supply current to operate the starter motor. The amperage requirements of a starter motor may be over 200 amperes. This requirement is also affected by temperature, engine size, and engine condition.

A typical 12-volt automotive battery is made up of six cells connected in series (Figure 3-2). This means the positive side of a cell element is connected to the negative side of the next cell

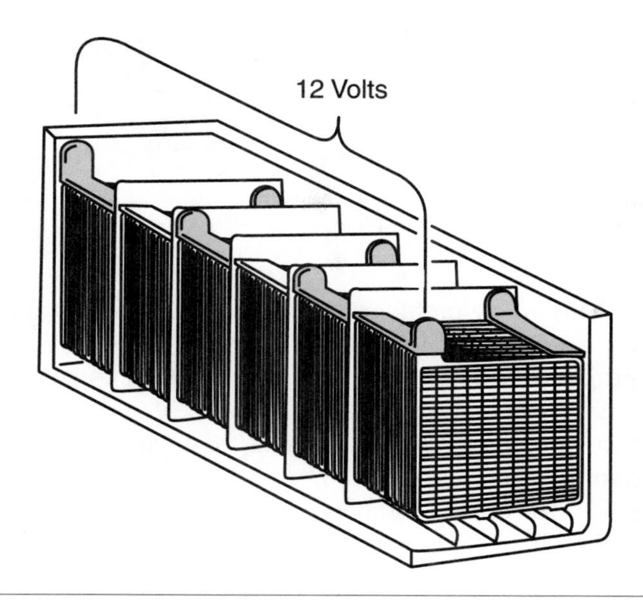


Figure 3-2 The 12-volt battery consists of six 2-volt cells that are wired in series.

element. This is repeated throughout all six cells. The six cells produce 2.1 volts each. Wiring the cells in series produces the 12.6 volts required by the automotive electrical system. The cell elements are submerged in a cell case that is filled with electrolyte solution.

All automotive batteries have two terminals. One terminal is a positive connection, the other is a negative connection. The **battery terminals** extend through the cover or the side of the battery case. Following are the most common types of battery terminals (Figure 3-3):

**Battery terminals** provide a means of connecting the battery plates to the vehicle's electrical system.

1. *Post or top terminals:* Used on most automotive batteries. The positive post is larger than the negative post to prevent connecting the battery in reverse polarity.
2. *Side terminals:* Positioned in the side of the container near the top. These terminals are threaded and require a special bolt to connect the cables. Polarity identification is by positive and negative symbols.
3. *L terminals:* Used on specialty batteries and some imports.

The condition of the battery is critical for proper operation of the starting system and all other vehicle electrical systems. The battery is often overlooked during diagnosis of many poor

**Shop Manual**
Chapter 3, page 101

Side terminal

Post or top terminal

"L" terminal

Figure 3-3 The most common types of automotive battery terminals.

| Temperature | % of Cranking Power |
|---|---|
| 80°F (26.7°C) | 100 |
| 32°F (0°C) | 65 |
| 0°F (-17.8°C) | 40 |

**Figure 3-4** Temperatures affect the cranking power of the battery.

starting or no-start problems. Many problems encountered within the starting system may be attributed to poor battery condition.

In addition, the proper battery selection for the vehicle is important. Some of the aspects that determine the battery rating required for a vehicle include engine size, engine type, climatic conditions, vehicle options, and so on. The requirements for electrical energy to crank the engine increase as the temperature decreases. Battery power drops drastically as temperatures drop below freezing (Figure 3-4). The engine also becomes harder to crank due to some oil's tendencies to thicken when cold, resulting in increased friction.

As a general rule, it takes 1 ampere of cold-cranking power per cubic inch of engine displacement. Therefore, a 200 CID engine should be fitted with a battery of at least 200 CCA. To convert this into metric measurement, it takes 1 ampere of cold-cranking power for every 16 cm of engine displacement. A 1.6-liter engine requires at least a battery rated at 100 CCA. This rule may not apply to vehicles equipped with several electrical accessories. The best method of determining the correct battery is to refer to the manufacturer's specifications.

**Shop Manual**
Chapter 3, page 102

The battery that is selected should fit the battery holding fixture and the hold-down must be able to be installed. It is also important that the height of the battery not allow the terminals to short across the vehicle hood when it is shut. Battery Council International (BCI) group numbers are used to indicate the physical size and other features of the battery (Figure 3-5). This group number does not indicate the current capacity of the battery.

All batteries must be secured in the vehicle to prevent damage and the possibility of shorting across the terminals if it tips. Normal vibrations cause the plates to shed their active materials. Hold-downs reduce the amount of vibration and help increase the life of the battery.

## The Starter Motor

The **armature** is the movable component of the motor that consists of a conductor wound around a laminated iron core and is used to create a magnetic field.

**Pole shoes** are made of high-magnetic permeability material to help concentrate and direct the lines of force in the field assembly.

Starter motors use the interaction of magnetic fields to convert electrical energy into mechanical energy. Figure 3-6 illustrates a simple electromagnet style of starter motor. The inside windings are called the **armature.** The armature rotates within the stationary outside windings, called the field, which has windings coiled around **pole shoes.** When current is applied to the field and the armature, both produce magnetic flux lines. The direction of the windings places the left pole at a south polarity and the right side at a north polarity. The lines of force move from north to south in the field. In the armature, the flux lines circle in one direction on one side of the loop and in the opposite direction on the other side. Current now sets up a magnetic field around the loop of wire, interacting with the north and south fields and causing a turning force on the loop. This force causes the loop to turn in the direction of the weaker field (Figure 3-7).

Worn starter bushings, shorted windings, or excessive electrical resistance can cause the starter to turn slower than required to start the engine. In addition, excessive mechanical friction can cause the engine to rotate slowly. Before condemning the engine, test the starting system first.

# Lubrication Systems

**Shop Manual**
Chapter 3, page 112

When the engine is operating, the moving parts generate heat due to friction. If this heat and friction were not controlled, the components would weld together. In addition, heat is created from

| Grp. Size | Vlt. | Cold cranking power—amps for 30 secs. at 0°F* | No. of mo. war-ran-ted | Size of battery container in inches (incl. terminals) | | |
|---|---|---|---|---|---|---|
| | | | | Lgth. | Wd. | Ht. |
| 17HF | 6 | 400 | 24 | 7 1/4 | 6 3/4 | 9 |
| 21 | 12 | 450 | 60 | 8 | 6 3/4 | 8 1/2 |
| 22F | 12 | 430 | 60 | 9 | 6 7/8 | 8 1/8 |
| | 12 | 380 | 55 | 9 | 6 7/8 | 8 1/8 |
| | 12 | 330 | 40 | 9 | 6 7/8 | 8 1/8 |
| 22NF | 12 | 330 | 24 | 9 1/2 | 5 1/2 | 8 7/8 |
| 24 | 12 | 525 | 60 | 10 1/4 | 6 7/8 | 8 5/8 |
| | 12 | 450 | 55 | 10 1/4 | 6 7/8 | 8 5/8 |
| | 12 | 410 | 48 | 10 1/4 | 6 7/8 | 8 5/8 |
| | 12 | 380 | 40 | 10 1/4 | 6 7/8 | 8 5/8 |
| | 12 | 325 | 36 | 10 1/4 | 6 7/8 | 8 5/8 |
| | 12 | 290 | 30 | 10 1/4 | 6 7/8 | 8 5/8 |
| 24F | 12 | 525 | 60 | 10 1/4 | 6 7/8 | 8 5/8 |
| | 12 | 450 | 55 | 10 1/4 | 6 7/8 | 8 5/8 |
| | 12 | 410 | 48 | 10 1/4 | 6 7/8 | 8 5/8 |
| | 12 | 380 | 40 | 10 1/4 | 6 7/8 | 8 5/8 |
| | 12 | 325 | 36 | 10 1/4 | 6 7/8 | 8 5/8 |
| | 12 | 290 | 30 | 10 1/4 | 6 7/8 | 8 5/8 |
| 27 | 12 | 560 | 60 | 12 | 6 7/8 | 8 5/8 |
| 27F | 12 | 560 | 60 | 12 | 6 7/8 | 9 |
| 41 | 12 | 525 | 60 | 11 9/16 | 6 13/16 | 6 15/16 |
| 42 | 12 | 450 | 60 | 9 5/8 | 6 7/8 | 6 3/4 |
| | 12 | 340 | 40 | 9 5/8 | 6 7/8 | 6 3/4 |
| 45 | 12 | 420 | 60 | 9 1/2 | 5 1/2 | 8 7/8 |
| 46 | 12 | 460 | 60 | 10 1/4 | 6 7/8 | 8 5/8 |
| 48 | 12 | 440 | 60 | 12 | 6 7/8 | 7 1/2 |
| 49 | 12 | 600 | 60 | 14 1/2 | 6 7/8 | 7 1/2 |
| 56 | 12 | 450 | 60 | 10 | 6 | 8 3/8 |
| | 12 | 380 | 48 | 10 | 6 | 8 3/8 |
| 58 | 12 | 425 | 60 | 9 1/4 | 7 1/4 | 6 7/8 |
| 71 | 12 | 450 | 60 | 8 | 7 1/4 | 8 1/2 |
| | 12 | 395 | 55 | 8 | 7 1/4 | 8 1/2 |
| | 12 | 330 | 36 | 8 | 7 1/4 | 8 1/2 |
| 72 | 12 | 490 | 60 | 9 | 7 1/4 | 8 1/4 |
| | 12 | 380 | 48 | 9 | 7 1/4 | 8 1/4 |
| 74 | 12 | 585 | 60 | 10 1/4 | 7 1/4 | 8 3/4 |
| | 12 | 525 | 60 | 10 1/4 | 7 1/4 | 8 3/4 |
| | 12 | 505 | 60 | 10 1/4 | 7 1/4 | 8 3/4 |
| | 12 | 450 | 55 | 10 1/4 | 7 1/4 | 8 3/4 |
| | 12 | 410 | 48 | 10 1/4 | 7 1/4 | 8 3/4 |
| | 12 | 380 | 40 | 10 1/4 | 7 1/4 | 8 3/4 |
| | 12 | 325 | 36 | 10 1/4 | 7 1/4 | 8 3/4 |

*Meets or exceeds Battery Council International rating standards.

**Figure 3-5** BCI battery group numbers indicate the size and features of the battery.

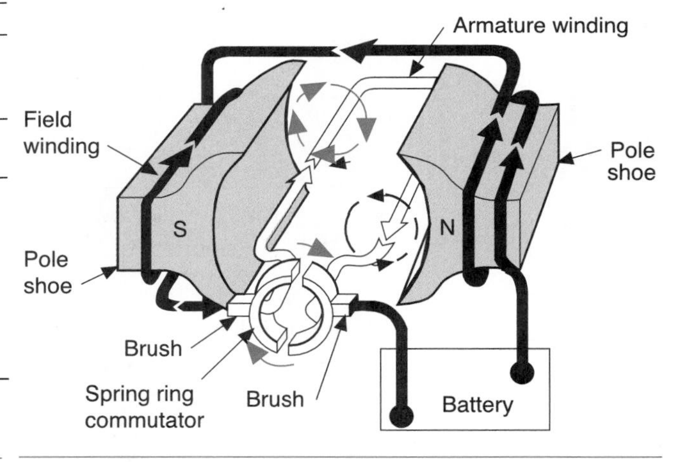


**Figure 3-6** Simple electromagnetic motor.

the combustion process. It is the function of the engine's **lubrication system** to supply oil to the high friction and wear locations and to dissipate heat away from them (Figure 3-8).

It is impossible to eliminate all of the friction within an engine, but a properly operating lubrication system will work to reduce it. Lubrication systems provide an oil film to prevent moving parts from coming in direct contact with each other (Figure 3-9). Oil molecules work as small

The engine's **lubrication system** supplies oil to high friction and wear locations.

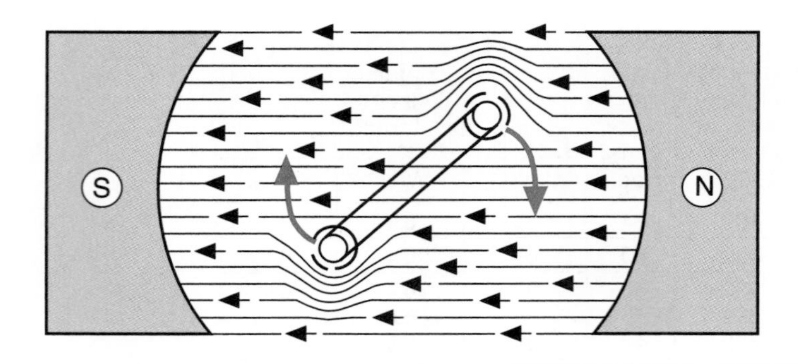


Figure 3-7 Rotation of the conductor is in the direction of the weaker field.

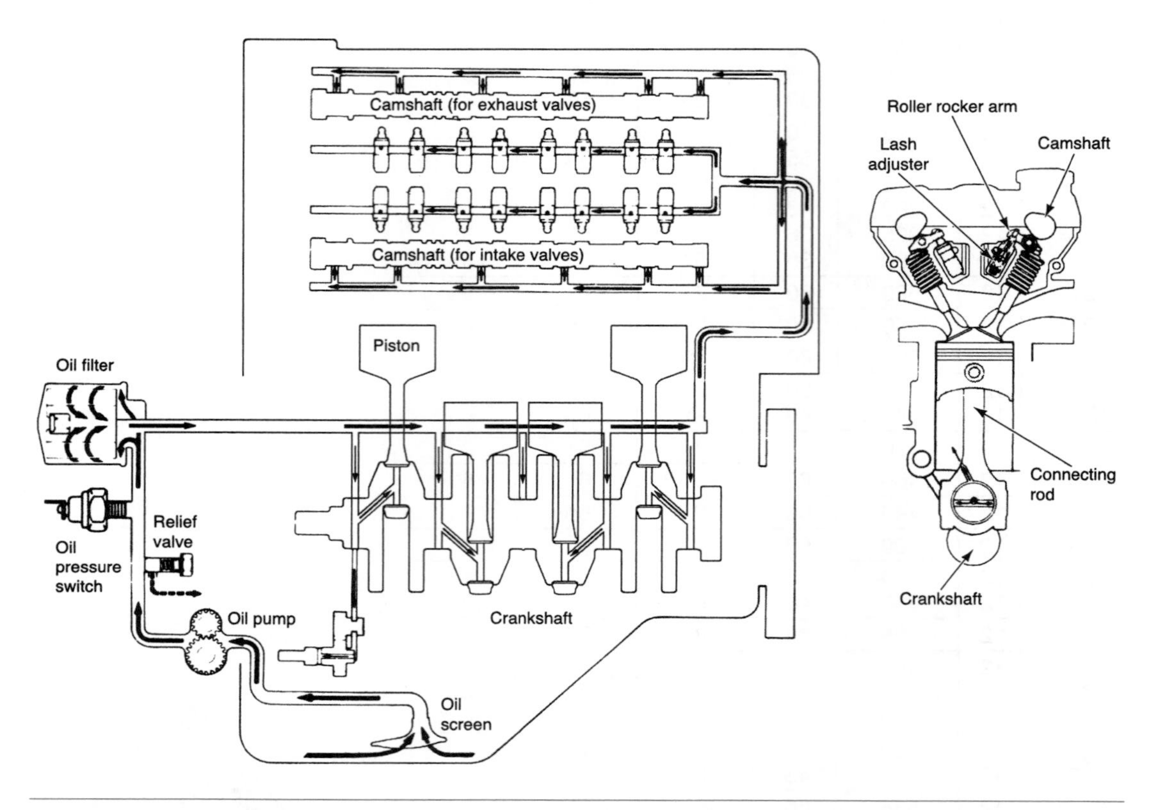


Figure 3-8 Oil flow diagram indicates the areas where oil is provided by the lubrication system. (Courtesy of Hyundai Motor America)

bearings rolling over each other to eliminate friction (Figure 3-10). Another function of the lubrication system is to act as a shock absorber between the connecting rod and crankshaft.

Besides providing reduction of friction, engine oil absorbs heat and transfers it to another area for cooling. As the oil flows through the engine, it conducts heat from the parts it comes in contact with. When the oil is returned to the oil pan, it is cooled by the air passing over the pan. Other purposes of the oil include sealing the piston rings and washing away abrasive metal and dirt. To perform these functions, the engine lubrication system includes the following components:

- Engine oil
- Oil pan or sump
- Oil filter
- Oil pump

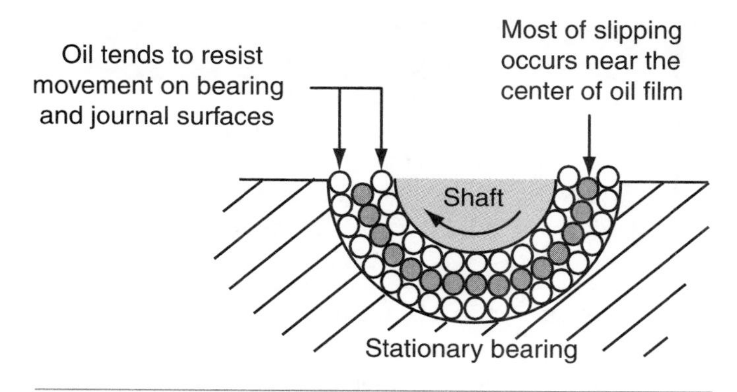


Figure 3-9 The oil develops a film to prevent metal-to-metal contact.

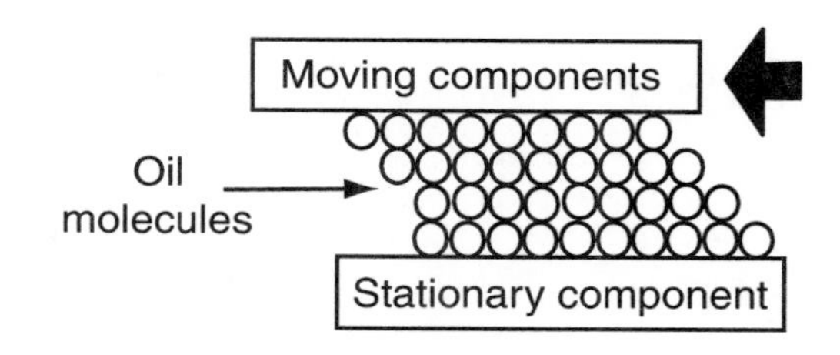


Figure 3-10 Oil molecules work as small bearings to reduce friction.

- Oil galleries
- Oil coolers (heavy-duty applications)

## Engine Oil

Engine oil must provide a variety of functions under all of the extreme conditions the engine will operate. To perform these tasks, additives are mixed with natural oil. A brief description of the types of additives used follows:

1. Antifoaming agents—These additives are included to prevent aeration of the oil. Aeration will result in the oil pump not providing sufficient lubrication to the parts and cause low oil pressure.
2. Antioxidation agents—Heat and oil agitation result in **oxidation.** These additives work to prevent the buildup of varnish and to prevent the oil from breaking down into harmful substances that can damage engine bearings.
3. Detergents and dispersants—These are added to prevent deposit buildup resulting from carbon, metal particles, and dirt. Detergents break up larger deposits and prevent smaller ones from grouping together. A dispersant is added to prevent carbon particles from grouping together.
4. Viscosity index improver—As the oil increases in temperature, it has a tendency to thin out. These additives prevent oil thinning.
5. Pour point depressants—Prevent oil from becoming too thick to pour at low temperatures.
6. Corrosion and rust inhibitors—Displace water from the metal surfaces and neutralize acid.
7. Cohesion agents—Maintain a film of oil in high-pressure points to prevent wear. Cohesion agents are deposited on parts as the oil flows over them and remain on the parts as the oil is pressed out.

**Oxidation** occurs when hot engine oil combines with oxygen and forms carbon.

Oil is rated by two organizations: the Society of Automotive Engineers (SAE) and the American Petroleum Institute (API). The Society of Automotive Engineers has standardized oil viscosity ratings, while the American Petroleum Institute rates oil to classify its service or quality limits. These rating systems were developed to make proper selection of engine oil easier.

The API classifies engine oils by a two-letter system (Figure 3-11). The prefix letter is either listed as **S-class** or **C-class** to classify the oil usage. The second letter denotes various grades of oil within each classification and denotes the oil's ability to meet the engine manufacturer's warranty requirements (Figure 3-12). At present, many manufacturers of gasoline engines demand an SJ oil classification for warranty requirements. Most gasoline engine manufacturers also demand an engine oil with the API Starburst symbol displayed on the oil container (Figure 3-13).

**S-class** is used for automotive gasoline engines. **C-class** is designed for commercial or diesel engines. The "S" stands for spark ignition, the "C" stands for compression ignition.

SAE ratings provide a numeric system of determining the oil's **viscosity.** The higher the number, the thicker or heavier the weight of the oil. For example, oil classified as SAE 50 is

**Viscosity** is the measure of oil thickness.

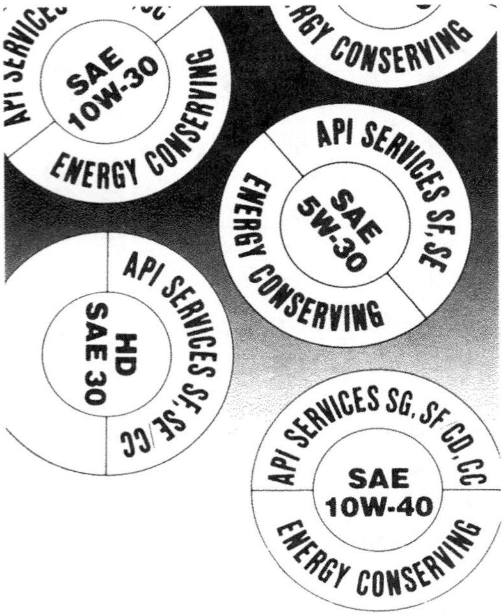

Figure 3-11 Oil containers have identification labels for API and SAE ratings. (Courtesy of Valvoline Company)

| API GASOLINE ENGINE DESIGNATION | |
|---|---|
| **SC** | Service typical of gasoline engines in 1964 through 1967. Oil designed for this service provides control of high and low temperature deposits, wear, dust, and corrosion in gasoline engines. |
| **SD** | Service typical of gasoline engines in 1968 through 1970. Oils designed for this service provide more protection against high and low temperature deposits, wear, rust, and corrosion in gasoline engines. SD oil can be used in engines requiring SC oil. |
| **SE** | Service typical of gasoline engines in automobiles and some trucks beginning in 1972. Oil designed for this service provides more protection against oil oxidation, high temperature engine deposits, rust, and corrosion in gasoline engines. SE oil can be used in engines requiring SC or SD oil. |
| **SF** | Service typical of gasoline engines in automobiles and some trucks beginning with 1980. SF oils provide increased oxidation stability and improved antiwear performance over oils that meet API designation SE. It also provides protection against engine deposits, rust, and corrosion. SF oils can be used in engines requiring SC, SD, or SE oils. |
| **SG** | Service typical of gasoline automobiles and lightduty trucks, plus CC classification diesel engines beginning in the late 1980s. SG oils provide the best protection against engine wear, oxidation, engine deposits, rust, and corrosion. It can be used in engines requiring SC, SD, or SF oils. |

Figure 3-12 The API rating indicates the quality of the oil and its ability to meet the manufacturer's requirements.

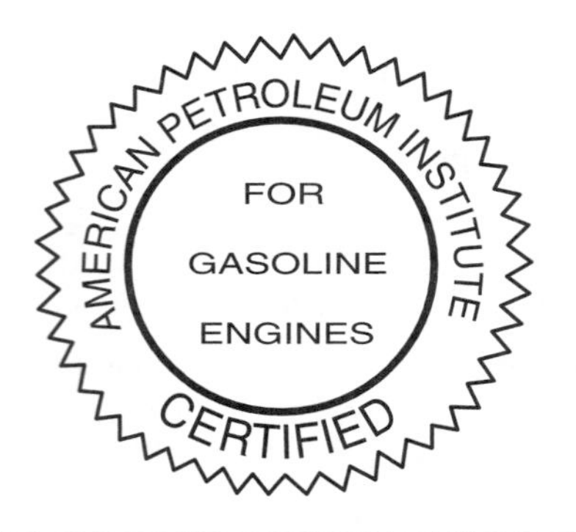


Figure 3-13 API Starburst symbol displayed on oil containers.

thicker than SAE 20. The thicker the oil, the slower it will pour. Thicker oils should be used when engine temperatures are high, but they can cause lubrication problems in cooler climates. Thinner oils will flow through the engine faster and easier at colder temperatures, but may **break down** under higher temperatures or heavy engine loads. To provide a compromise between these conditions, multiviscosity oils have been developed. For example, an SAE 10W-30 oil has a weight of 10W at lower temperatures to provide easy flow. As the engine temperature increases, the weight of the oil actually increases to a weight of 30 to prevent damage resulting from too thin oil. The "W" in the designation refers to winter, which means the viscosity is determined at 0°F (–18°C). If there is no "W," the viscosity is determined at 210°F (100°C).

**Oil breakdown** results from high temperatures for extended periods of time. The oil combines with oxygen and can cause carbon deposits in the engine.

Oils have recently been developed that may be labeled "energy conserving." These oils use friction modifiers to reduce friction and increase fuel economy.

Selection of oil is based upon the type of engine and the conditions it will be running in. The API ratings must meet the requirements of the engine and provide protection under the normal expected running conditions. A main concern in selecting SAE ratings is ambient temperatures (Figure 3-14). Always select oil that meets or exceeds the manufacturer's recommendations.

In recent years, synthetic oils have become increasingly popular. These oils are produced in a laboratory instead of being refined from crude. The advantage of synthetic oils are their stability over a wide temperature range. Before using these oils during an engine service, refer to the vehicle's warranty information to confirm use will not void the warranty.

## Oil Pumps

There are two basic types of **oil pumps:** rotor and gear. Both types are **positive displacement pumps.** The rotary pump generally has a four-lobe inner rotor with a five-lobe outer rotor (Figure 3-15). The outer rotor is driven by the inner rotor. As the lobes come out of mesh, a vacuum is

**Shop Manual**
Chapter 3, page 112

**SAE GRADES OF MOTOR OIL**

| Lowest Atmospheric Temperature Expected | Single-Grade Oils | Multigrade Oils |
|---|---|---|
| 32°F (0°C) | 20, 20W, 30 | 10W-30, 10W-40, 15W-40, 20W-40, 20W-50 |
| 0°F (–18°C) | 10W | 5W-30, 10W-30, 10W-40, 15W-40 |
| –15°F (–26°C) | 10W | 10W-30, 10W-40, 5W-30 |
| Below –15°F (–26°C) | 5W* | *5W-20, 5W-30 |

*SAE 5W and 5W-20 grade oils are not recommended for sustained high-speed driving.

**Figure 3-14** The typical temperature the engine will be operating in determines the selection of the proper SAE grade.

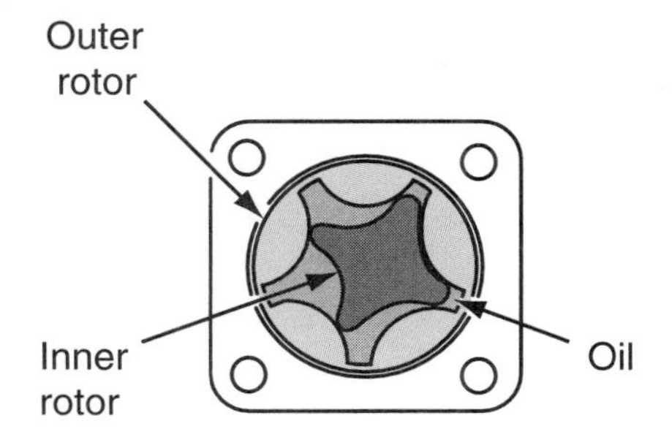


**Figure 3-15** Typical rotor-type oil pump.

The **oil pump** creates a vacuum so atmospheric pressures can push oil from the sump. The pump then pressurizes the oil and delivers it throughout the engine by use of galleries.

**Positive displacement pumps** deliver the same amount of oil with every revolution, regardless of speed.

created, and atmospheric pressure on the oil pushes it into the pickup tube. The oil is trapped between the lobes as it is directed to the outlet. As the lobes come back into mesh, the oil is expelled from the pump.

Gear pumps can use two gears riding in mesh with each other or use two gears and a crescent design (Figure 3-16). Both types operate in the same manner as the rotor-type pump. The advantage of the rotor-type pump is its capability to deliver a greater volume of oil since the cavities are larger.

In the past, most oil pumps were driven off of the camshaft by a drive shaft fitting into the bottom of the distributor shaft (Figure 3-17). This is still a popular method for engines using distributor ignition systems. Many of today's engines do not use a distributor and drive the oil pump by the front of the crankshaft (Figure 3-18).

Since oil pumps are positive displacement types, output pressures must be regulated to prevent excessive pressure buildup. A pressure relief valve opens to return oil to the sump or pump inlet if the specified pressure is exceeded (Figure 3-19). A calibrated spring holds the valve closed, allowing pressure to increase. Once the oil pressure is great enough to overcome the spring pressure, the valve opens and returns the oil to the sump or pump inlet.

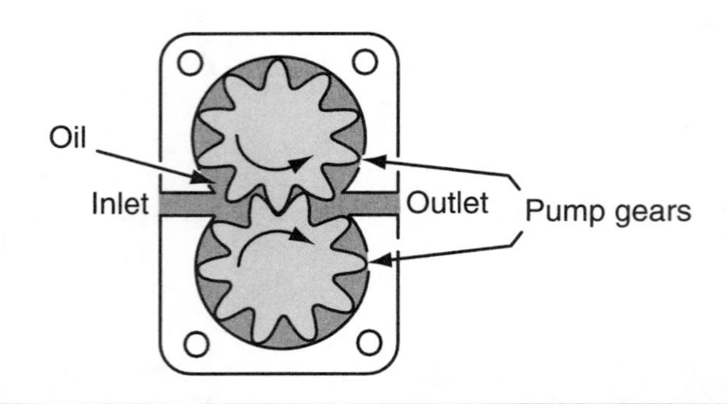


**Figure 3-16** Typical gear-type oil pump.

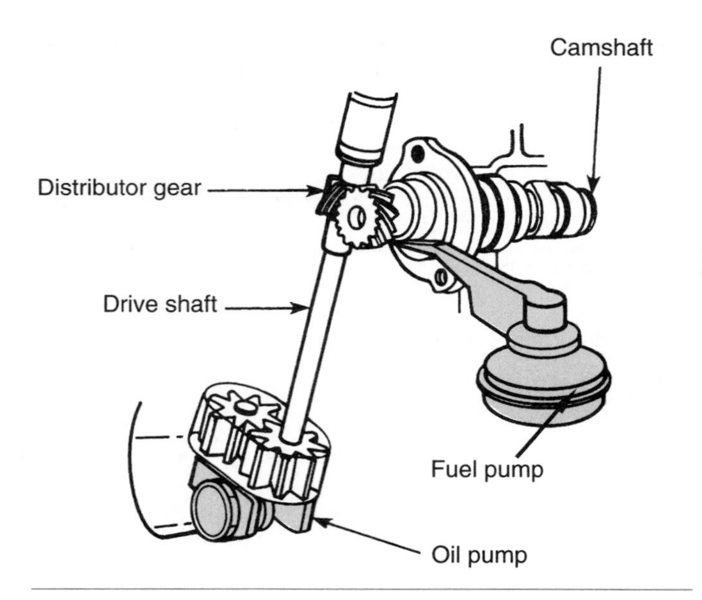


**Figure 3-17** Many oil pumps are driven by a gear on the camshaft.

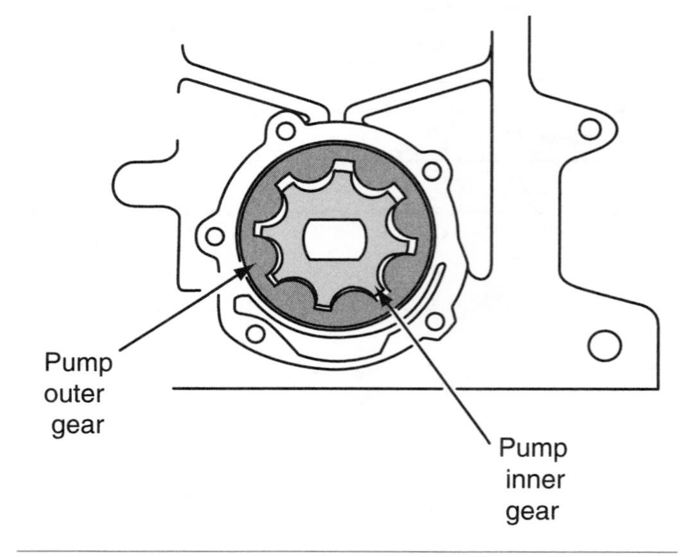


**Figure 3-18** Many engines have their oil pump located at the front of the engine block and drive it directly off of the crankshaft.

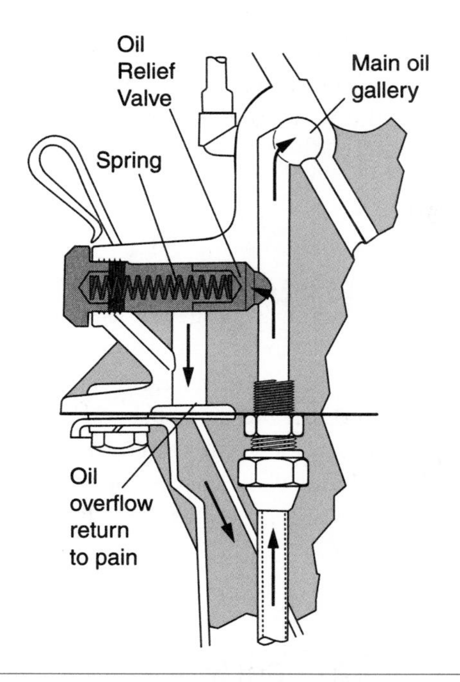


Figure 3-19 The oil pressure relief valve opens to return the oil to the sump if pressures are excessive.

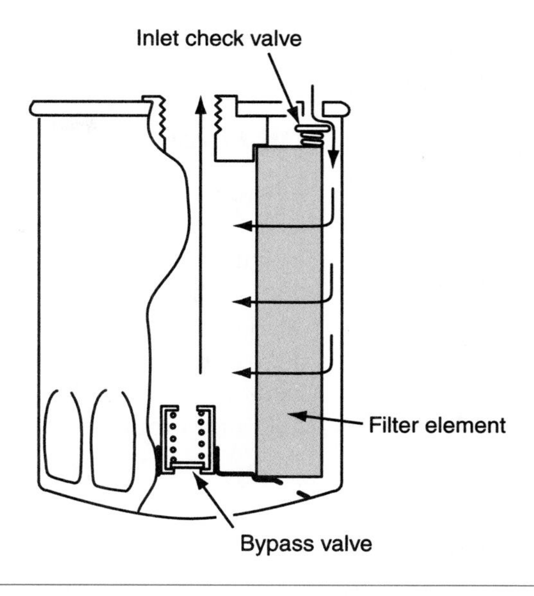


Figure 3-20 Oil flow through the oil filter.

## Oil Filter

As the oil flows through the engine, it works to clean dirt and deposits from the internal parts. These contaminants are deposited into the oil pan or sump with the oil. Since the oil pump pickup tube syphons the oil from the sump, it can also pick up these contaminants. The pickup tube has a screen mesh to prevent larger contaminants from being picked up and sent back into the engine. The finer contaminants must be filtered from the oil to prevent them from being sent with the oil through the engine. Oil filter elements have a pleated paper or fibrous material designed to filter out particles between 20 and 30 **microns** (Figure 3-20). Most oil filters use a **full-filtration system.** Under pressure from the pump, oil enters the filter on the outer areas of the element and works its way toward the center. In the event the filter becomes plugged, a bypass valve will open and allow the oil to enter the galleries. If this occurs, the oil will no longer be filtered (Figure 3-21).

A micron is a thousandth of a millimeter or about 0.0008 inch.

A full-filtration system means all of the oil is filtered before it enters the oil galleries.

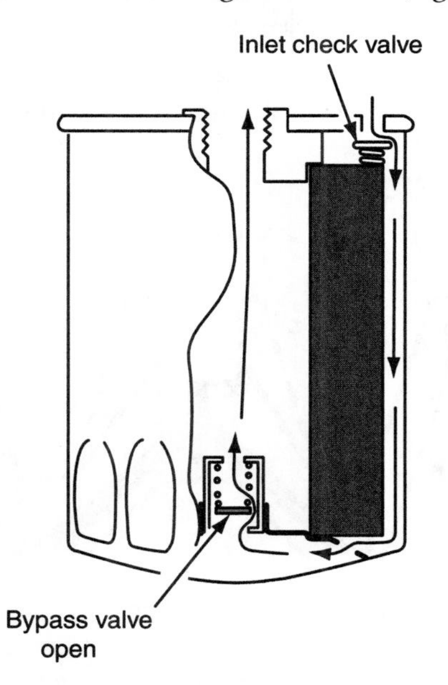


Figure 3-21 If the oil filter becomes plugged, the bypass valve opens to protect the engine.

The **inlet check valve** keeps the oil filter filled at all times so when the engine is started, an instantaneous supply of oil is available.

The **saddle** is the portion of the crankcase bore that holds the bearing half in place.

The connecting rod journal is also referred to as the crank pin.

An **inlet check valve** is used to prevent oil drainback from the oil pump when the engine is shut off. The check valve is a rubber flap covering the inside of the inlet holes. When the engine is started and the pump begins to operate, the valve opens and allows oil to flow to the filter.

## Oil Flow

After the oil leaves the filter, it is directed through galleries to various parts of the engine. Oil galleries are drilled passages in the cylinder block and head. A main oil gallery is usually drilled the length of the block. From the main gallery, pressurized oil branches off to upper and lower portions of the engine (Figure 3-22). Oil is directed to the crankshaft main bearings through the main bearing **saddles.** Passages drilled in the crankshaft then direct the oil to the rod bearings (Figure 3-23). Some manufacturers drill a small spit or squirt hole in the connecting rod to spray pressurized oil delivered to the bearings out and onto the cylinder walls (Figure 3-24). When the spit hole aligns with the oil passage in the rod journal, it squirts the oil onto the cylinder wall.

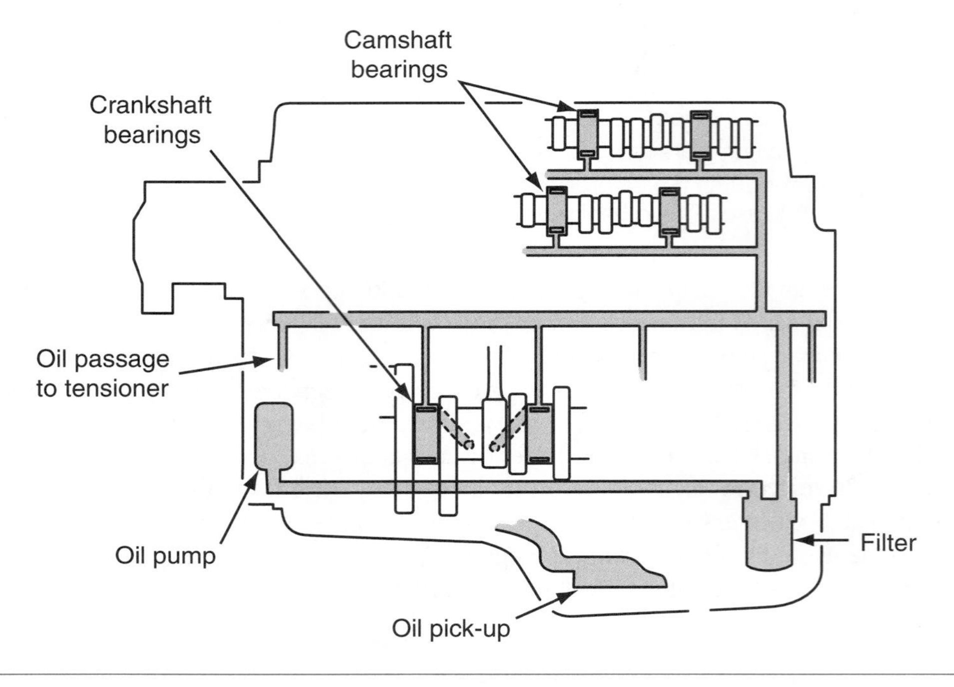


**Figure 3-22** The lubrication system delivers oil to the high friction and wear areas of the engine.

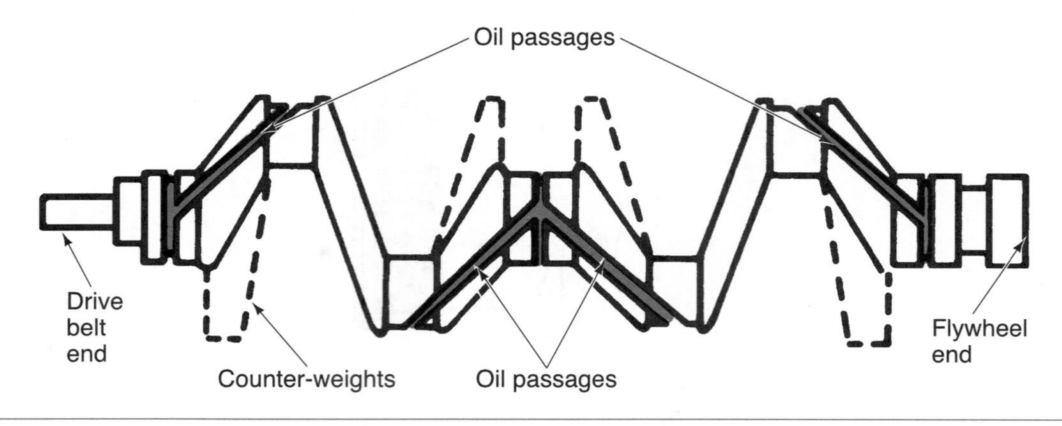


**Figure 3-23** Oil passages drilled into the crankshaft provide lubrication to the main bearings.

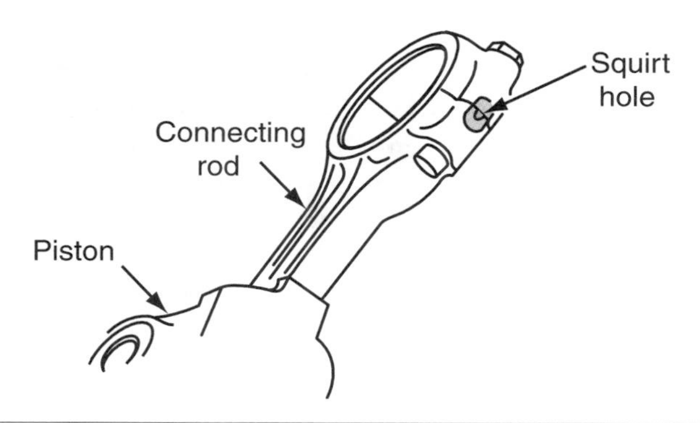


**Figure 3-24** A spit hole is used to lubricate the cylinder wall and keep the piston head cool.

Each camshaft bearing (on OHV engines) also receives oil from passages drilled in the block (Figure 3-25). The valve train can receive oil through passages drilled in the block and then through a hollow rocker arm shaft. Engines using pushrods usually pump oil from the lifters through the pushrods to the valve train (Figure 3-26). Oil sent to the cylinder head is returned to the sump through drain passages cast into the head and block.

Not all areas of the engine are lubricated by pressurized oil. The crankshaft rotates in the oil sump, creating some splash-and-throw effects. Oil that is picked up in this manner is thrown throughout the crankcase. This oil splashes onto the cylinder walls below the pistons. This provides lubrication and cooling to the piston pin and piston rings. In addition, oil that is forced out of the connecting rod bearings is thrown to lubricate parts that are not fed pressurized oil. Valve timing chains are often lubricated by oil draining from the cylinder head being dumped onto the chain and gears.

**AUTHOR'S NOTE:** During my experience as an automotive technician and instructor, I encountered a significant number of vehicle owners who attempted to save money on car maintenance by extending engine oil and filter changes beyond the manufacturer's recommended interval. My experience also indicated that if these individuals kept their vehicles very long, they spent a lot more on premature, expensive engine repairs than they saved on maintenance costs. Therefore, proper lubrication system maintenance is extremely important to provide long engine life!

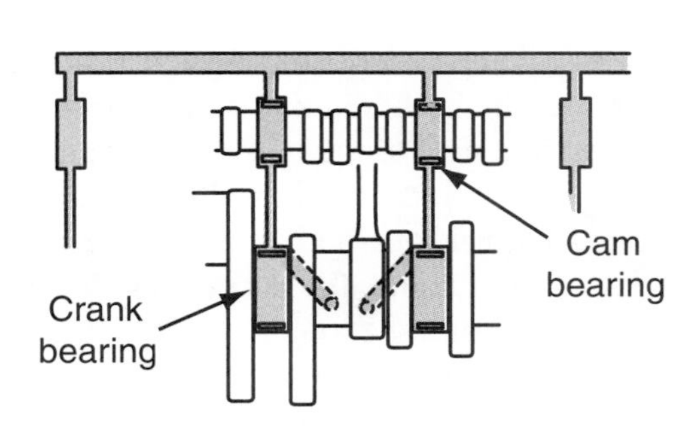


**Figure 3-25** OHV engines have branches from the main oil gallery to direct pressurized oil to the camshaft journals.

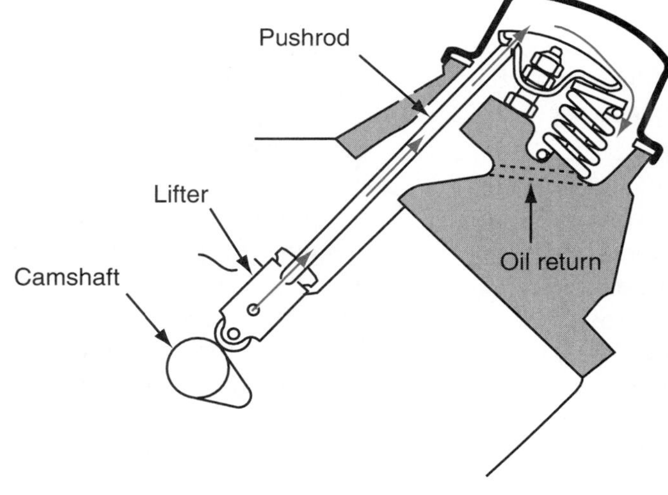


**Figure 3-26** OHV engines often use a hollow pushrod to deliver pressurized oil to the upper engine.

# Cooling Systems

**Shop Manual**
Chapter 3, page 118

The **cooling system** is used to disperse the heat of engine operation to the atmosphere.

The heat created when the combustion process is completed would quickly increase to a point where engine damage would occur. The **cooling system** is used to disperse the heat to the atmosphere. The components of the cooling system include:

- Coolant
- Radiator
- Radiator cap
- Recovery system
- Coolant jackets
- Hoses
- Water pump
- Thermostat
- Cooling fan
- Heater system
- Transmission cooler
- Temperature warning system

**Shop Manual**
Chapter 3, page 120

The temperatures used for boiling points are at sea level.

**Electrolysis** is the chemical and electrical process of decomposition that occurs when two dissimilar metals are joined with the presence of moisture. The lesser of the two metals is eaten away.

**Nucleate boiling** is the process of maintaining the overall temperature of a coolant to a level below its boiling point, but allowing the portions of the coolant actually contacting the surfaces (the nuclei) to boil into a gas.

## Coolant

Water by itself does not provide the proper characteristics needed to protect an engine. It works quite well to transfer heat, but does not protect the engine in colder climates. The boiling point of water is 212°F (100°C), and it freezes at 32°F (0°C). Temperatures around the cylinders and in the cylinder head can reach above 500°F (250°C), and the engine will often be exposed to ambient temperatures below 32°F (0°C). In addition, water reacts with metals to produce rust, corrosion, and **electrolysis.** Most engine manufacturers use a solution of water and ethylene glycol as engine coolants.

The proper coolant mixture will protect the engine under most conditions. Water expands as it freezes. If the coolant in the engine block freezes, the expansion could cause the block or cylinder head to crack. If the coolant boils, liquid is no longer in contact with the cylinder walls and combustion chamber. The vapors are not capable of removing the heat, and the pistons may collapse or the head may warp from the excessive heat.

Ethylene glycol by itself has a boiling point of 330°F (165°C), but it does not transfer heat very well. The freezing point of ethylene glycol is –8°F (–20°C). To improve the transfer of heat and lower the freezing point, water is added in a mix of about 50/50. At a 50/50 mix, the boiling point is 226°F (108°C), and the freezing point is –34°F (–37°C). These characteristics can be altered by changing the mixture (Figure 3-27).

It may appear that lowering the boiling point by adding water is not desirable. In fact, this is required for proper engine cooling. Under pressure, the boiling point of the coolant mix is about 263°F (128°C). As was stated earlier, the temperature next to the cylinder or cylinder head may be in excess of 500°F (250°C). As coolant droplets touch the metal walls, they are turned into a gas as they boil. The superheated gas bubbles are quickly carried away into the middle of the coolant flow, where they cool and condense back into a liquid. If this **nucleate boiling** did not occur, the coolant would not be capable of removing heat fast enough to protect the engine.

The engine coolant should always be replaced whenever an engine is rebuilt. Special agents are added to antifreezes used in aluminum engines or with aluminum radiators. Always refer to the service manual for the type and mixture recommended. In addition, proper maintenance of the cooling system is required to maintain good operation. Over time, the coolant can become slightly acidic because of the minerals and metals in the cooling system. A small electri-

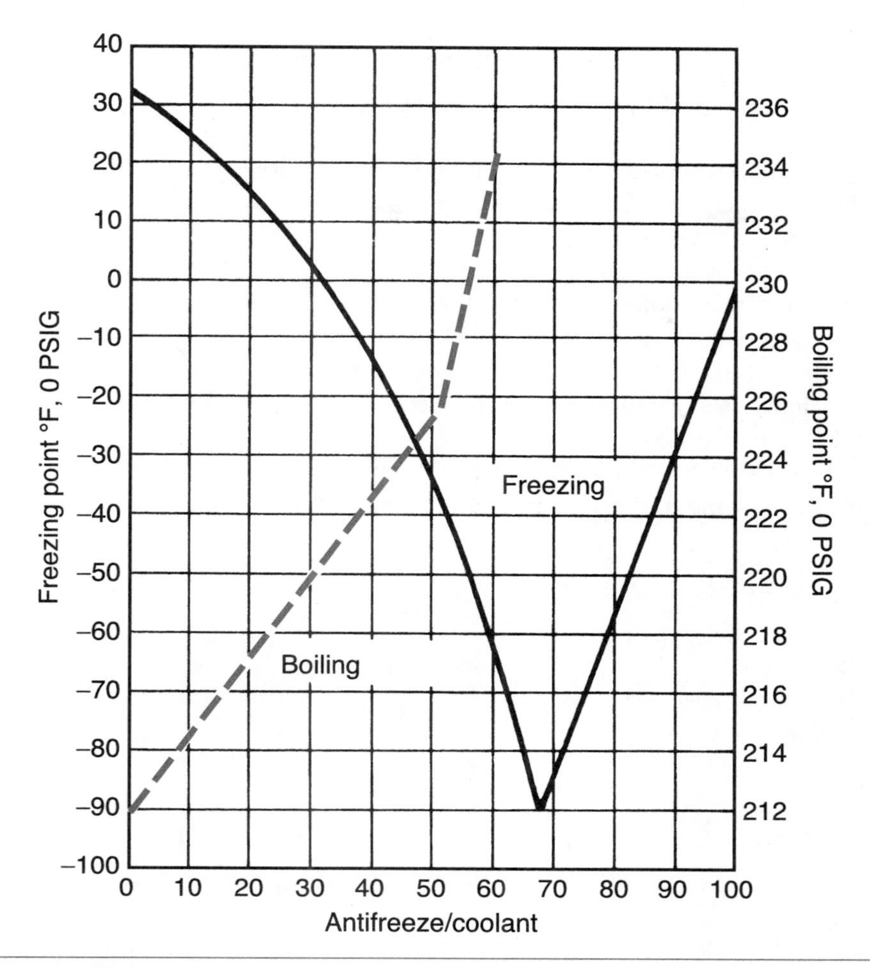


**Figure 3-27** Changing the strength of the antifreeze solution changes its boiling and freezing points.

cal current may flow between metals through the acid. The electrical current has a corrosive effect on the metal used in the cooling system. Anticorrosion agents are mixed into the antifreeze; however, these agents may be used up in a year's time. All vehicle manufacturers provide a maintenance schedule recommending cooling system flushing and refill based on time and mileage.

## A BIT OF HISTORY

In the early years of the automobile, water was used as engine coolant. The water worked well to remove heat from the engines, but it would freeze, resulting in cracked blocks. Motorists used several different substances to prevent the water from freezing. Most of these substances prevented freezing but had other adverse effects to the engine. Some of these early substances included salt, calcium chloride, soda, sugar, honey, engine oil, and coal oil. The first successful antifreeze was made from wood or grain alcohol. However, these substances lowered the boiling point of the water and evaporated at higher engine temperatures. The first permanent antifreeze (meaning it would not evaporate) was made with a glycerine base.

**Shop Manual**
Chapter 3, page 128

The **radiator** is a heat exchanger used to transfer heat from the engine to the air passing through it.

## Radiator

The **radiator** is a series of tubes and fins that transfers the heat in the coolant to the air. As the coolant is circulated throughout the engine, it attracts and absorbs the heat within the engine, then flows into the radiator intake tank. The coolant then flows through the tubes to the outlet tank. As the heated coolant flows through the radiator tubes, the heat is dissipated into the air through the fins.

There are two basic radiator designs: downflow and crossflow. Downflow radiators have vertical fins that direct coolant flow from the top inlet tank to the bottom outlet tank (Figure 3-28). Crossflow radiators use horizontal fins to direct coolant flow across the core (Figure 3-29). The core of the radiator can be constructed using a tube-and-fin or a cellular-type design (Figure 3-30). The material used for the core is usually copper, brass, or aluminum. Aluminum-core radiators generally use nylon-constructed tanks.

For heat to be transferred to the air effectively, there must be a difference in temperature between the coolant and the air. The greater the temperature difference, the more effectively heat

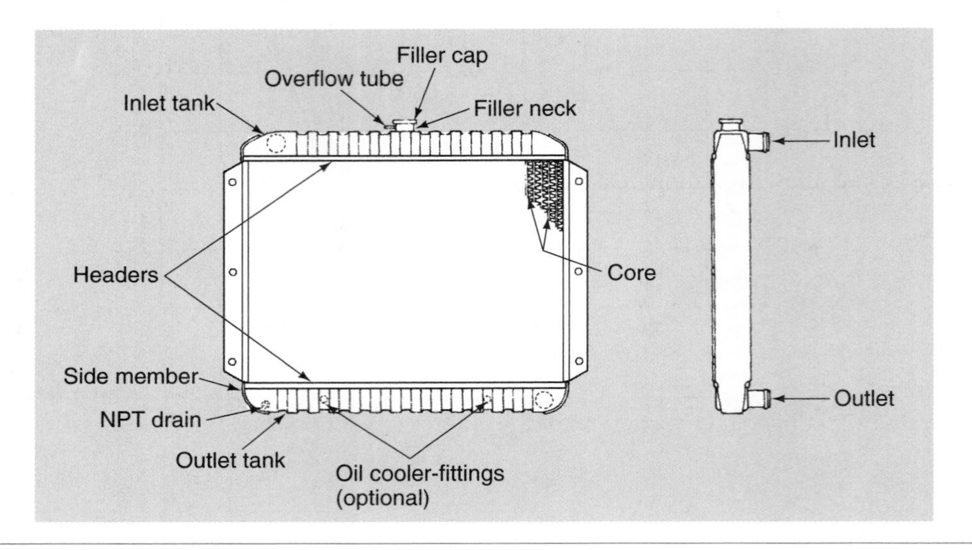


Figure 3-28 A typical downflow radiator. (Reprinted with the permission from SAE 1983 Handbook, Volume 3, Society of Automotive Engineers, Inc.)

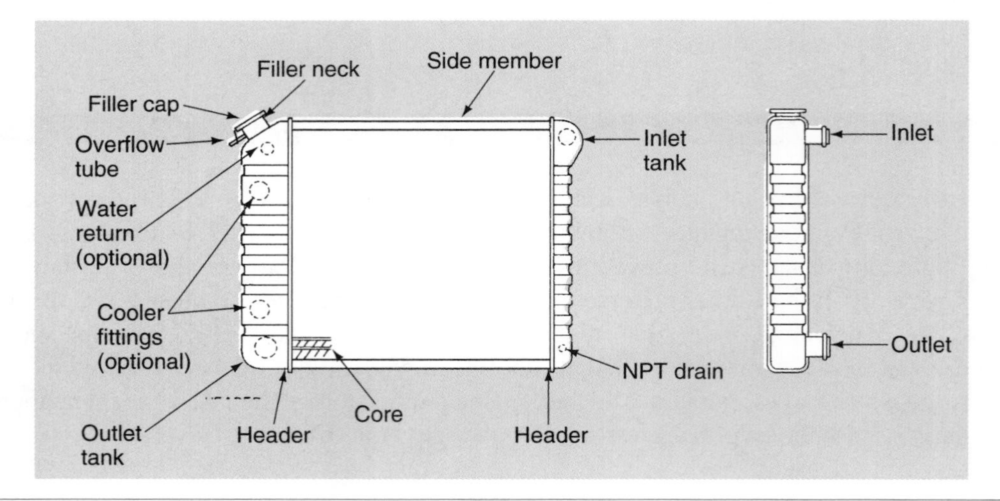


Figure 3-29 A typical crossflow radiator. (Reprinted with the permission from SAE 1983 Handbook, Volume 3, Society of Automotive Engineers, Inc.)

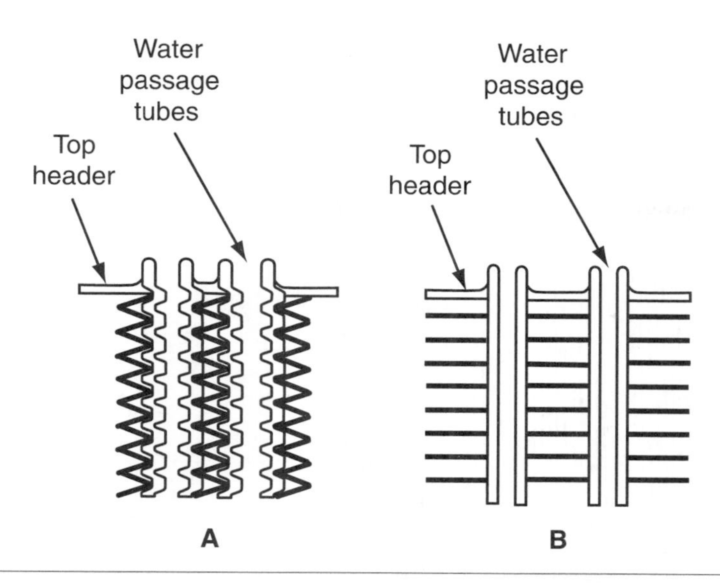


Figure 3-30 (A) Construction of a cellular radiator core. (B) Construction of a tube-and-fin radiator core.

is transferred. Manufacturers increase the temperature of the coolant by pressurizing the radiator. The radiator cap allows for an increase of pressure within the cooling system, increasing the boiling point of the coolant; for example, if the pressure is increased to 15 psi (103 kPa), the boiling point of the 50/50 mix would be increased to 266°F (134°C). Since the boiling point is increased, the operating temperature of the engine can also be increased.

For every pound of pressure increase, boiling point of water is raised about $3\frac{1}{4}$°F.

The radiator cap uses a pressure valve to pressurize the radiator to between 14 and 18 pounds (Figure 3-31). If the pressure increases over the setting of the cap, the cap's seal will lift and release the pressure into a recovery tank (Figure 3-32).

The **coolant recovery system** contains a reservoir that is connected to the radiator by a small hose. The coolant expelled from the radiator during a high-pressure condition is sent to this reservoir. When the engine cools, a void is created in the radiator, and the vacuum valve in the radiator cap opens (Figure 3-33). Under this condition, atmospheric pressure on the coolant in the overflow reservoir pushes it back into the radiator.

The **coolant recovery system** prevents loss of coolant if the engine overheats, and it keeps air from entering the system.

Whenever an engine is rebuilt, the radiator should be thoroughly cleaned and then inspected by pressure testing it. The radiator cap should be replaced with a new one.

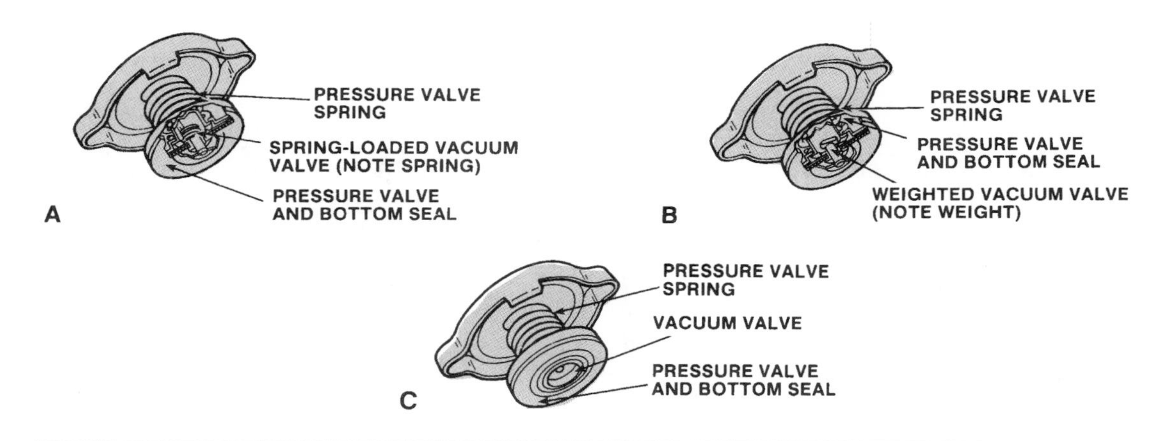


Figure 3-31 Three typical types of radiator caps: (A) constant pressure type; (B) pressure vent type; (C) closed system type.

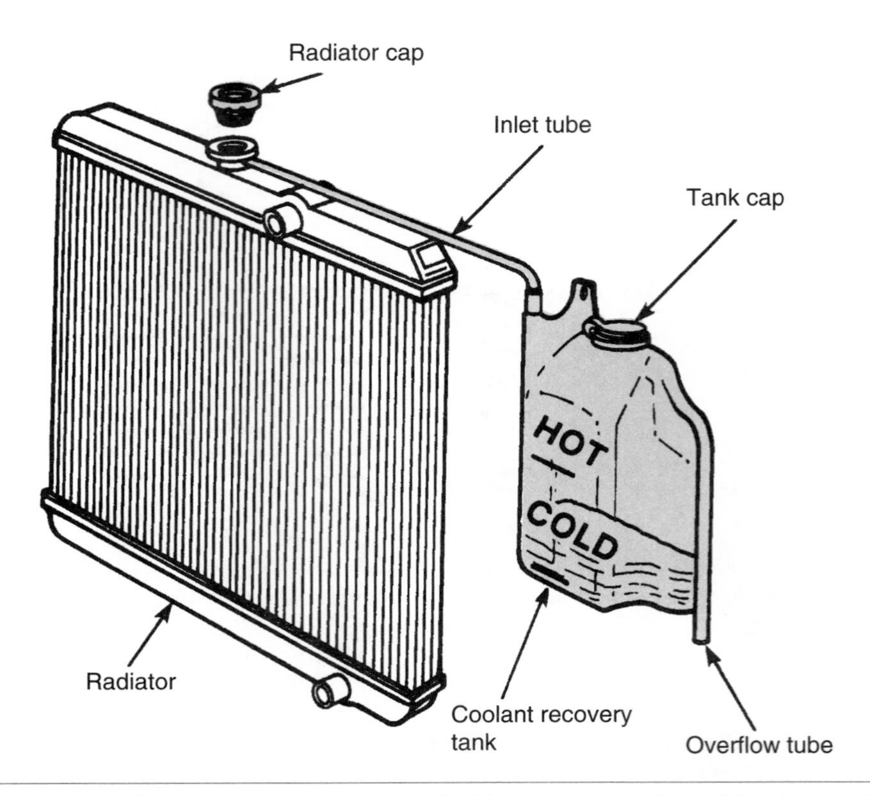


Figure 3-32 A typical coolant recovery system holds the coolant released from the radiator.

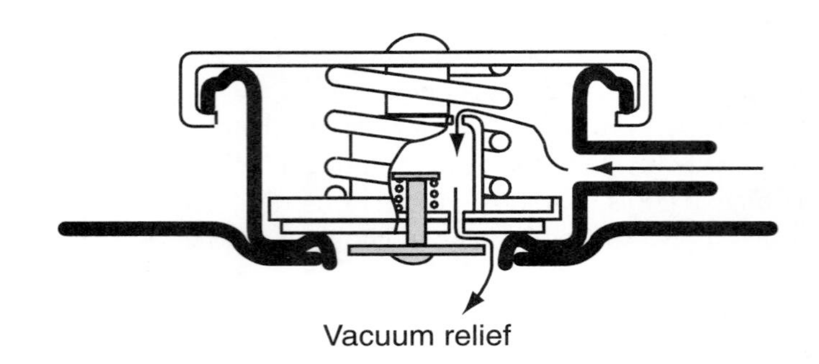


Figure 3-33 When the vacuum valve opens, coolant from the recovery reservoir enters the radiator.

The **heater core** dissipates heat like a radiator. The radiated heat is used to warm the passenger compartment.

## Heater Core

The **heater core** is similar to a small version of the radiator. The heater core is located in a housing, usually in the passenger compartment of the vehicle (Figure 3-34). Some of the hot engine coolant is routed to the heater core by hoses. The heat is dispersed to the air inside the vehicle, thus warming the passenger compartment. To aid in quicker heating of the compartment, a heater fan blows the radiated heat into the compartment.

## Hoses

Hoses are used to direct the coolant from the engine into the radiator, and back to the engine. In addition, hoses are used to direct coolant from the engine into the heater core, and (on some engines) to bypass the thermostat.

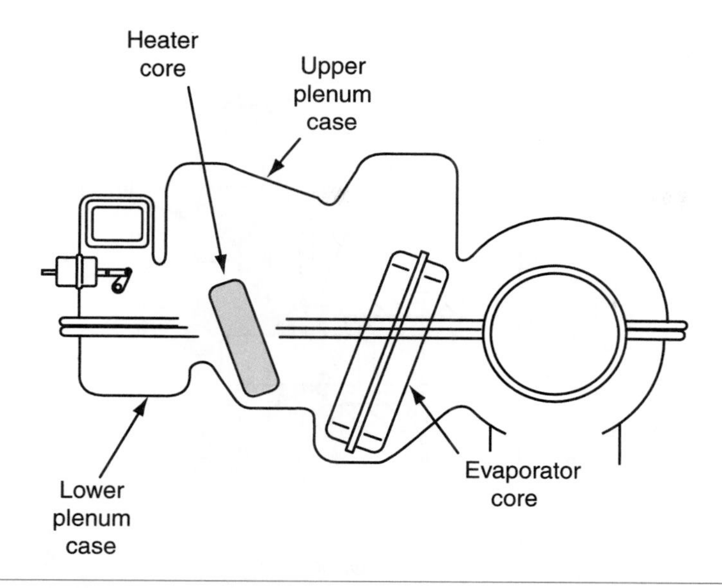


Figure 3-34 The heater core uses hot engine coolant to warm the passenger compartment.

Since radiator and heater hoses deteriorate from the inside out, most manufacturers recommend periodic replacement of the hoses as preventive maintenance. Whenever the engine is rebuilt, the hoses should always be replaced.

## Water Pump

The water pump is the heart of the engine's cooling system. It forces the coolant through the engine block and into the radiator and heater core (Figure 3-35). The water pump is driven by an accessory belt, the timing belt, or directly from the camshaft. Most water pumps are centrifugal

**Shop Manual**
Chapter 3, page 129

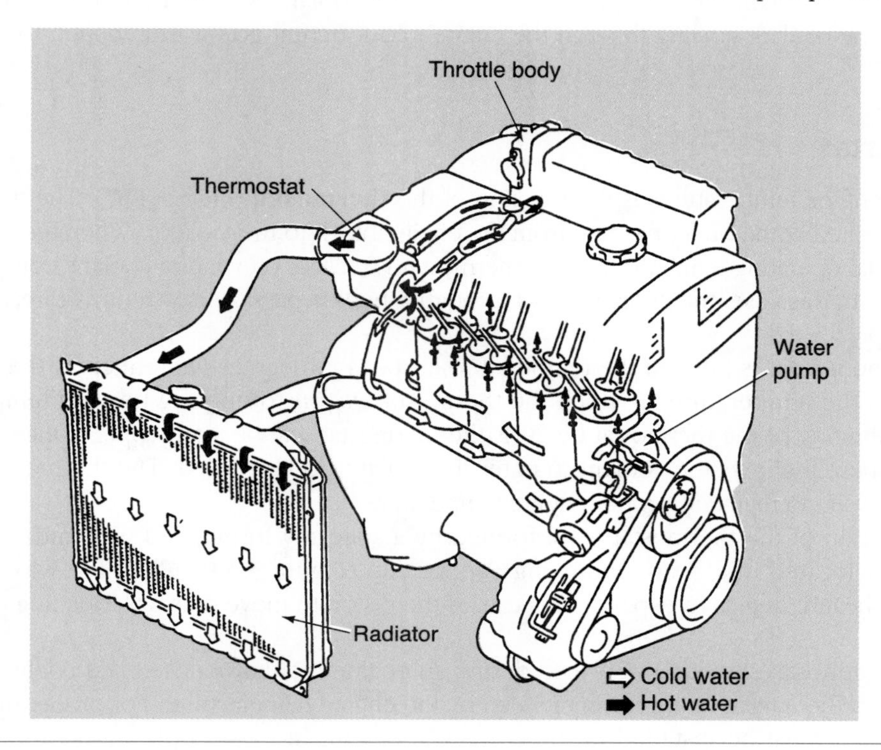


Figure 3-35 Coolant flow through the engine. (Courtesy of Hyundai Motor America)

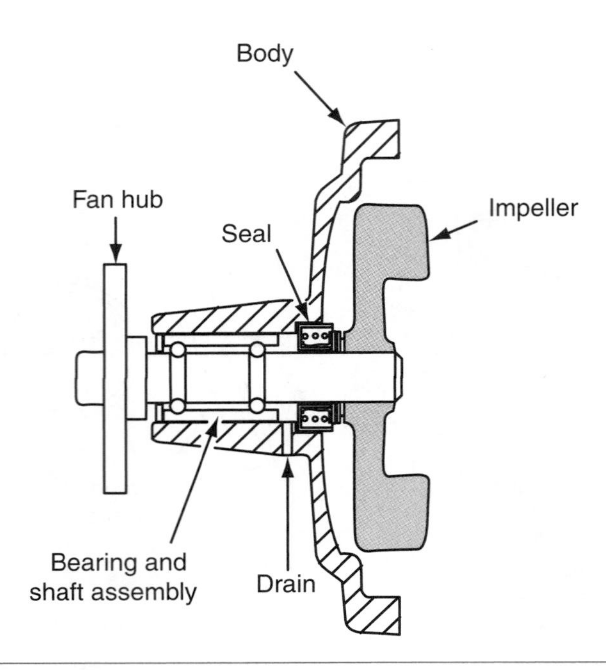


**Figure 3-36** Most water pumps use an impeller to move the coolant through the system.

design, using a rotating impeller to move the coolant (Figure 3-36). When the engine is running, the impeller rotates, forcing coolant from the inside of the cavity outward toward the tips by centrifugal force. Once inside the block, the coolant flows around the cylinders and into the cylinder heads, absorbing the heat from these components. If the thermostat is open, the coolant will then enter the radiator. The void created by the empty impeller cavity allows the pressurized coolant to be pushed from the radiator to fill the cavity and repeat the cycle. When the thermostat is closed because coolant temperatures are too cold, the coolant is circulated through a bypass. This keeps the coolant circling through the engine block until it gets warm enough to open the thermostat.

## Thermostat

**Shop Manual**
Chapter 3, page 124

The **thermostat** allows the engine to quickly reach normal operating temperatures and maintains the desired temperature.

Control of engine temperatures is the function of the **thermostat** (Figure 3-37). The thermostat is usually located at the outlet passage from the engine block to the radiator. When the coolant is below normal operating temperatures, the thermostat is closed, preventing coolant from entering the radiator. In this case, the coolant flows through a bypass passage and returns directly to the water pump.

The thermostat is rated at the temperature it opens in degrees Fahrenheit. If the rating is 195°, this is the temperature at which the thermostat *begins* to open. When the temperature exceeds the rating of the thermostat by 20°F, the thermostat will be totally open. Once the thermostat is open, it allows the coolant to enter the radiator to be cooled. The thermostat cycles open and closed to maintain proper engine temperature.

Operation of the thermostat is performed by a specially formulated wax and powdered metal pellet located in a heat-conducting copper cup (Figure 3-38). When the wax pellet is exposed to heat, it begins to expand. This causes the piston to move outward, opening the valve (Figure 3-39).

The thermostat should be replaced during an engine rebuild or scheduled cooling system service. Use a thermostat rating recommended by the engine manufacturer. For proper operation, the thermostat must be installed in the correct direction. In most applications, the pellet is installed toward the engine block, but always refer to the appropriate service manual.

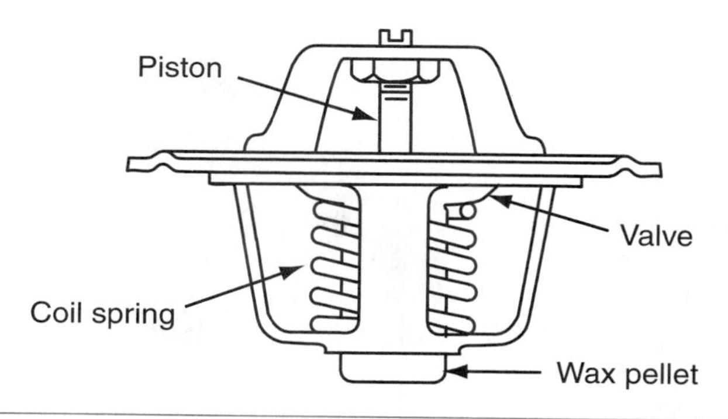

Figure 3-37 The thermostat controls the temperature of the engine.

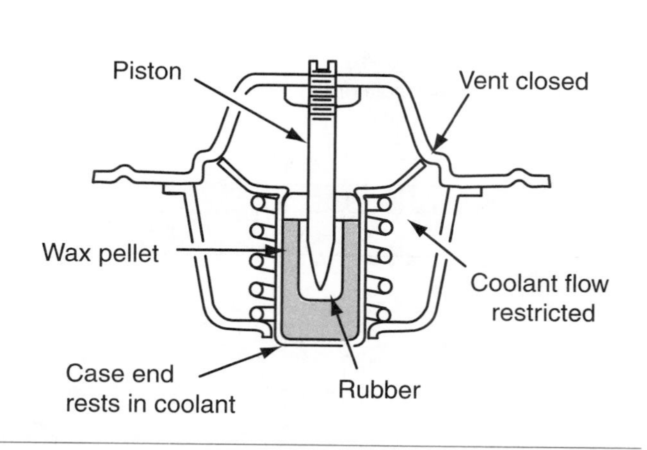

Figure 3-38 The wax pellet controls the flow of coolant into the radiator.

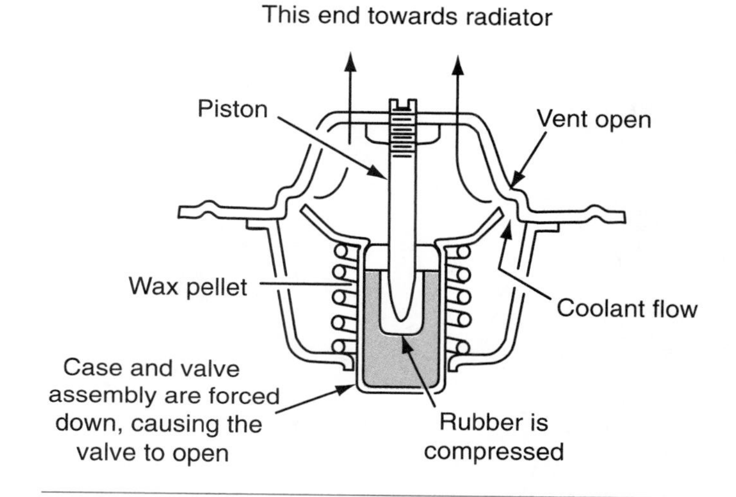

Figure 3-39 When the pellet expands, it opens the thermostat.

On many engines, the thermostat is located on the outlet side of the water pump. The thermostat is mounted in a housing on top of the engine, and the upper radiator hose is connected to this housing. A number of engines designed in the 1990s, such as the General Motors Gen III V8 engine used in light-duty trucks, Corvette, Camaro, and Firebird, have the thermostat mounted in a housing on the inlet side of the water pump (Figure 3-40). When the thermostat is located on the outlet side of the water pump, cold coolant from the radiator is supplied to the water pump when the thermostat begins to open. This action tends to close the thermostat, resulting in thermostat cycling and a more gradual engine warmup. When the thermostat is mounted on the inlet side of the pump, thermostat cycling is reduced, and engine temperature increases faster with fewer fluctuations.

## Coolant Flow

Coolant flow through the engine can be one of two ways: parallel or series. In a parallel flow system, the coolant flows into the engine block and into the cylinder head through passages beside each cylinder (Figure 3-41). In a series flow system, the coolant enters the engine block and flows around each cylinder. It then flows to the rear of the block and through passages to the cylinder head. The coolant then flows through the head to the front of the engine (usually the highest point of the cooling system) and out to the radiator (Figure 3-42).

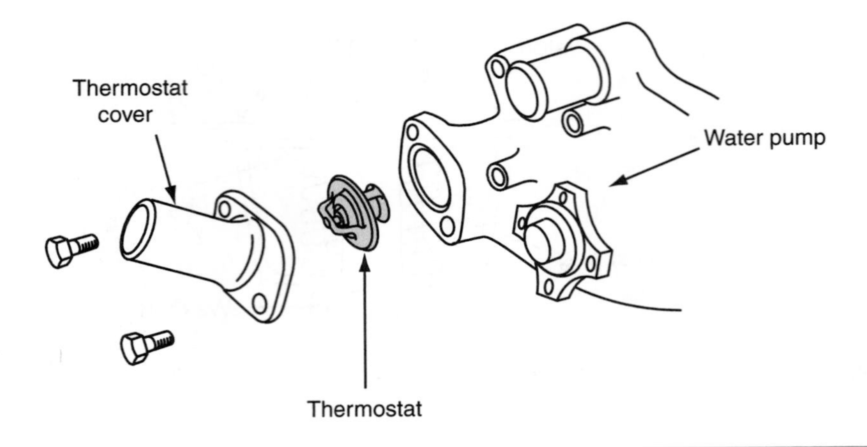


Figure 3-40 When the pellet expands, it opens the thermostat.

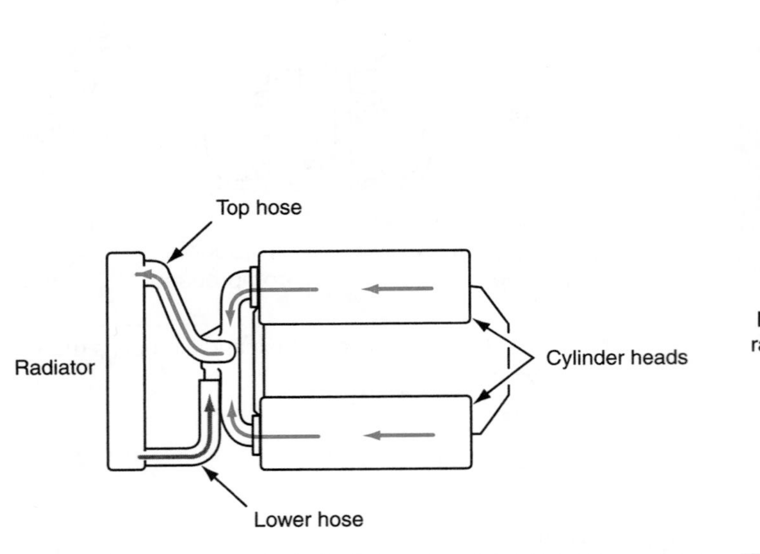


Figure 3-41 The parallel flow sends the coolant around each cylinder and up to the cylinder head. There is more than one path for coolant flow to the cylinder head.

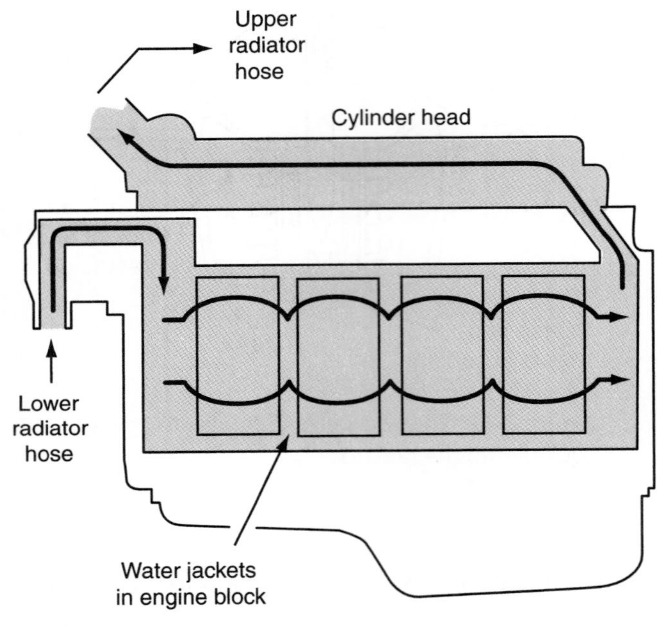


Figure 3-42 A series flow sends the coolant around all of the cylinders before sending it to the cylinder head.

The **head gasket** is used to prevent compression pressures, gases, and fluids from leaking. It is located on the connection between the cylinder head and engine block.

The cooling system passages are more than hollow portions in a casting. They are designed and located so no steam pockets can develop. Any steam will go to the top of the radiator. This is accomplished by bleed holes among the cylinder block, **head gasket,** and cylinder head.

Corrosion or contaminants can plug the coolant passages. This results in reduced heat transfer from the cylinder to the coolant. To prevent the formation of deposits, the cooling system should be flushed and coolant changed at regular intervals as specified by the manufacturer. Anytime the engine is removed and rebuilt, it is a good practice to thoroughly clean the coolant passages in the block and cylinder head.

## Reverse Flow Cooling Systems

A **reverse flow cooling system** is used in some engines, such as the General Motors LT1 5.7L V8 engine in some model years of Camaro, Firebird, and Corvette. In these engines, the water pump forces coolant through the cylinder heads first and then through the block passages (Figure 3-43). The water pump is driven from a gear and shaft mounted on the front of the engine. The water pump drive gear is rotated by gear teeth on the back of the camshaft sprocket (Figure 3-44). Coolant flows from the block through other water pump passages and through the upper radiator hose to the radiator. Coolant returns through the lower radiator hose to the inlet side of the water pump. On a conventional flow cooling system, the coolant flows through the block first and then through the cylinder heads. Circulating coolant through the cylinder heads first improves combustion chamber cooling, and allows the engine to

In a **reverse flow cooling system** coolant flows from the water pump through the cylinder heads and then through the engine block.

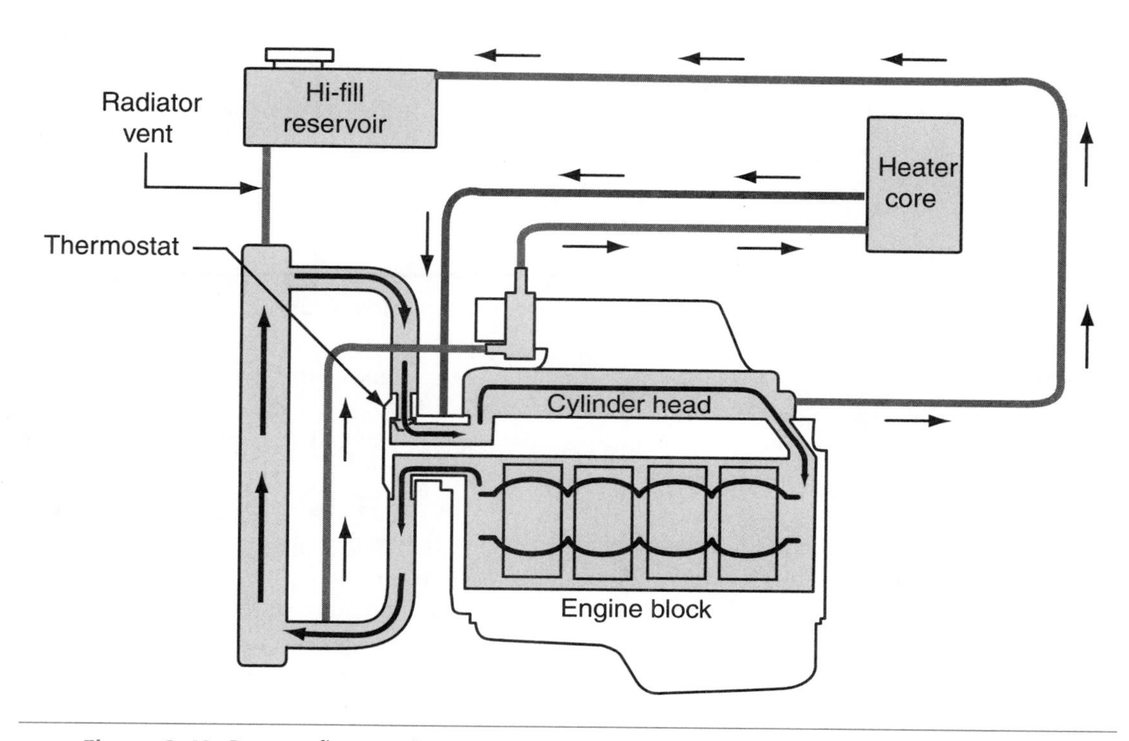


**Figure 3-43** Reverse flow cooling system.

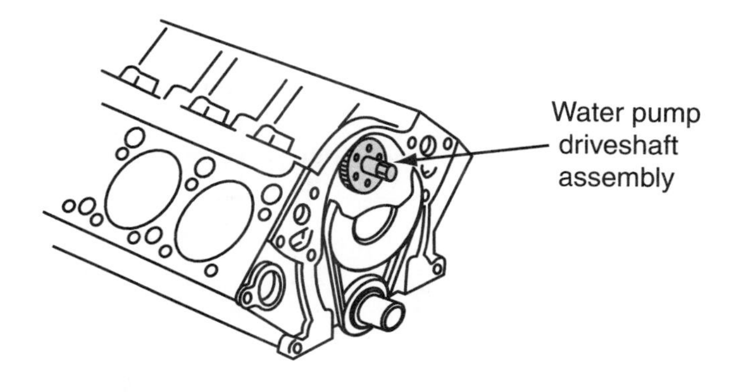


**Figure 3-44** Water pump drive.

operate with a higher compression ratio. The compression ratio in the LT1 engine is 10.25:1. Improved combustion chamber cooling also allows the engine to operate with increased spark advance without encountering detonation problems. This action improves fuel economy and engine performance.

The thermostat is located in a housing on the inlet side of the water pump. A smaller diameter hose connected to the upper radiator hose supplies coolant to the throttle body and heater core. This coolant is returned to the inlet side of the water pump. A radiator vent tube is connected from the top of the radiator to the hi-fill reservoir. The pressure cap is installed on the hi-fill reservoir rather than on the radiator, and the hi-fill reservoir is designed to withstand cooling system pressure. The pressure cap contains conventional pressure and vacuum valves. When filling the cooling system, coolant must be added through the hi-fill reservoir. A bleed pipe assembly is connected to the back of both cylinder heads and a hose connects this pipe to the hi-fill reservoir. The bleed pipe and hose bleed any air from the cylinder heads into the hi-fill reservoir. If air is trapped in the aluminum cylinder heads, hot spots may be created in the aluminum cylinder heads that result in cylinder head damage. When filling the cooling system, bleed valves in the thermostat housing and throttle body coolant passage must be opened to bleed air from the cooling system. The bleed valve in the throttle body may be located in a different location depending on the vehicle application. When coolant comes out of the bleed valves, they should be closed.

## Cooling Fans

**Shop Manual**
Chapter 3, page 125

To increase the efficiency of the cooling system, a fan is mounted to direct flow of air through the radiator core or past the radiator tubes. In the past, at high vehicle speeds air flow through the grill and past the radiator core was sufficient to remove the heat. The cooling fan was really only needed for low-speed conditions when air flow was reduced. As modern cars have become more aerodynamic, air flow through the grill has declined; thus proper operation of the cooling fan has become more critical.

The cooling fan can be either driven mechanically by the engine or by an electric motor. Electric drive fans are common on today's vehicles because they only operate when needed, thus reducing engine loads. Some of the earlier designs of electric cooling fans use a temperature switch in the radiator or engine block that closes when the temperature of the coolant reaches a predetermined value. With the switch closed, the electrical circuit is completed for the fan motor relay control circuit and the fan turns on (Figure 3-45). The feed to the motor is supplied through the ignition switch or directly from the battery. If it is supplied directly from the battery, the cooling fans can come on even though the engine is not running.

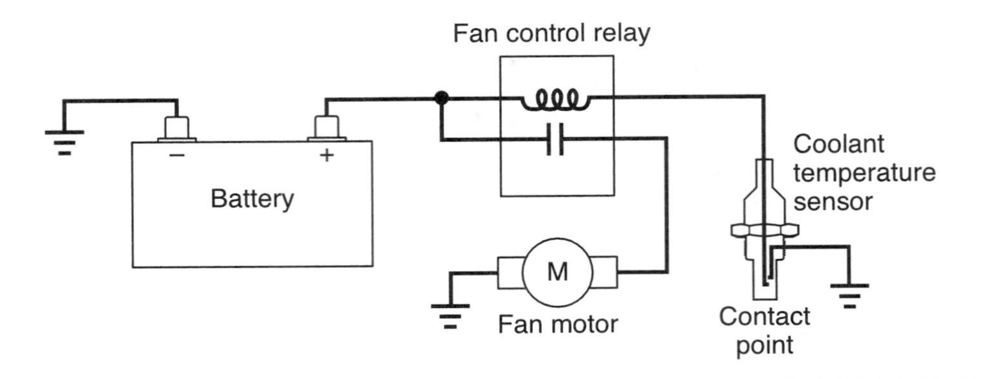


**Figure 3-45** Simplified electric fan circuit.

In recent years, most cooling fans are controlled by the powertrain control module (PCM). The PCM receives engine coolant temperature input signals from the thermistor-type engine coolant temperature (ECT) sensor. When the thermistor value indicates the temperature is hot enough to turn on the fans, the PCM activates the fan control relay (Figure 3-46). On some vehicles, the PCM also turns on the fans whenever air-conditioning is turned on, regardless of the engine temperature.

Belt-driven fans usually attach to the water pump pulley (Figure 3-47). Some belt-driven fans use a viscous-drive fan clutch to enhance performance (Figure 3-48). The clutch operates the fan in relation to engine temperature by a silicone oil (Figure 3-49). If the engine is hot, the silicone oil in the clutch expands and locks the fan to the pump hub. The fan now rotates at the same speed as the water pump. When the engine is cold, the silicone oil contracts and the fan rotates at a reduced speed. Some fan clutches use a thermostatic coil that winds and unwinds in response to the engine temperatures (Figure 3-50). The coil operates a valve that prevents or allows silicone to leave the reservoir and flow into the working chamber.

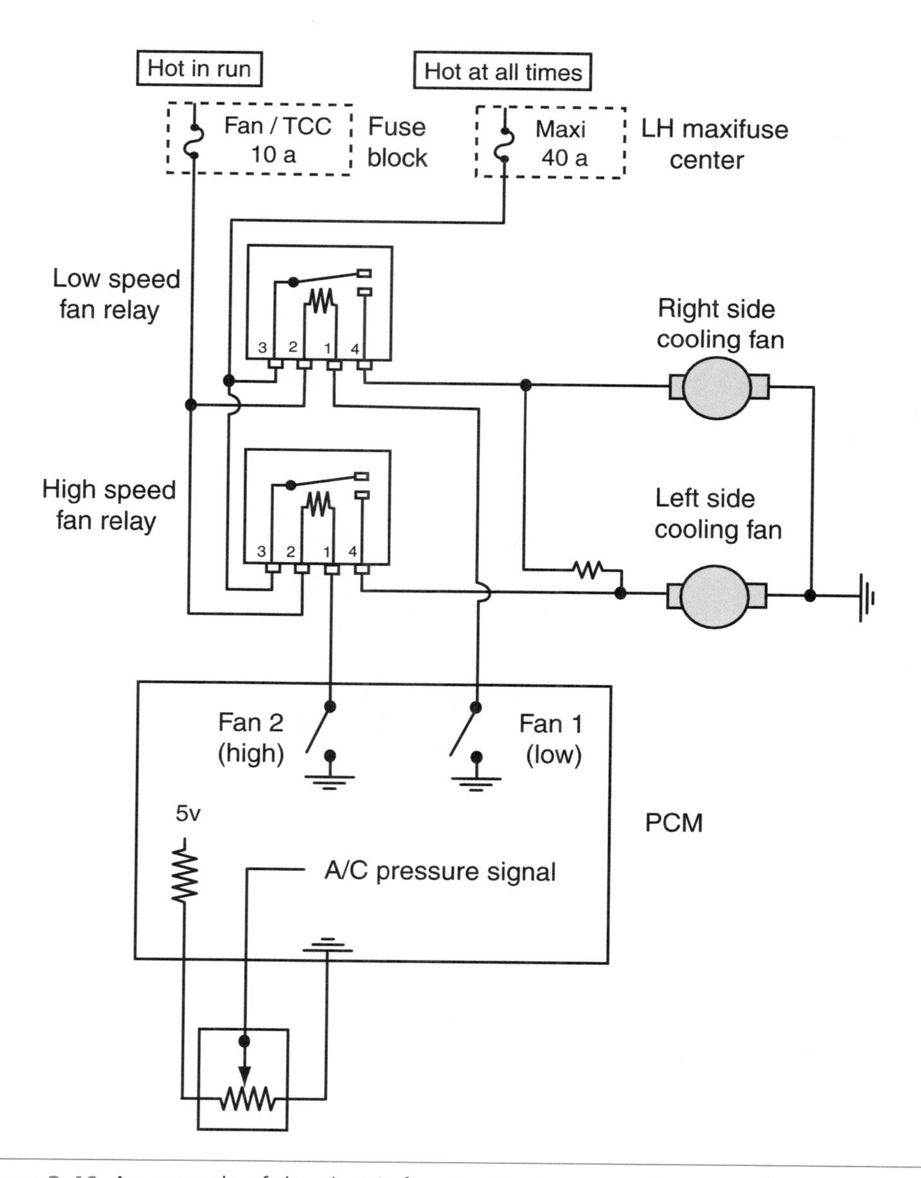


Figure 3-46 An example of the electric fan circuit using an engine controller.

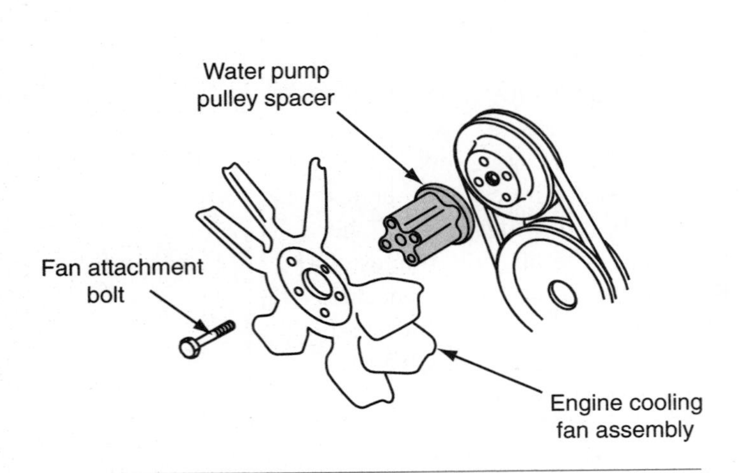


Figure 3-47 Belt-driven cooling fans are usually mounted to the water pump pulley.

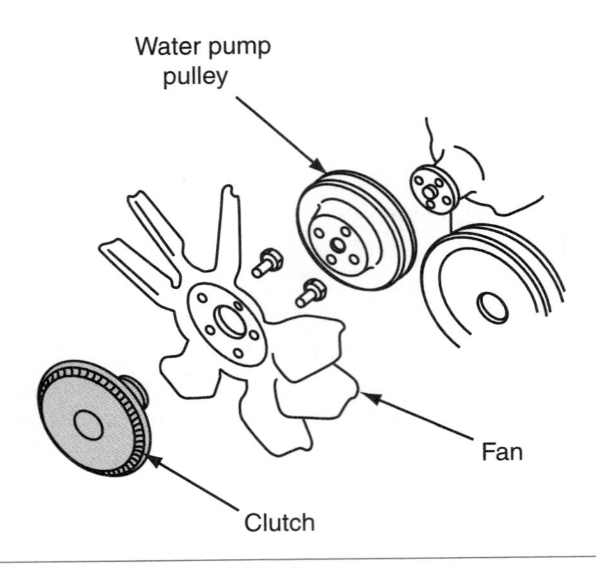


Figure 3-48 Some fans use a clutch to reduce noise and load on the engine.

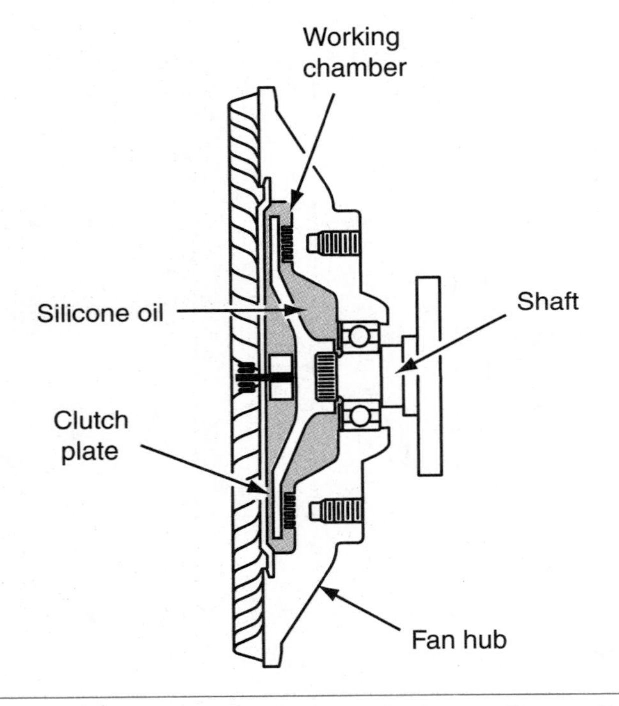


Figure 3-49 The viscous-drive clutch uses silicone oil to lock the fan to the hub.

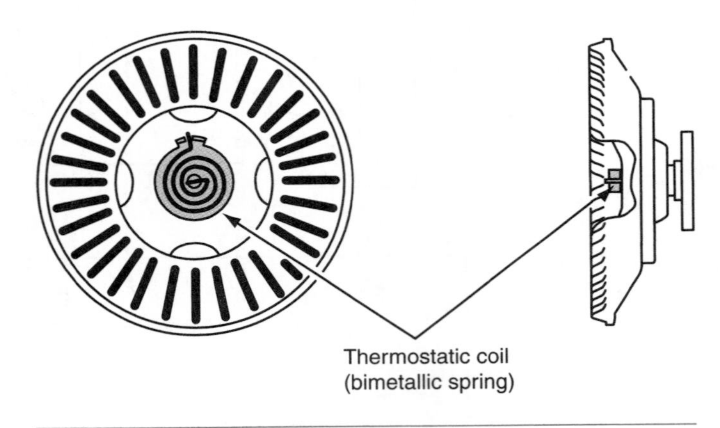


Figure 3-50 A thermostatic spring connected to a piston is another common type of clutch fan.

## Lubrication and Cooling System Warning Indicators

Vehicle manufacturers provide some method of indicating problems with the lubrication or cooling system to the driver. This is done by the use of an indicating gauge or by the illumination of a light. Some manufacturers control the operation of the gauge or light by computers. Regardless of the type of control, sensors or switches provide the needed input.

## Gauge Sending Units

The most common sensor type used for monitoring the cooling system is a **thermistor.** The lubrication system uses piezoresistive sensors to measure oil pressure.

In a simple coolant temperature-sensing circuit, current is sent from the gauge unit into the top terminal of the sending unit, through the variable resistor (thermistor), and to the engine block (ground). The resistance value of the thermistor changes in proportion to coolant temperature (Figure 3-51). As the temperature rises, the resistance decreases, and the current flow through the gauge increases. As the coolant temperature lowers, the resistance value increases, and the current flow decreases.

The **piezoresistive sensor** sending unit is threaded into the oil delivery passage of the engine and the pressure that is exerted by the oil causes the flexible diaphragm to move (Figure 3-52). The diaphragm movement is transferred to a contact arm that slides along the resistor. The position of the sliding contacts on the arm in relation to the resistance coil determines the resistance value and the amount of current flow through the gauge to ground.

The **thermistor** is a resistor whose resistance changes in relation to changes in temperature. It is often used as a coolant temperature sensor.

A **piezoresistive sensor** is sensitive to pressure changes. The most common use of this type of sensor is to measure the engine oil pressure.

## Warning Lamps

A **warning light** may be used to warn of low oil pressure or high coolant temperature. Unlike gauge sending units, the sending unit for a warning light is nothing more than a simple switch. The style of switch can be either normally open or normally closed, depending on the monitored system.

A **warning light** is a lamp that is illuminated to warn the driver of a possible problem or hazardous condition.

Most oil pressure warning circuits use a normally closed switch (Figure 3-53). A diaphragm in the sending unit is exposed to the oil pressure. The switch contacts are controlled by the movement of the diaphragm. When the ignition switch is turned to the on position with the engine not running, the oil warning light turns on. Since there is no pressure to the diaphragm, the contacts remain closed, and the circuit is complete to ground. When the engine is started, oil pressure builds and the diaphragm moves the contacts apart. This opens the circuit, and the warning light goes off. The amount of oil pressure required to move the diaphragm is about 3 psi. If the oil warning light comes on while the engine is running, it indicates that the oil pressure has dropped below the 3 psi limit.

Most coolant temperature warning light circuits use a normally open switch (Figure 3-54). The temperature sending unit consists of a fixed contact and a contact on a bimetallic strip. As the coolant temperature increases, the bimetallic strip bends. As the strip bends, the contacts move closer to each other. Once a predetermined temperature level has been exceeded, the contacts are closed, and the circuit to ground is closed. When this happens, the warning light is turned on.

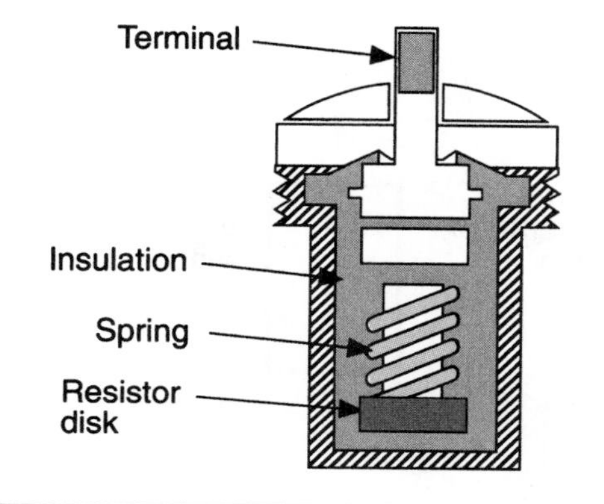


Figure 3-51 A thermistor used to sense engine temperature.

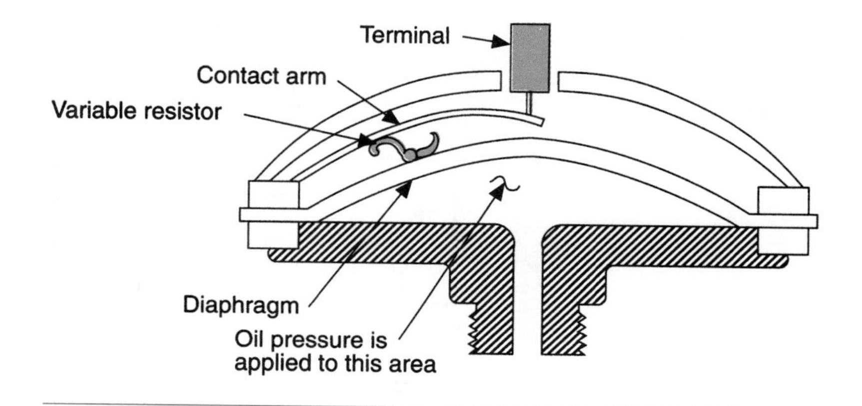


Figure 3-52 Piezoresistive sensor used for measuring engine oil pressure.

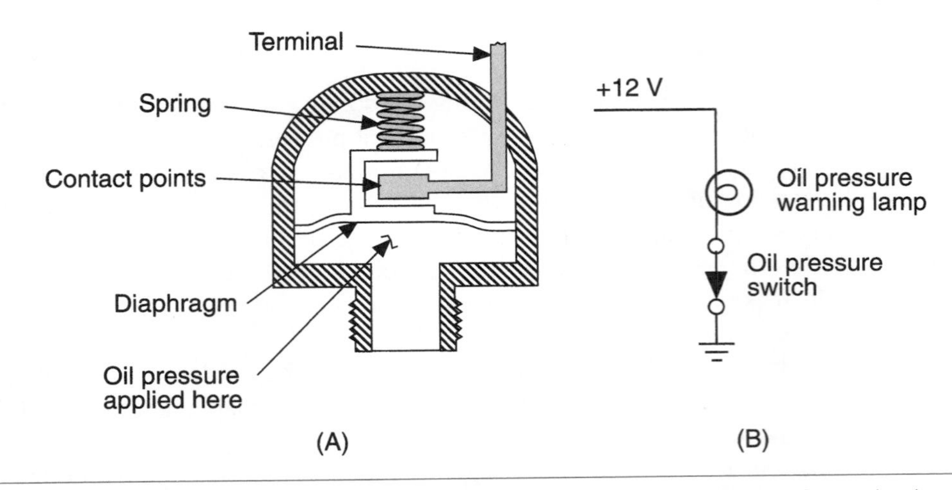


**Figure 3-53** (A) Oil pressure light sending unit. (B) Oil pressure warning lamp circuit.

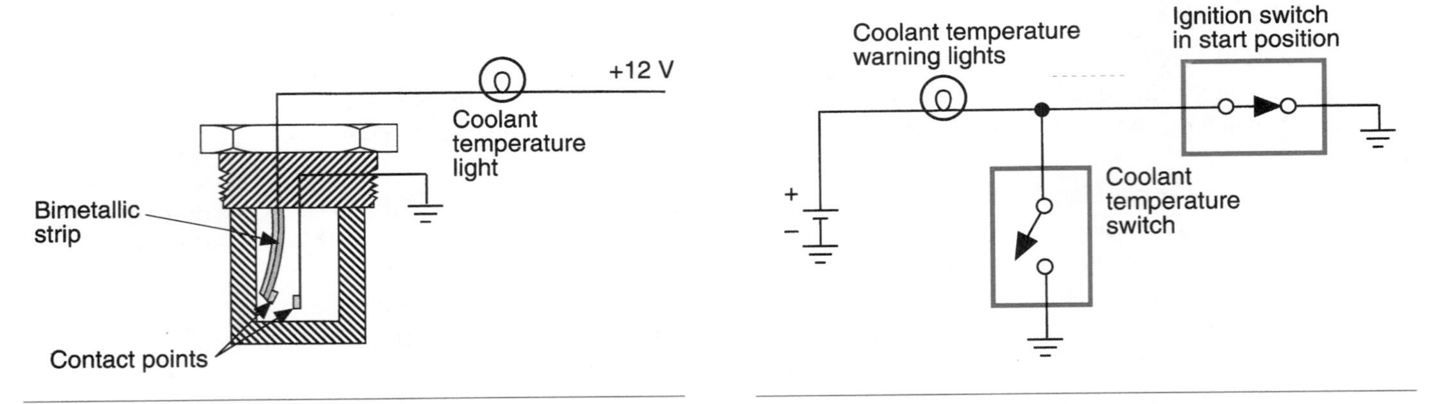


**Figure 3-54** Temperature indicator light circuit.

**Figure 3-55** Prove-out circuit included in normally open (NO) coolant temperature light system.

A **prove-out circuit** completes the warning light circuit to ground through the ignition switch when it is in the START position. The warning light will be on during engine cranking to indicate to the driver that the bulb is working properly.

With normally open switches, the contacts are not closed when the ignition switch is turned to ON. In order to perform a bulb check on normally open switches, a **prove-out circuit** is included (Figure 3-55).

It is possible to have more than one sender unit connected to a single bulb (Figure 3-56). The light will come on whenever oil pressure is low or coolant temperature is too high.

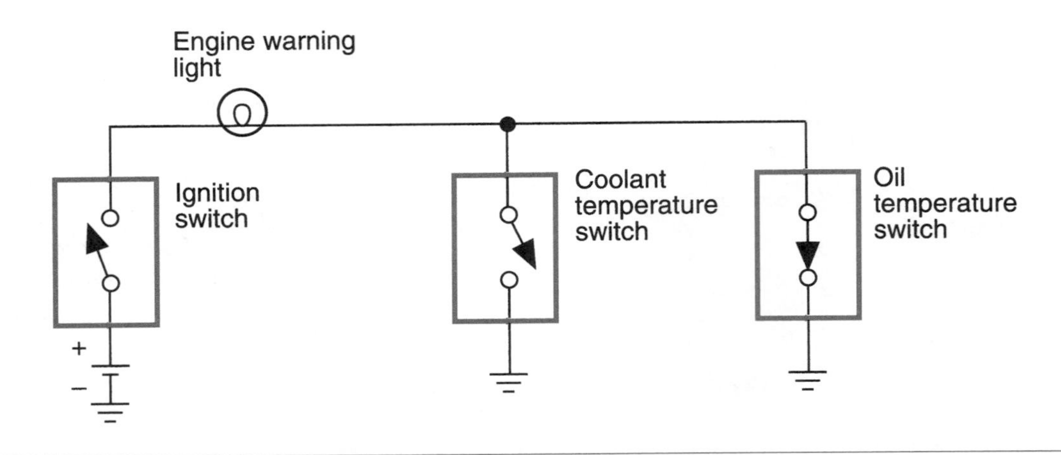


**Figure 3-56** One warning lamp used with two sensors.

### Message Center

Some vehicles manufactured in the late 1990s have a message center that displays a number of warning messages to alert the driver regarding dangerous vehicle operating conditions. Warning messages related to the lubrication and cooling system include: LOW COOLANT, CHECK COOLANT TEMPERATURE, ENGINE OVERHEATED, CHECK ENGINE OIL PRESSURE, CHECK ENGINE OIL LEVEL, and CHANGE ENGINE OIL. The messages are displayed in the message center by the PCM. The PCM receives input signals from the ECT sensor. If the engine coolant reaches a specific temperature above the normal operating temperature, the PCM illuminates the CHECK COOLANT TEMPERATURE. If the engine coolant temperature increases a specific number of degrees, the PCM turns on the ENGINE OVERHEATED message. Special sensors are required to send a low coolant signal or a low engine oil signal to the PCM. The PCM looks at several different engine operating parameters to illuminate the CHANGE OIL MESSAGE. If the engine operating parameters such as engine temperature and rpm are continually in the normal operating range, the PCM will probably turn on the CHANGE OIL MESSAGE at the approximate interval recommended by the vehicle manufacturer for oil change intervals. However, under severe vehicle operating conditions such as trailer towing, the engine temperature may be above normal, and the PCM may illuminate the CHANGE ENGINE OIL message at a lower mileage interval. On some vehicles, the CHANGE ENGINE OIL message comes on for 10 to 25 seconds each time the engine is started. After this time period, the message is turned off. After the oil is changed, the CHANGE ENGINE OIL message must be reset. On some vehicles, this is done by pushing the accelerator pedal wide open three times in a 5-second interval with the ignition switch on and the engine not running. A different procedure may be required to reset the CHANGE ENGINE OIL message on other vehicles. Always consult the vehicle manufacturer's service manual for the proper procedure.

Refer to *Today's Technician: Automotive Engine Performance* for a detailed explanation of fuel system operation.

## Fuel System

The **fuel system** includes the intake system that is used to bring air into the engine and the components used to deliver the fuel to the engine. The fuel delivery system includes the fuel tank, fuel pump, lines, fuel filters, and carburetor or fuel injectors (Figure 3-57).

The **fuel system** includes the intake system that is used to bring air into the engine and the components used to deliver the fuel to the engine.

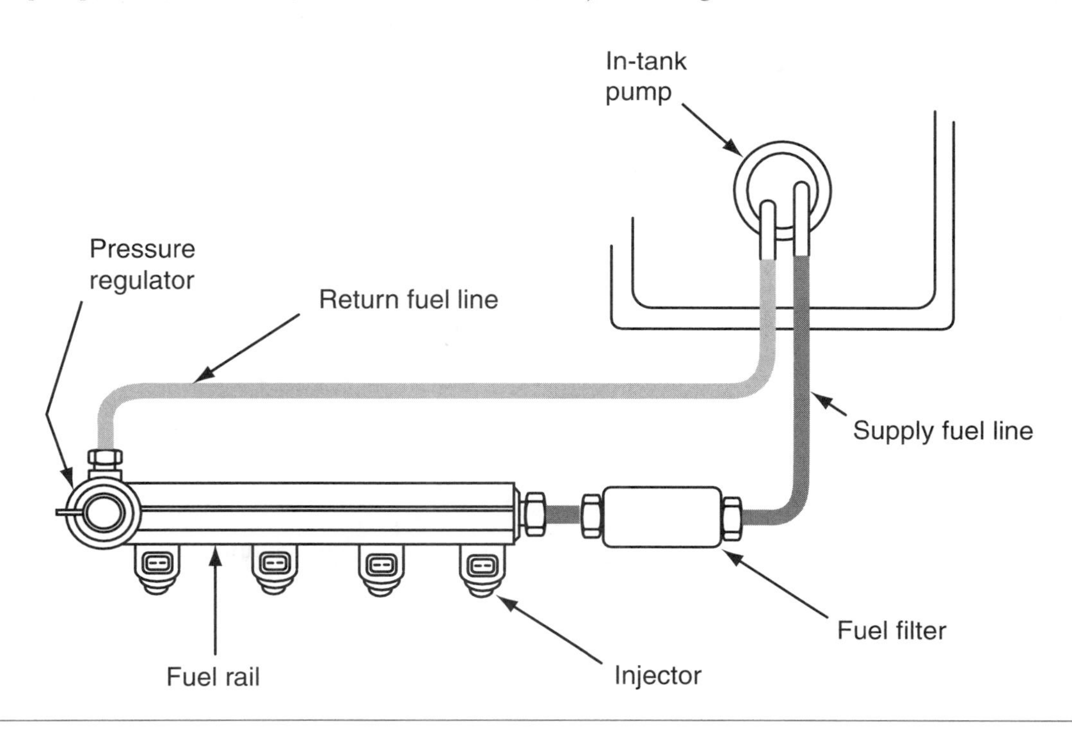


**Figure 3-57** The fuel delivery system.

The **intake manifold** (see Chapter 5) delivers the air or air/fuel mixture to each cylinder.

The air delivery system includes the air intake duct, the air filter assembly, and the **intake manifold** (Figure 3-58).The ultimate goal of the fuel system is to provide the most efficient air/fuel mixture for the running condition of the engine.

## Ignition System

Refer to *Today's Technician: Automotive Engine Performance* for a detailed explanation of ignition system operation.

The delivery of the spark used to ignite the compressed air/fuel mixture is the function of the **ignition system.**

**Burn time** is the amount of time from the instant the mixture is ignited until the combustion is complete.

A **distributor ignition (DI)** system has a distributor that distributes spark to each spark plug.

Making the spark arrive earlier is referred to as **spark advance.**

The compressed air/fuel mixture must be ignited at the correct time. The delivery of the spark is the function of the **ignition system.** If the spark is not delivered at the correct time, poor engine performance and fuel economy will result. **Burn time** of the fuel plays into the calculation of spark delivery. Combustion should be completed by the time the piston is at 10 degrees after top dead center (ATDC) during the power stroke. If the spark occurs too early, when the piston is moving up during the compression stroke, the piston will have to overcome the combustion pressures. If the spark occurs too late, the combustion pressures are not as effective in pushing the piston down during the power stroke. In either case, power output is reduced.

Engine speed is another consideration in spark delivery. As engine speed increases, the amount of piston movement also increases for the same amount of time. Burn time of the gasoline remains constant, but the piston will travel more as engine speed increases. To compensate for this, the spark must be delivered at an earlier time in the piston movement. Advance weights in combination with a vacuum advance unit or the engine computer will advance the spark timing as required based on engine speed and load.

Engines with computer-controlled carburetors and some engines with electronic fuel injection (EFI) systems have **distributor ignition (DI)** systems. In these DI systems, the distributor cap and rotor distribute the spark to the proper spark plug wire and spark plug in the engine firing order. A pickup coil is mounted in the distributor, and a reluctor is fastened to the distributor shaft so the reluctor rotates with the shaft. The reluctor has a high point for each engine cylinder. Each time a reluctor high point rotates past the pickup coil, a voltage is induced in this coil. If the engine is cranking, the pickup coil signal goes directly to the ignition module. When the engine is running, this voltage signal is sent from the ignition module to the PCM on the ignition control (IC) reference wire (Figure 3-59). Several sensors inform the PCM regarding engine operating conditions, such as engine temperature, intake air temperature, engine rpm, throttle opening, engine load, and the amount of air entering the engine. The PCM continually scans the input sensor signals and determines the **spark advance** required by the engine. The PCM sends a voltage signal back to the ignition module on the IC EST wire, and this signal informs the module to open the primary ignition circuit and fire the proper spark plug at the right instant. The IC bypass wire contains a quick disconnect connector that must be disconnected when checking ignition

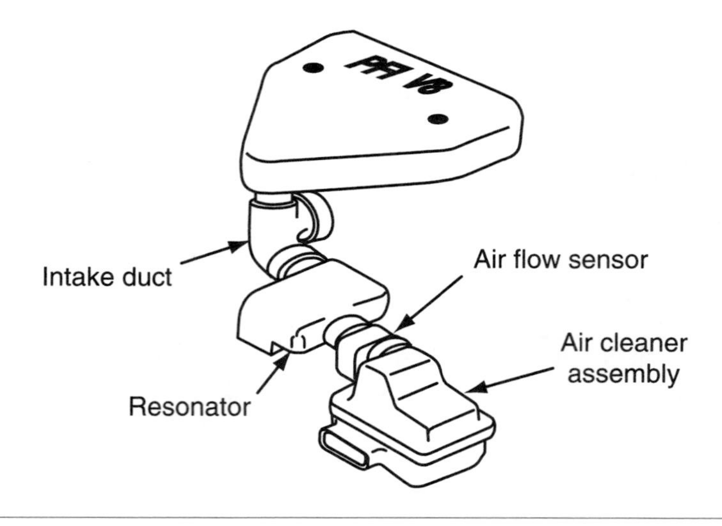


**Figure 3-58** Air intake system components.

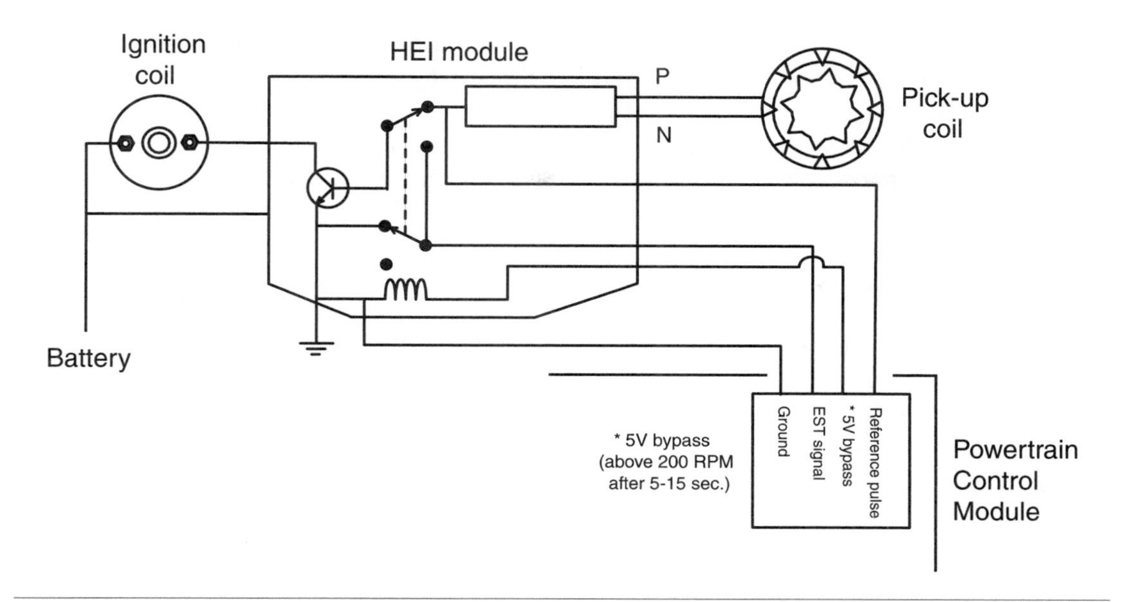


Figure 3-59 Distributor ignition (DI) system.

timing. Disconnecting the quick disconnect connector prevents the PCM from providing any spark advance. The ignition module may be mounted internally in the distributor housing or externally from the distributor. Some ignition modules are contained in the PCM.

In an **electronic ignition (EI)** system, the distributor is no longer required. Some EI systems have an ignition coil for each pair of spark plugs (Figure 3-60). Each coil fires two spark plugs at the same instant. One of the spark plugs being fired is in a cylinder that is on the compression stroke, while the other spark plug is in a cylinder on the exhaust stroke. This type of ignition system may be called a "waste spark" system because firing a cylinder on the exhaust stroke has no effect on engine operation. In most EI systems, a crankshaft position (CKP) sensor and a camshaft position (CMP) sensor send voltage signals to the ignition module and the PCM. These signals inform the PCM regarding the crankshaft and camshaft position. On the basis of the input signals that it receives, the PCM sends voltage signals to the ignition module on the elec-

An **electronic ignition (EI)** system uses an electronic module to open and close the primary circuit.

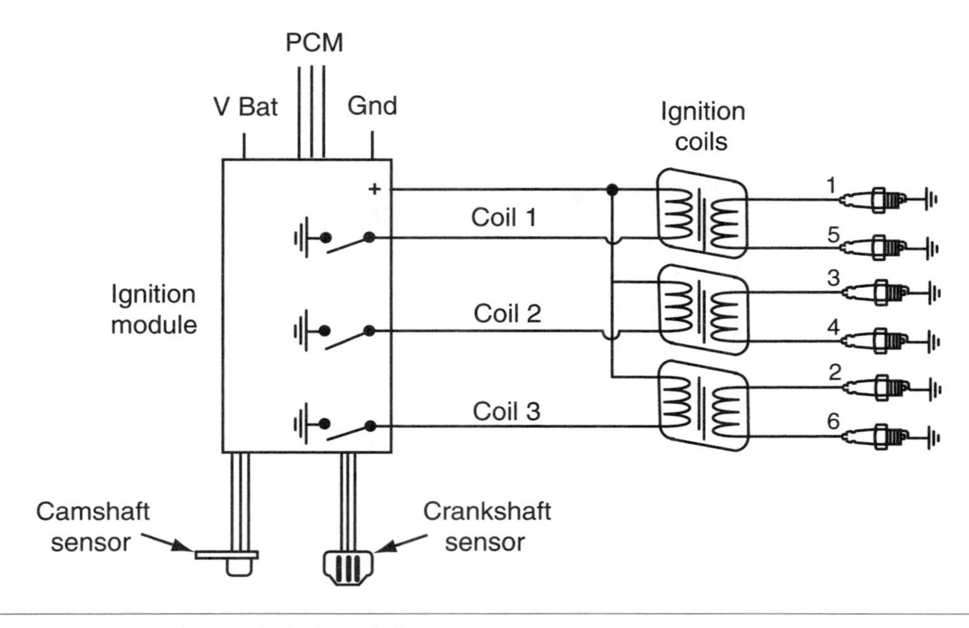


Figure 3-60 Electronic ignition (EI) system.

tronic spark timing (EST) wiring. These voltage signals inform the ignition module to open the primary circuit on the appropriate coil and fire the spark plugs at the proper instant. Many engines now have EI systems with an individual coil connected directly to each spark plug or connected through a short spark plug wire to the spark plug (Figure 3-61). These EI systems may be called coil-on-spark-plug, or coil-near-spark-plug systems. These ignition systems have CKP and CMP sensors and the operating principles are similar to other EI systems.

## Exhaust System

The **exhaust manifold** (see Chapter 5) collects and then directs engine exhaust gases from the cylinders.

The exhaust system removes the byproducts of the combustion process from the cylinders. In order for a full charge of air/fuel mixture to fill the cylinder, all of the burned fuel must be expelled. **Exhaust manifolds** are designed to effectively direct the burned fuel from the cylinders to the exhaust pipe.

Common exhaust systems include the following components:

- Exhaust manifold
- Crossover pipe
- Muffler
- Exhaust pipe
- Catalytic converter
- Tail pipe

Refer to *Today's Technician: Automotive Engine Performance* for a detailed explanation of emission system operation.

## Emission Control System

The **emission control system** includes various devices connected to the engine or exhaust system to reduce harmful emissions of hydrocarbons (HC), carbon monoxide (CO), and oxides of nitrogen ($NO_x$).

The **emission control system** helps to reduce the harmful emissions resulting from the combustion process. The actual controls on an engine depend upon the design of the engine. Most vehicles have these emission control systems:

- Catalytic converter
- Secondary air injection system
- Positive crankcase ventilation valve (PCV) system
- Exhaust gas recirculation (EGR) system
- Evaporative or enhanced evaporative system

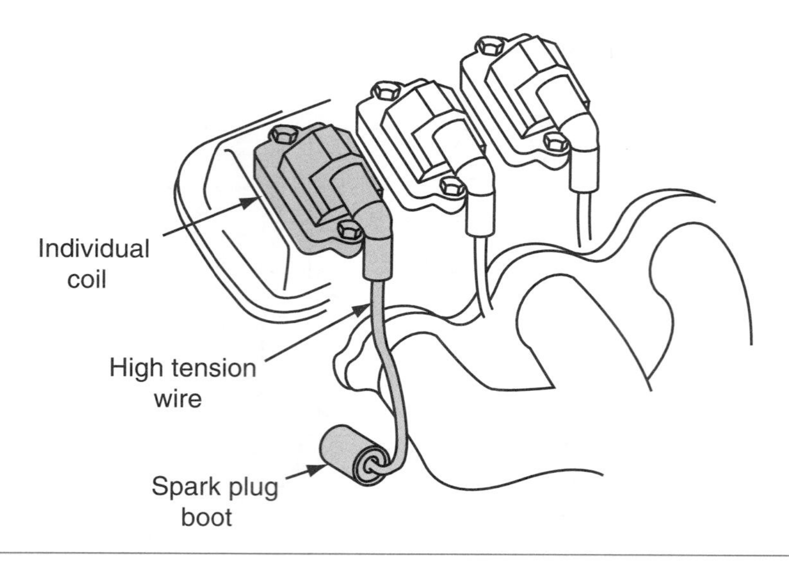


**Figure 3-61** Coil-near-spark-plug ignition system.

## Summary

- The starting system is a combination of mechanical and electrical parts that work together to start the engine.
- The starting system components include the battery, cable and wires, the ignition switch, the starter solenoid or relay, the starter motor, the starter drive and flywheel ring gear, and the starting safety switch.
- An automotive battery is an electrochemical device that provides for and stores electrical energy.
- Electrical energy is produced in the battery by the chemical reaction that occurs between two dissimilar plates that are immersed in an electrolyte solution.
- The lubrication system provides an oil film to prevent moving parts from coming in direct contact with each other. Oil molecules work as small bearings rolling over each other to eliminate friction. Another function is to act as a shock absorber between the connecting rod and crankshaft.
- Many different types of additives are used in engine oils to formulate a lubricant that will meet all of the demands of today's engines.
- Oil is rated by two organizations: the Society of Automotive Engineers (SAE) and the American Petroleum Institute (API). The Society of Automotive Engineers has standardized oil viscosity ratings, while the American Petroleum Institute rates oil to classify its service or quality limits.
- Oil filter elements have a pleated paper or fibrous material designed to filter out particles between 20 and 30 microns.
- There are two basic types of oil pumps: rotor and gear. Both types are positive displacement pumps.
- The radiator is a series of tubes and fins that transfer heat from the coolant to the air.
- In order for heat to be effectively transferred to the air, there must be a difference between the temperature of the coolant and the temperature of the air. Manufacturers increase the temperature of the coolant by pressurizing the radiator.
- The radiator cap allows for an increase of pressure within the cooling system, increasing the boiling point of the coolant.
- The water pump is the heart of the engine's cooling system. It forces the coolant through the engine block and into the radiator and heater core.
- Control of engine temperatures is the function of the thermostat. When the coolant is below normal operating temperature, the thermostat is closed, preventing coolant from entering the radiator. When normal operating temperature is obtained, the thermostat opens, allowing the coolant to enter the radiator to be cooled.
- In a reverse flow cooling system, coolant flows from the water pump through the cylinder heads and then through the engine block.
- In some instrument panels, a message center displays specific warning messages to alert the driver regarding dangerous vehicle operating conditions.
- Some engines have electronic distributor ignition (DI) systems, while most engines manufactured at present have electronic ignition (EI) systems which do not have a distributor.

### Terms to Know

Armature
Battery terminals
Burn time
C-class oil
Coolant recovery system
Cooling system
Distributor ignition (DI)
Electrolysis
Electronic ignition (EI)
Emission control system
Exhaust manifold
Fuel system
Full-filtration system
Head gaskets
Heater core
Ignition system
Inlet check valve
Intake manifold
Lubrication system
Micron
Nucleate boiling
Oil breakdown
Oil pump
Oxidation
Piezoresistive sensor
Pole shoes
Positive displacement pump
Prove-out circuit
Radiator
Reverse flow cooling system
Saddle
S-class oil
Spark advance
Thermistor
Thermostat
Viscosity
Warning light

# Review Questions

## Short Answer Essays

1. What are the purposes of the engine's lubrication system?
2. Explain the purpose of the Society of Automotive Engineers' (SAE) classifications of oil.
3. Explain the purpose of the American Petroleum Institute's (API) classifications of oil.
4. Describe the two basic types of oil pumps: rotary and gear.
5. What is the purpose of the starting system?
6. Explain the operation of the thermostat.
7. Describe the function of the radiator.
8. Explain the function of the water pump.
9. Explain the purpose of the pressure and vacuum valves used in the radiator cap.
10. Describe the purpose of antifreeze and explain its characteristics.

## Fill-in-the-Blanks

1. It is the function of the engine's lubrication system to supply oil to the ________________ ________________ and ________________ locations and to remove heat from them.
2. The Society of Automotive Engineers has standardized oil ________________ ratings, while the American Petroleum Institute rates oil to classify its ________________ or quality limits.
3. Oil pumps are ________________ displacement pumps.
4. A pressure relief valve opens to return oil to the sump if the pressure is ________________ .
5. If the oil filter becomes plugged, a ________________ ________________ will open and allow the oil to enter the galleries.
6. Coolant can become slightly acid because of the minerals and metals in the cooling system. A small ________________ ________________ may flow between metals through the acid and have a corrosive effect on the metals.
7. For heat to be effectively transferred to the air, there must be a ________________ in temperature between the coolant and the air.
8. The radiator cap allows for an increase of ________________ within the cooling system.
9. Operation of the thermostat is performed by a specially formulated wax and powdered metal ________________ located in a heat-conducting copper cup.
10. Viscous-drive fan clutches operate the fan in relation to engine ________________ by a silicone oil.

## Multiple Choice

1. Lubrication systems are being discussed.
   *Technician A* says the oil bypass valve opens when pressures are exceeded in the oil galleries.
   *Technician B* says the relief valve opens when the filter is plugged.
   Who is correct?
   **A.** A only **C.** Both A and B
   **B.** B only **D.** Neither A nor B

2. *Technician A* says the oil pump is a positive displacement pump.
   *Technician B* says the lubrication system uses both pressurized oil and splash to lubricate the engine.
   Who is correct?
   **A.** A only **C.** Both A and B
   **B.** B only **D.** Neither A nor B

3. *Technician A* says some lubrication systems use an inlet check valve to prevent oil drainback from the oil pump.
   *Technician B* says the oil filter is located after the oil pump.
   Who is correct?
   **A.** A only **C.** Both A and B
   **B.** B only **D.** Neither A nor B

4. An engine is experiencing low oil pressure.
   *Technician A* says the relief valve may be stuck closed.
   *Technician B* says there may be excessive clearance between the pump gears.
   Who is correct?
   **A.** A only **C.** Both A and B
   **B.** B only **D.** Neither A nor B

5. Engine coolant is being discussed.
   *Technician A* says a solution of pure antifreeze provides the best protection against overheating.
   *Technician B* says electrolysis in the cooling system causes metal corrosion.
   Who is correct?
   **A.** A only **C.** Both A and B
   **B.** B only **D.** Neither A nor B

6. *Technician A* says downflow radiators have vertical fins directing coolant flow from the top inlet tank to the bottom outlet tank.
   *Technician B* says crossflow radiators use horizontal fins to direct coolant flow through the core.
   Who is correct?
   **A.** A only **C.** Both A and B
   **B.** B only **D.** Neither A nor B

7. *Technician A* says the radiator cap pressurizes the cooling system so heat can be better transferred to the atmosphere.
   *Technician B* says as the pressure is increased, the boiling point decreases.
   Who is correct?
   **A.** A only **C.** Both A and B
   **B.** B only **D.** Neither A nor B

8. *Technician A* says when the thermostat is closed because coolant temperatures are too cold, the coolant is circulated through a bypass.
   *Technician B* says the pellet of the thermostat is installed facing toward the radiator.
   Who is correct?
   **A.** A only **C.** Both A and B
   **B.** B only **D.** Neither A nor B

9. *Technician A* says the temperature rating of the thermostat is the point at which it is completely open.
   *Technician B* says the thermostat opens when the pellet expands.
   Who is correct?
   **A.** A only **C.** Both A and B
   **B.** B only **D.** Neither A nor B

10. Cooling fans are being discussed.
    *Technician A* says electric fans may be controlled by a temperature switch.
    *Technician B* says viscous-drive fan clutches do not rotate if the engine temperature is hot.
    Who is correct?
    **A.** A only **C.** Both A and B
    **B.** B only **D.** Neither A nor B

# Engine Materials and Fasteners

Upon completion and review of this chapter, you should be able to:

- ❒ Describe and identify the various types of materials used in engine construction, including iron, steel, aluminum, plastics, ceramics, and composites.
- ❒ Explain common usages of aluminum alloys in engine construction.
- ❒ Describe the different manufacturing processes, including casting, forging, machining, stamping, and powdered metal.
- ❒ Describe various treatment methods, including heat treating, tempering, annealing, case hardening, and shot peening.
- ❒ Describe the methods used to locate cracks in castings.
- ❒ Properly identify, inspect, and select the correct fasteners required to assemble engine components to the block.

## Introduction

This chapter covers the materials and machining processes used to construct the components of the engine. Today's technician should be able to identify these characteristics properly in order to perform repairs or machining operations on the components.

**Metallurgy** is the science of extracting metals from their ores and refining them for various uses.

## Engine Materials

Today's engines are constructed with the use of several different types of materials. The same engine can have iron, steel, aluminum, plastic, ceramic, and composite components. Although a complete understanding of **metallurgy** is not a prerequisite for performing engine repairs, today's technician is challenged to perform repairs requiring the proper handling, machining, and service of these materials.

Metals are divided into two basic groups: **ferrous** and **nonferrous.** All metals have a grain structure that can be seen if the metal is fractured (Figure 4-1). Grain size and position determine the strength and other characteristics of the metal.

**Ferrous metals** contain iron. Cast iron and steel are examples of ferrous metals. These metals will attract a magnet.

**Nonferrous metals** contain no iron. Aluminum, magnesium, and titanium are examples of nonferrous metals. These metals will not attract a magnet.

Figure 4-1 The properties and strength of a metal can be seen in its grain.

Coke is a very hot-burning fuel formed when coal is heated in the absence of air. Coal becomes coke above 1,022°F (550°C). It is produced in special coke furnaces.

Alloys are mixtures of two or more metals. For example, brass is an alloy of copper and zinc.

Gray cast iron is easy to cast and machine and absorbs vibration and resists corrosion.

Nodular iron contains a specific carbon content plus magnesium and other additives.

Iron containing very low carbon (between 0.05 and 1.7 percent) is called steel.

Carbon steel has a specific carbon content.

Stainless steel is an alloy that is highly resistant to rust and corrosion.

Tensile strength is the metal's resistance to being pulled apart.

**Shop Manual**
Chapter 4, page 192

## Iron

Iron is the most common metal used to produce engine components. It comes from iron ore, which is retrieved from the earth. The ore is heated in blast furnaces to burn off impurities. **Coke** is used to fuel the furnaces, resulting in some of the carbon from the coke being deposited into the iron. When the liquid iron is poured from the coke furnace, the resultant billet (called pig iron) will have about a 5 percent carbon content. This makes the iron brittle.

## Cast Iron

Once the iron cools, it can be shipped for several uses. When shipped to an engine manufacturing plant, the pig iron is remelted and poured into a cast or mold. The resultant form, when the iron cools, is referred to as cast iron. During the process of remelting the iron, the amount of carbon content can be controlled, **alloys** can be added, or special heat treatments can be performed to achieve the desired strength and characteristics of the cast iron. The most common type of cast iron used in automotive engine construction is **gray cast.** Gray cast is easy to cast and machine. It also absorbs vibrations and resists corrosion.

Some engine applications require additional strength above that provided by gray cast. **Nodular iron** is used in some engines for the construction of crankshafts, camshafts, and flywheels. Nodular iron contains between 2 and 2.65 percent carbon with small amounts of magnesium and other additives. It is also heat treated, causing the carbon to form as small balls or nodules. The result is a cast iron with reduced brittleness.

## Steel

Steel is produced by heating the iron at a controlled temperature to burn off most of the carbon, phosphorous, sulfur, silicon, and manganese. As the process continues, the correct amount of carbon is then readded. This type of **steel** is referred to as **carbon steel.**

Low carbon steel (carbon content between 0.05 and 0.30 percent) can be easily bent and formed. As the carbon content of the steel increases, so does its resistance to denting and penetration. However, the higher the carbon content, the more brittle the steel. Medium carbon steel (carbon content between 0.30 and 0.60 percent) is used for connecting rods, crankshafts, and camshafts in some engines. High carbon steel (carbon content between 0.70 and 1.70 percent) has limited use in engine applications, but is used to make drill bits, files, hammers, and other tools.

The quality of steel can be improved through the addition of alloys. Some of the most common alloys include nickel, molybdenum, tungsten, vanadium, silicon, manganese, and chromium. Steel with 11 to 26 percent content of chromium is referred to as **stainless steel.** Nickel is often added to make the steel able to withstand sudden shock loads. If increased **tensile strength** is required, vanadium can be added.

The addition of chromium and molybdenum (sometimes called chrome-moly) produces a very strong steel. This alloy is often used to make crankshafts in high-performance engines. Chrome-moly is also used in the construction of tube frames used in racing cars.

## Magnesium

Magnesium is a very lightweight metal, about 2/3 the weight of aluminum. Pure magnesium is expensive and has a low tensile strength; however, it can be alloyed with other metals (such as aluminum). Magnesium alloy is used in valve covers on some engines; for example, the Porsche 911 used magnesium alloy for the engine block.

## Aluminum

Aluminum is a silver-white ductile metallic element. It is the most abundant metal in the earth's crust. There is no pure aluminum found in the earth though. It is always mixed with other elements. The aluminum we use comes from an ore called bauxite, containing 50 percent alumina (aluminum oxide). The bauxite is crushed and then ground into a fine powder. Then the impurities are separated from the powdered ore by mixing it with a hot caustic soda solution. It is then pumped into large pressure tanks called digesters. The temperature in the tanks, which is maintained at 300°F (149°C), dissolves the alumina, but the impurities remain solid. The impurities can then be filtered out. The remaining liquid is pumped into precipitator tanks where it is allowed to cool slowly. The alumina comes out of the liquid as crystals when it cools. The crystals are then heated until white hot to remove any water. The result is a dry, white powder. To complete the process of turning alumina into aluminum, it is dissolved in a substance called cryolite. An electric current is passed through the pot containing the mixture from carbon anodes that hang in the liquid from overhead bars. The current flows through the liquid and causes the liquid to break up. Pure aluminum then falls to the bottom of the pot.

Although aluminum is not as strong as steel, it weighs about one-third less than steel. Pure aluminum is not strong enough to be used for most engine applications, but the attractiveness of weight reduction has led to the development of many types of aluminum alloys that are proving acceptable. Today, most pistons used in automotive engines are constructed of aluminum alloys.

The light weight of aluminum pistons allows for higher engine speed and increased engine responsiveness. The lighter pistons also reduce the amount of load on the connecting rods and crankshaft.

Many manufacturers are now using aluminum alloys for engine block construction. Aluminum is too soft to be used by itself and is unable to withstand the wear caused by the piston rings traveling in the cylinder. Most aluminum blocks of today's engines are either fitted with special cylinder liners (made of iron, steel, or composite) or special casting.

Most manufacturers use an aluminum alloy to construct cylinder heads, engine mounts, water pump housings, air-conditioning generator housings, intake manifolds, and valve covers.

## Titanium

Titanium alloys offer the benefits of light weight and high strength. Titanium alloys are almost as strong as steel, at about half the weight. The most common use for titanium is in racing applications for the construction of connecting rods and valves. The high cost of titanium alloys, coupled with its difficulty in being welded and machined, have limited their use in production vehicles. However, the NSX from Acura was introduced in 1990 with titanium connecting rods.

## Plastics

Plastic is made from petroleum.

Plastic is a substance made from petroleum by a special method that joins atoms together to make long chains of atoms. The word "plastic" means something that can be pressed into a new shape. After pure plastic is made, it can be improved by adding other material, to give it additional strength, stiffness, or density.

Plastics were first introduced in the automotive engine when American Motors used plastic to make valve covers on their 258 CID 6-cylinder engine. In subsequent years, other manufacturers have used plastics to help reduce the overall weight of the engine. Today some manufacturers are using forms of plastics to make intake manifolds, pulleys, and valve timing train covers. Plastic intake manifolds transfer less heat to the intake air compared to cast iron or aluminum intake manifolds. This action reduces intake air temperature and engine detonation.

**Composites** are man-made materials using two or more different components tightly bound together. The result is a material that has characteristics that neither component possesses on its own.

## Composites

Manufacturers are experimenting with increased use of **composite** materials. These man-made materials are showing great promise in engine applications. Graphite reinforced fiber or nylon materials have been successfully used as connecting rods, pushrods, rocker arms, intake manifolds, and cylinder liners.

Ceramic is a combination of nonmetallic powdered materials fired in special kilns. The end product is a new product.

**Shop Manual**
Chapter 4, page 101

## Ceramics

**Ceramics** also show good promise in automotive engine uses. They are lightweight, provide good frictional reduction, are heat resistant, isolate sound and vibrations, and are brittle. Engine components constructed from ceramics require special handling due to being so brittle. Ceramic components that are in use today include the compressor and turbine wheels of some General Motors' turbochargers and the rocker arm pads on some Mitsubishi engines. In addition, the aftermarket parts suppliers are making ceramic valves, valve seats, valve spring retainers, and wrist pins available.

Typical uses of ceramics today also include the rotor of some turbochargers, the liner for exhaust ports, intake and exhaust valves, valve seats, and piston pins.

# Manufacturing Processes

Not only do different types of irons, steels, and aluminum have different qualities, the manufacturing process also determines their strength and properties. The most common processes are casting, forging, machining, and stamping. In addition to the manufacturing process, the material may be specially treated to increase strength.

## Casting

As discussed earlier, casting requires heating the metal to a liquid, then pouring it into a mold. When the metal cools, it returns to a solid state. The molds can be made of foundry sand. After the metal cools, the sand is broken away to expose the part.

The mold is made by packing the foundry sand around a wooden pattern. The pattern is removed prior to the liquid metal being poured in. This type of casting usually leaves a rough, grainy appearance.

Other casting methods incorporate the use of a polystyrene foam pattern. This pattern is left in the mold when the molten metal is poured in. The heat of the molten metal vaporizes the foam.

Cast iron is commonly used for engine block, camshaft, connecting rod, and crankshaft construction.

Aluminum is usually cast into permanent, reusable molds. The molten aluminum alloys are forced into the mold under pressure or through the use of centrifugal force. The pressure is used to help eliminate any air pockets that may affect the machining process.

Cast aluminum is commonly used for cylinder head, bellhousing, piston, and intake manifold construction.

## Forging

Forging heats the metal to a state in which it can be worked and reshaped. It is not heated until it becomes a liquid as it is when casting. When the metal is hot enough to be worked, a forging die is forced onto the metal under great pressure. The metal then assumes the shape of the forging die. If the metal must be worked into complex shapes, several forging dies may be used. Each die will alter the shape until the desired results are obtained.

When the forging die is closed around the hot metal, some of the metal is forced into the parting lines of the die. This excess metal is called flash and is usually removed after the part is removed from the die.

Forged parts are very strong because the high pressure used to force the die onto the component causes the grain structure to follow the shape of the part. The pressure also causes the molecules of the metal to become very compact and tightly bound.

Most engines equipped with turbochargers use forged pistons and some use forged connecting rods and crankshafts. High-performance engines often use a forged crankshaft and camshaft to increase the endurance of the engine.

## Machining

To construct a component from a piece of steel or iron billet is very time consuming and very expensive. Despite this, many engine components are constructed in this manner. Machined components are stronger than cast, but not as strong as forging. Machined components used in today's engines include rocker arms, piston pins, lifters, and followers. High-performance engines use machined steel crankshafts and camshafts.

## Stamping

Some engine components of simple design, and not requiring much thickness, can be stamped out of a sheet of metal. The metal is not heated during this process. Depending on the final design of the component, it can be stamped on a press punch. This process is used if the piece is to remain flat. The punch is in the shape of the piece to be cut. The press forces the punch through the sheet of metal, much like a cookie cutter through dough. If the finished product must have some bends to it, it is bent by the stamping process, then trimmed to final specifications. Common components made from stamped steel include oil pans, valve covers, timing train covers, and heat shields.

## Powder Metallurgy

Metal powder may be derived by cooling melted metal very quickly. Other methods include reducing the metal oxide, electrolysis, and crushing. The powder can then be blended with other metals to produce an alloy. The powder is then poured into a die and compacted by a cold press. Next it is **sintered** to make the powder bond together.

This process is currently being used to construct some connecting rods and valve seats. After the connecting rod is removed from the die, it is forged into final shape, then shot peened to increase its strength. This process provides a strong component at a light weight. It also eliminates many of the machining processes forged or machined connecting rods must undergo.

Powder metallurgy is the manufacture of metal parts by compacting and sintering metal powders. **Sintering** is done by heating the metal to a temperature below its melting point in a controlled atmosphere. The metal is then pressed to increase its density.

## Treatment Methods

To obtain the desired result of the component, engine manufacturers may have the component treated by heat, chemical, or shot-peening processes. Heating the metal will change its grain structure. By heat treating a metal component, its properties can be altered. To heat treat a metal properly, it is heated to a desired temperature depending on the metal used and the desired results. Once the temperature is reached, it is maintained for a specific amount of time. The last step is to cool the metal at a controlled rate.

To harden a carbon steel, it is heated to 1,400°F (760°C), then quickly cooled. Heating the steel causes its grain structure to become finer. When it is cooled very fast, the grain structure does not have time to change and remains very fine. The harder the steel becomes, the more brittle it becomes.

To counteract the effects of hardening steel, **tempering** of the metal may be the next step. The steel is heated to a temperature between 300 and 1,100°F (150 to 600°C), then cooled slowly. The higher the temperature used to temper the steel, the more hardening is lost, yet the toughness of the metal is increased.

Hard metals must be softened in order to be machined. This process is called **annealing.** Annealing is much like tempering except the temperatures are increased to above 1,000°F (550°C). The metal is then allowed to cool at a slower rate than used for tempering. This makes the grain structure of the metal coarse. If desired, the metal can be hardened and tempered again after the machining process is completed.

To help protect the shell or outer surfaces of a component, the manufacturer may require it to be **case hardened.** This process does not alter the core structure of the metal. Case hardening

In **tempering**, metal is heated to a specific temperature to reduce the brittleness of hardened carbon steel.

**Annealing** is a heat-treatment process to reduce metal hardness or brittleness, relieve stresses, improve machinability, or facilitate cold working of the metal.

**Case hardening** is a heating and cooling process for hardening metal. Because the case hardening is only on the surface of the component, it is usually removed during machining procedures performed at the time the engine is being rebuilt. It is not necessary to replace the case hardening in most applications.

**Shot peening** is performed by blasting the component's surface with steel or glass shot to prestress the material.

**Shop Manual**
Chapter 4, page 198

is performed after all machining processes are completed. Case hardening of crankshafts protects the journals from wear and fractures. Most manufacturers now use a process called ion nitriding to case harden their components. The component is placed into a pressurized chamber filled with hydrogen and nitrogen gases. An electrical current is then applied through the component. This changes the molecular structure and allows the induction of the gases onto the surface area of the component.

Another form of metal treatment is **shot peening.** Shot peening is performed by blasting the component's surface with steel or glass shot. When the balls hit the component, a small dent is made. Thousands of these dents are applied to the component's surface. The dents overlap each other and compress the component's surface. This process of prestressing means any tension forces applied to the part must overcome the compression forces before the part will crack.

## Cracks

Cracks are the result of stress in the casting. A crack can cause fluid or compression leakage. When performing engine service requiring the removal of the intake manifold, exhaust manifold, cylinder heads, and so on, check the component for cracks. The following is a list of some of the most common causes for this stress:

1. Fatigue
2. Excessive flexing
3. Impact damage
4. Extreme temperature changes in a very short time
5. Freezing of the engine coolant
6. Excessive overheating
7. Detonation
8. Defects during the casting process

**AUTHOR'S NOTE:** It has been my experience that one of the common causes of cracked cylinder heads is adding cold water to the cooling system on an overheated engine. This action subjects the cylinder head to a very sudden change in temperature resulting in a cracked head. If the engine overheats, do not loosen the radiator cap, and do not add water and/or antifreeze to the cooling system until the engine cools down.

## Detecting Cracks

Many cracks are detected by a thorough visual inspection. However, very small stress cracks may not be detected in this manner. There are several different methods of crack detection. Following is a sample of the four most common:

1. *Magnetic particle inspection (MPI).* Uses an electromagnet to create a magnetic field in the casting. Because all magnets have north and south poles, the magnetic field runs from one pole to the other through the casting. A crack in the casting causes a break in the field, creating opposite poles (Figure 4-2). The magnetic powder is attracted to this area.
2. *Magnetic fluorescent.* This method uses a fluorescent paste dripped onto the casting. The casting is then placed in a magnetic field and observed under a black light. Cracks are visible by white, gray, or yellow streaks.
3. *Penetrant dye.* Using dye penetrant is a three-step process. First, the special dye is sprayed onto the casting surface and allowed to dry. The excess dye is then wiped away. Next, a special remover is sprayed over the surface and the casting is rinsed with water. Third, the developer is sprayed onto the casting. As it dries, any dye left in the cracks seeps through the developer. The crack shows as a red line against a white background.

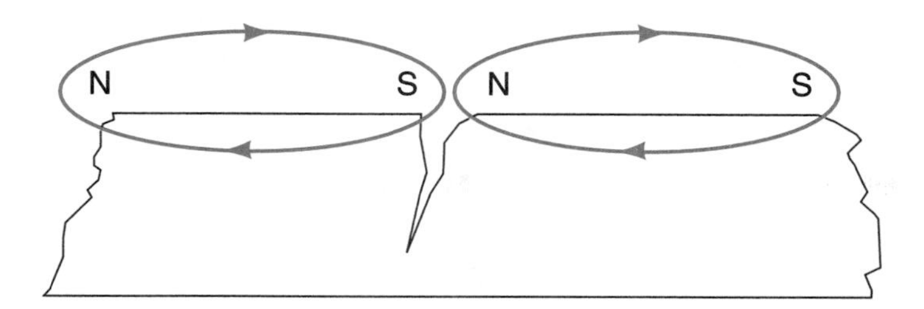


Figure 4-2 A crack causes a break in the magnetic field, creating opposite magnetic poles.

4. *Pressurizing the block or head.* The casting can be pressurized with air or water. When air is used, the casting is attached to a special fixture and 40 to 60 psi of air pressure is applied. A soapy solution is then sprayed onto the casting while looking for bubbles. Water pressure testing uses hot water to cause casting expansion. If the casting is leaking, water will be found on the outside of it.

## Fasteners

There are many different types of fasteners used throughout the engine. The most common are threaded fasteners, including bolts, studs, screws, and nuts. These fasteners must be inspected for thread damage, fillet damage, and stretch before they can be reused (Figure 4-3). In addition, many threaded fasteners are not designed for reuse; the service manual should be referenced for the manufacturer's recommendations. If a threaded fastener requires replacement, there are some concerns the technician must be aware of:

- Select a fastener of the same diameter, thread pitch, strength, and length as the original.
- All bolts in the same connection must be of the same grade.
- Use nut grades that match their respective bolts.
- Use the correct washers and pins as originally equipped.
- Torque the fasteners to the specified value.
- Use torque-to-yield bolts where specified.

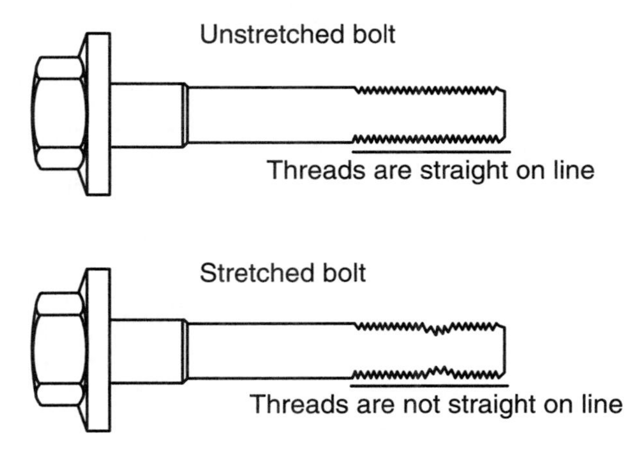


Figure 4-3 Check all bolts for stretch and other damage before reusing them.

Unified National Fine-thread or Unified National Course-thread bolts may be referred to as fine-thread or coarse-thread bolts.

**Bolt diameter** indicates the diameter of an imaginary line running through the center of the bolt threaded area.

**Pitch** is the number of threads per inch on a bolt in the USC system.

**Bolt length** indicates the distance from the underside of the bolt head to the tip of the bolt threaded end.

Threaded fasteners used in automotive applications are classified by the Unified National Series using four basic categories:

1. Unified National Coarse (UNC or NC)
2. Unified National Fine (UNF or NF)
3. Unified National Extrafine (UNEF or NEF)
4. Unified National Pipe Thread (UNPT or NPT)

In recent years, the automotive industry has switched to the use of metric fasteners. Metric threads are classified as course or fine, as denoted by an SI or ISO lettering.

The most common type of threaded fastener used on the engine is the bolt (Figure 4-4). To understand proper selection of a fastener, terminology must be defined (Figure 4-5). The head of the bolt is used to torque the fastener. Several head designs are used, including hex, torx, slot, and spline. **Bolt diameter** is the measure across the threaded area or shank. The **pitch** (used in the English system) is the number of threads per inch. In the metric system, thread pitch is a measure of the distance (in millimeters) between two adjacent threads. **Bolt length** is the distance from the bottom of the head to the end of the bolt. The **grade** of the bolt denotes its strength and is used to designate the amount of stress the bolt can withstand. The grade of the bolt depends upon the material it is constructed from, bolt diameter, and **thread depth. Grade marks** are placed on the top of the head (in the English system) to identify the bolt's strength (Figure 4-6). In the metric system, the strength of the bolt is identified by a property class number on the head (Figure 4-7). The larger the number, the greater the tensile strength.

Like bolts, nuts are graded according to their tensile strength (Figure 4-8). As discussed earlier, the nut grade must be matched to the bolt grade. The strength of the connection is only as strong as the lowest grade used; for example, if a grade 8 bolt is used with a grade 5 nut, the connection is only a grade 5.

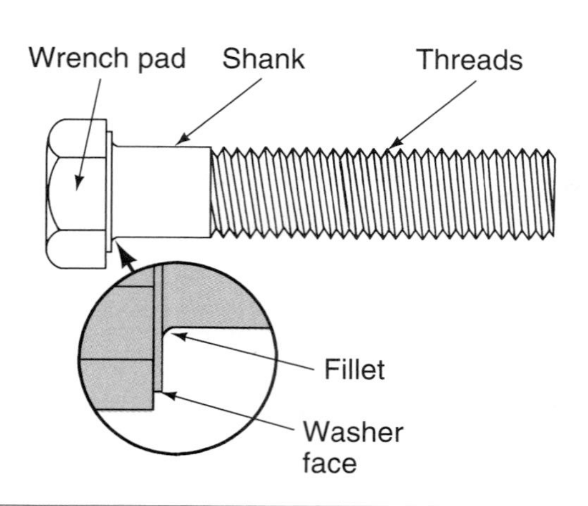


**Figure 4-4** Typical bolt.

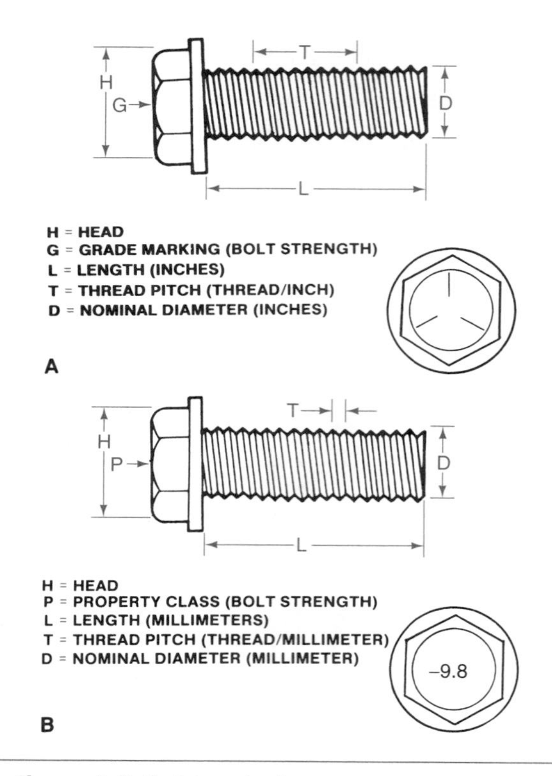


**Figure 4-5** Bolt terminology.

| SAE grade markings | 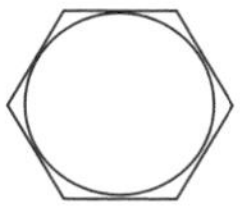 | 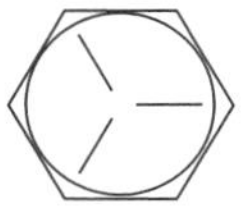 | 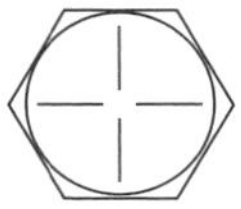 |  |  |
|---|---|---|---|---|---|
| Definition | No lines: unmarked indeterminate quality SAE grades 0-1-2 | 3 lines: common commercial quality Automotive and AN bolts SAE grade 5 | 4 lines: medium commercial quality Automotive and AN bolts SAE grade 6 | 5 lines: rarely used SAE grade 7 | 6 lines: best commercial quality NAS and aircraft screws SAE grade 8 |
| Material | Low carbon steel | Med. carbon steel tempered | Med. carbon steel quenched and tempered | Med. carbon alloy steel | Med. carbon alloy steel quenched and tempered |
| Tensile strength | 65,000 psi | 120,000 psi | 140,000 psi | 140,000 psi | 150,000 psi |

**Figure 4-6** Bolt grade identification marks.

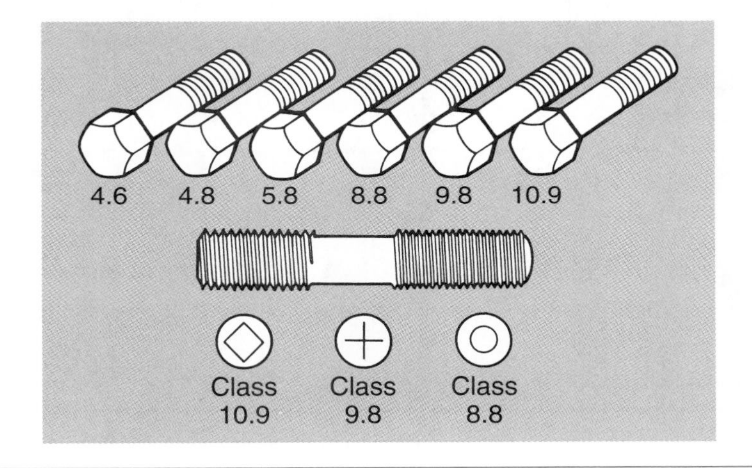


**Figure 4-7** Property class numbers.

Proper fastener torque is important to prevent thread damage and to provide the correct clamping forces. The service manual provides the manufacturer's recommended torque value and tightening sequence for most fasteners used in the engine. The amount of torque a fastener can withstand is based on its tensile strength (Figure 4-9). In order to obtain proper torque, the fastener's threads must be cleaned and may require light lubrication.

**Grade** is a classification of material.

**Thread depth** is the height of the thread from its base to the top of its peak.

**Grade marks** are radial lines on the bolt head.

## Torque-to-Yield

Modern engines are designed with very close tolerances. These tolerances require an equal amount of clamping forces at mating surfaces. Normal head bolt torque values have a calculated 25 percent safety factor; that is, they are torqued to only 75 percent of the bolt's maximum proof load (Figure 4-10). Using the chart, it can be seen that a small difference between torque values at the bolt head can result in a large difference in clamping forces. Because torque is actually force used to turn a fastener against friction, the actual clamping forces can vary even at the same torque value. Up to about 25 ft-lb (35 Nm) of torque, the clamping force is fairly constant; however, above this point, variation of actual clamping forces at the same torque value can be as

| Inch system | | Metric system | |
|---|---|---|---|
| Grade | Identification | Class | Identification |
| Hex nut grade 5 | 3 dots | Hex nut property grade 9 | Arabic 9 |
| Hex nut grade 8 | 6 dots | Hex nut property grade 10 | Arabic 10 |
| Increasing dots represent increasing strength. | | Can also have blue finish or paint dab on hex flat. Increasing numbers represent increasing strength. | |

Figure 4-8 Nut grade markings.

| STANDARD BOLT AND NUT TORQUE SPECIFICATIONS | | | | | |
|---|---|---|---|---|---|
| **Size Nut or Bolt** | **Torque (foot-pounds)** | **Size Nut or Bolt** | **Torque (foot-pounds)** | **Size Nut or Bolt** | **Torque (foot-pounds)** |
| 1/4–20 | 7–9 | 7/16–20 | 57–61 | 3/4–10 | 240–250 |
| 1/4–28 | 8–10 | 1/2–13 | 71–75 | 3/4–16 | 290–300 |
| 5/16–18 | 13–17 | 1/2–20 | 83–93 | 7/8–9 | 410–420 |
| 5/16–24 | 15–19 | 9/16–12 | 90–100 | 7/8–14 | 475–485 |
| 3/8–16 | 30–35 | 9/16–18 | 107–117 | 1–8 | 580–590 |
| 3/8–24 | 35–39 | 5/8–11 | 137–147 | 1–14 | 685–695 |
| 7/16–14 | 46–50 | 5/8–18 | 168–178 | | |

Figure 4-9 Standard bolt and nut torque specifications.

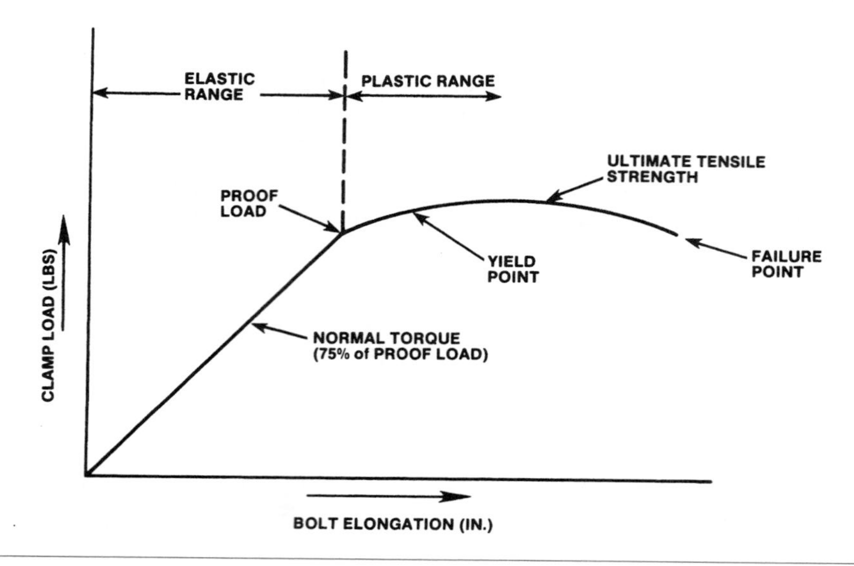


Figure 4-10 Relationship between proper clamp load and bolt failure.

high as 200 percent. This is due to variations in thread conditions or dirt and oil in some threads. Up to 90 percent of the torque is used up by friction, leaving 10 percent for the actual clamping. The result could be that some bolts have to provide more clamping force than others, distorting the cylinder bores.

To compensate and correct for these factors, many manufacturers use torque-to-yield bolts. The yield point of identical bolts does not vary much. A bolt that has been torqued to its yield point can be rotated an additional amount without any increase in clamping force. When a set of **torque-to-yield** fasteners is used, the torque is actually set to a point above the yield point of the bolt. This assures the set of fasteners will have an even clamping force.

**Torque-to-yield** is a stretch-type bolt that must be tightened to a specific torque and then rotated a certain number of degrees.

Manufacturers vary on specifications and procedures for securing torque-to-yield bolts. Always refer to the service manual for exact procedures. In most instances, a torque wrench is first used to tighten the bolts to their yield point. Next, the bolt is turned an additional amount as specified in the service manual.

**Shop Manual**
Chapter 4, page 97

The graph in Figure 4-10 indicates that a bolt can be elongated considerably at its yield point before it reaches its failure point. Also notice, the clamp load is consistent between the proof load and the failure point of the bolt. Bolts that are torqued to their yield points have been stretched beyond their elastic limit and require replacement whenever they are removed or loosened.

### A BIT OF HISTORY

In the 1940s Ralph H. Miller, an American, designed an engine with a compressor that forced more air-fuel mixture into the cylinder during the compression stroke. This design may be called a Miller-cycle engine. This type of engine is used in marine and industrial applications in addition to late-model Mazda Millenias.

## Summary

- ❒ Metals are divided into two basic groups: ferrous and nonferrous. Ferrous metals are those containing iron. Nonferrous metals contain no iron.
- ❒ Alloys are mixtures of two or more metals.
- ❒ Iron containing very low carbon (between 0.05 and 1.7 percent) is called steel.
- ❒ Tensile strength is the metal's resistance to being pulled apart.
- ❒ Composites are man-made materials using two or more different components tightly bound together. The result is a material consisting of characteristics that neither component possesses on its own.
- ❒ A ceramic is a combination of nonmetallic powdered materials fired in special kilns. The end product is a new product.
- ❒ Powder metallurgy is the manufacture of metal parts by compacting and sintering metal powders.
- ❒ Forging heats the metal to a state in which it can be worked and reshaped.
- ❒ Annealing is the process of heating the metal to remove its hardness.

**Terms to Know**

Alloys
Annealing
Bolt diameter
Bolt length
Carbon steel
Case hardening
Ceramics
Coke
Composite
Ferrous metals
Grade
Grade marks
Gray cast iron
Metallurgy
Nodular iron
Nonferrous metals
Pitch
Shot peening
Sintered
Stainless steel
Steel
Tempering
Tensile strength
Thread depth
Torque-to-yield

- ❒ Case hardening is the process of hardening the surface of the metal component.
- ❒ Cracks are the result of stress in the casting.
- ❒ Thread depth is the height of the thread from its base to the top of its peak.

# Review Questions

## Short Answer Essays

1. Briefly describe the following types of materials used in engine construction: iron, steel, aluminum, plastics, ceramics, and composites.
2. Briefly describe the following manufacturing processes: casting, forging, machining, stamping, and powdered metal.
3. Describe the following treatment methods: heat treating, tempering, annealing, case hardening, and shot peening.
4. Describe the advantage of aluminum alloy engine components compared to cast iron or steel components.
5. Describe aluminum alloy block and cylinder design.
6. List the methods available to locate cracks in castings.
7. Describe the purpose of torque-to-yield bolts.
8. Explain the bolt grade marking system.
9. Whenever a fastener is replaced, what must the technician consider?
10. Define the following terms used with bolts:

    Head

    Diameter

    Pitch

    Length

    Grade

## Fill-in-the-Blanks

1. ____________________ metals are those containing iron. ____________________ metals contain no iron.
2. ____________________ ____________________ is the metal's resistance to being pulled apart.
3. ____________________ is the process of heating the metal to remove its hardness.
4. Cracks are the result of ____________________ in the casting.
5. Low carbon steel can be easily ____________ and ____________ .
6. Thread depth is the height of the thread from its ____________________ to the top of its peak.
7. Most aluminum blocks are fitted with cylinder ____________.

8. Bolt diameter is the measure across the ____________________ area.

9. The ____________________ of the bolt denotes its strength and is used to designate the amount of stress the bolt can withstand.

10. When a set of torque-to-yield fasteners is used, the torque is actually set to a point ____________________ the yield point of the bolt.

## Multiple Choice

1. Treatment of metals is being discussed.
   *Technician A* says case hardening alters the structure of the metal throughout the component.
   *Technician B* says shot peening is performed by blasting the component's surface with steel or glass shot to compress the surface.
   Who is correct?
   **A.** A only **C.** Both A and B
   **B.** B only **D.** Neither A nor B

2. While discussing engine materials:
   *Technician A* says plastic intake manifolds are heavier than cast iron intake manifolds.
   *Technician B* says plastic intake manifolds do not transfer as much heat to the intake air.
   Who is correct?
   **A.** A only **C.** Both A and B
   **B.** B only **D.** Neither A nor B

3. Ceramic engine components have all these characteristics EXCEPT:
   **A.** Light weight
   **B.** Heat resistance
   **C.** Non-brittleness
   **D.** Ability to isolate sound and vibration.

4. While discussing engine materials:
   *Technician A* says gray cast iron blocks are easy to cast and machine.
   *Technician B* says nodular iron may be used in crankshafts.
   Who is correct?
   **A.** A only **C.** Both A and B
   **B.** B only **D.** Neither A nor B

5. *Technician A* says stainless steel is made by adding chromium to steel.
   *Technician B* says vanadium added to steel increases the tensile strength.
   Who is correct?
   **A.** A only **C.** Both A and B
   **B.** B only **D.** Neither A nor B

6. Forged engine components have these advantages and applications:
   **A.** Intake manifolds may be forged.
   **B.** Crankshafts may be forged.
   **C.** Rocker arm covers may be forged.
   **D.** Provide a softer metal that is more easily machined.

7. *Technician A* says crankshaft journals are case hardened.
   *Technician B* says case hardening protects the outer surfaces of a component.
   Who is correct?
   **A.** A only **C.** Both A and B
   **B.** B only **D.** Neither A nor B

8. Fastener selection is being discussed.
   *Technician A* says to select a fastener with the same or higher grade than originally installed.
   *Technician B* says the bolt size is determined by the size of the head.
   Who is correct?
   **A.** A only **C.** Both A and B
   **B.** B only **D.** Neither A nor B

9. *Technician A* says torque-to-yield bolts are first torqued to their yield point, then turned an additional amount.
   *Technician B* says torque-to-yield bolts are used to provide uniform clamping forces.
   Who is correct?
   **A.** A only **C.** Both A and B
   **B.** B only **D.** Neither A nor B

10. Engine components may be cracked by:
    **A.** Extreme temperature changes in a very short time.
    **B.** Extremely low, uniform temperatures.
    **C.** An excessive amount of antifreeze in the coolant.
    **D.** Engine operating temperature below normal.

# Intake and Exhaust Systems

Upon completion and review of this chapter, you should be able to:

- ❒ Describe the purpose of the air filter.
- ❒ Describe air filter design.
- ❒ Explain the operation of an airflow restrictor in the air cleaner.
- ❒ List three different materials used to manufacture intake manifolds.
- ❒ Describe the purpose of the intake manifold.
- ❒ Explain the advantages of aluminum and plastic intake manifolds compared to cast iron.
- ❒ Describe the operation and advantages of intake manifolds with dual runners.
- ❒ Describe two different methods for operating the valves that open and close the intake manifold runners.
- ❒ Explain how the engine creates vacuum.
- ❒ Describe how vacuum is used to operate and control many automotive devices.
- ❒ Explain the operation of exhaust system components, including exhaust manifold, gaskets, exhaust pipe and seal, catalytic converter, muffler, resonator, tailpipe, and clamps, brackets, and hangers.
- ❒ Properly perform an exhaust system inspection, and service and replace exhaust system components.

## Introduction

An internal combustion engine requires airflow, fuel, and spark to provide combustion in the cylinders. This air supply is drawn into the engine by the vacuum created during the intake stroke of the pistons. The air is mixed with fuel and delivered to the combustion chambers. Controlling the flow of air and the air/fuel mixture is the purpose of the induction system.

## Air Induction System

On carbureted engines, the induction system was quite simple. It consisted of an air cleaner housing mounted on top of the engine with a filter inside the housing. Its function was to filter dust and grit from the air being drawn into the carburetor. The carburetor supplied the fuel and the intake manifold delivered the air/fuel mixture to the cylinders.

The air intake system on a modern fuel-injected engine is rather complicated (Figure 5-1). Ducts channel cool air from outside the engine compartment to the throttle body assembly. The air filter is placed below the top of the engine to allow for aerodynamic body designs. Electronic sensors measure airflow, temperature, and density. On some engines, pulse air systems provide fresh air to the exhaust stream to oxidize unburned hydrocarbons in the exhaust. These components allow the air induction system to perform the following functions:

- Filter the air to protect the engine from wear
- Silence air intake noise
- Heat or cool the air as required
- Provide the air the engine needs to operate

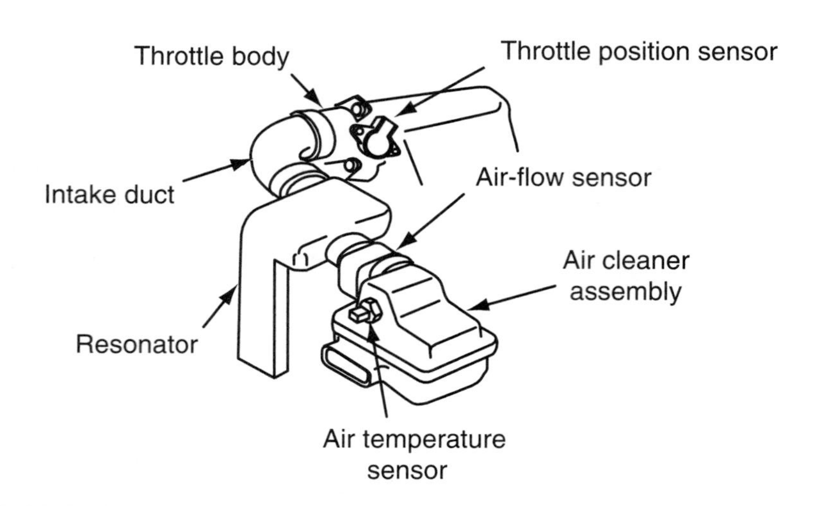


Figure 5-1 A late-model intake air distribution system.

- Monitor airflow temperature and density for more efficient combustion and a reduction of hydrocarbon (HC) and carbon monoxide (CO) emissions
- Operate with the PCV system to burn the crankcase fumes in the engine
- Provide air for some air injection systems

## Air Intake Ductwork

Figure 5-2 shows an air induction system for an older engine equipped with a carburetor. This system uses a **fresh air tube** to take in cool air from outside the engine compartment. It also uses a tube to pull in warm air from around the exhaust manifold.

This warm air tube warms the intake air during engine warmup in cold weather. As technology and body designs changed, air cleaner assemblies were placed away from the engine. Ductwork was and is used to direct the air into the throttle body. Cool outside air is drawn into the air cleaner assembly and, on some engines, warm air from around the exhaust is also brought in for cold engine operation (Figure 5-3).

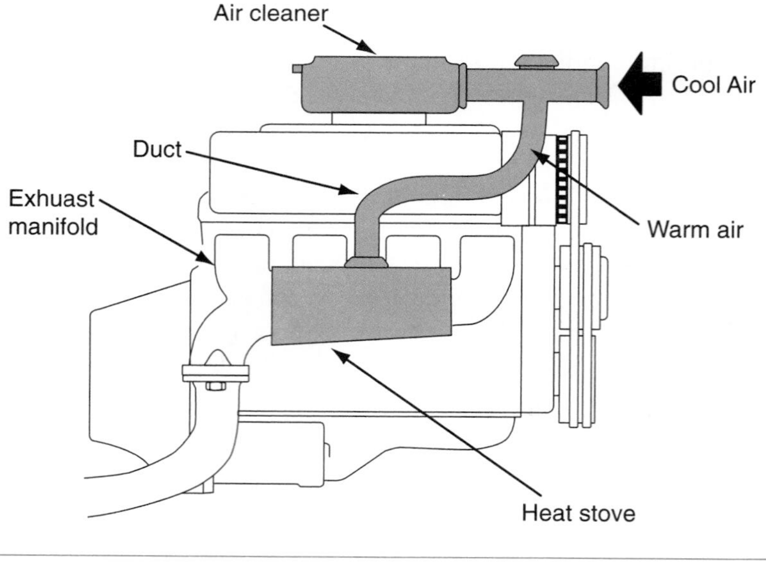

A **fresh air tube** in a heated air inlet system may be called a zip tube.



Figure 5-2 Typical air induction system for carburetor-equipped engines.

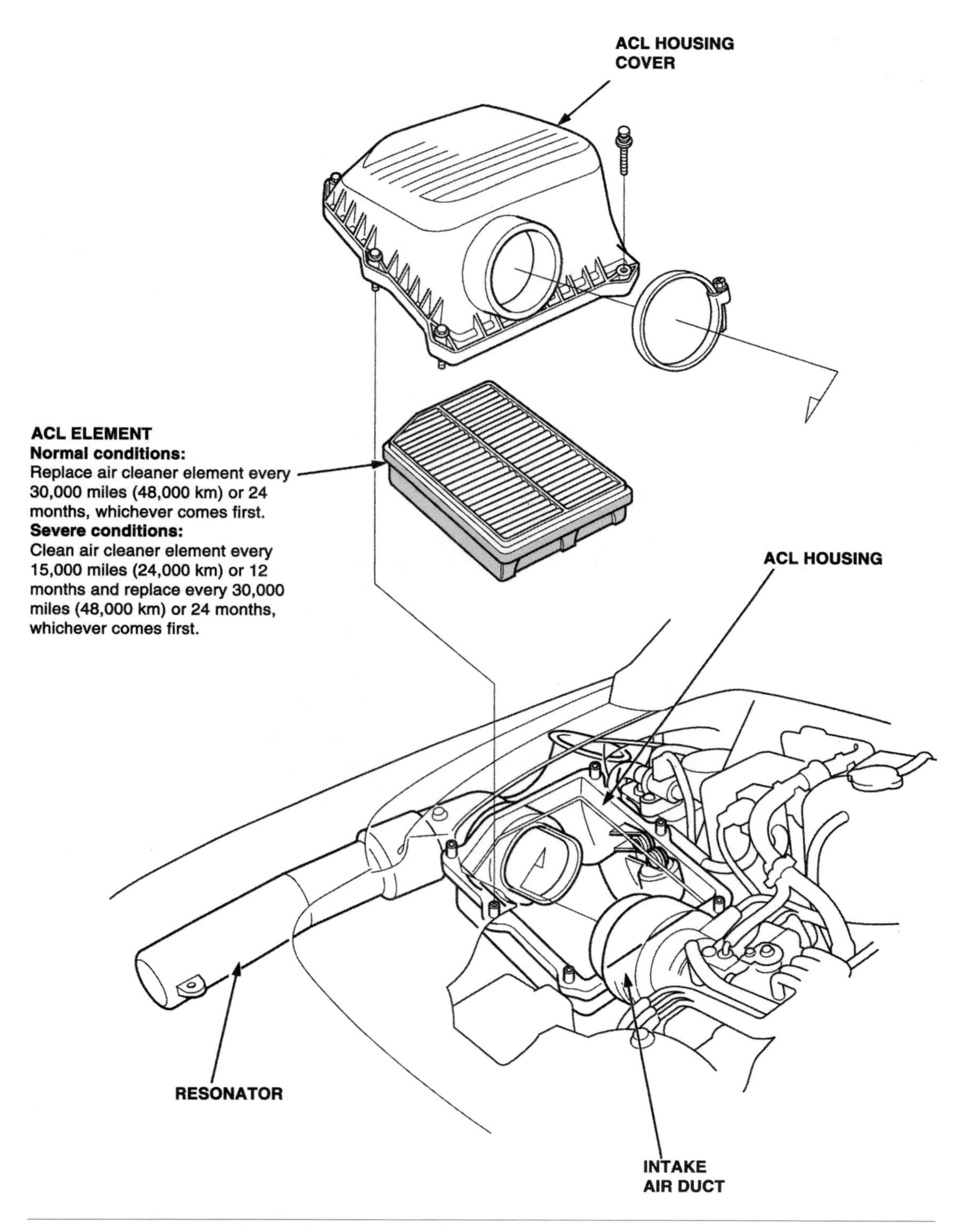


Figure 5-3 Typical air induction system for engines equipped with fuel injection. (Courtesy of American Honda Motor Co., Inc.)

The most recent designs have remote air cleaner assemblies with a **mass airflow (MAF) sensor** installed in the ductwork (Figure 5-4).

Other sensors may also be installed in the air cleaner assembly (Figure 5-5) or in the ductwork leading to the throttle body assembly. The air cleaner assembly also provides filtered air to the PCV system.

Be sure that the intake ductwork is properly installed and all connections are airtight, especially those between an airflow sensor or remote air cleaner and the throttle plate assembly.

In many fuel injection systems a **mass airflow (MAF) sensor** performs the same function as an airflow meter.

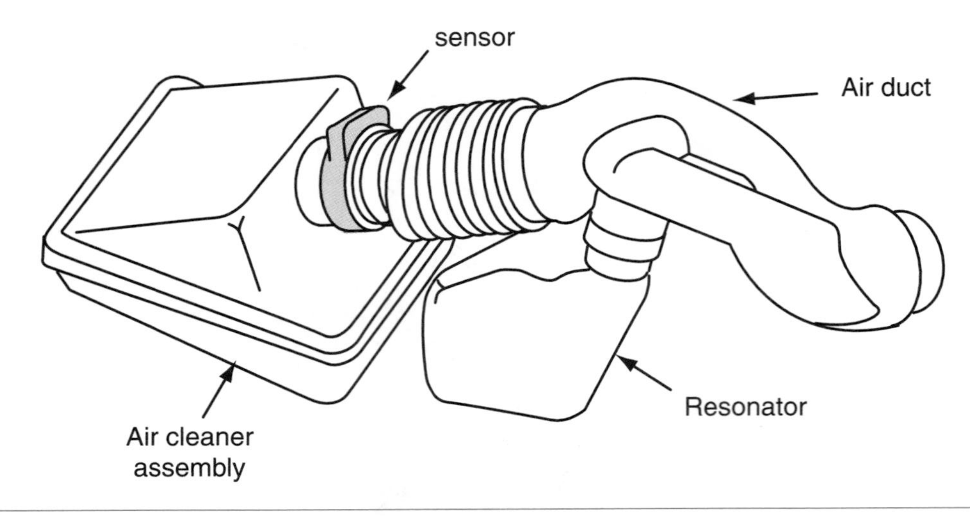


Figure 5-4 Ducts are used on remote air cleaners.

Figure 5-5 An intake air temperature sensor mounted inside the air cleaner.

Generally, metal or plastic air ducts are used when engine heat is not a problem. Special paper-metal ducts are used when they will be exposed to high engine temperatures.

## Air Cleaner/Filter

The **air filter** cleans all air entering the engine.

The primary function of the **air filter** is to prevent airborne contaminants and abrasives from entering into the engine. Without proper filtration, these contaminants can cause serious damage and appreciably shorten engine life. All incoming air should pass through the filter element before entering the engine.

### Air Filter Design

Air filters are basically assemblies of pleated paper supported by a layer of fine mesh wire screen. The screen gives the paper some strength and also filters out large particles of dirt. A thick plastic-like gasket material normally surrounds the ends of the filter. This gasket adds strength to the filter and serves to seal the filter in its housing. If the filter does not seal well in the housing, dirt and dust can be pulled into the air stream to the cylinders. In most air filters the air flows from the outside of the element to the inside as it enters the intake system. On some air intake systems, the air flows from the inside of the element to the outside.

The shape and size of the air filter element depends on its housing; the filter must be the correct size for the housing or dirt will be drawn into the engine. On today's engines, air filters are either flat (Figure 5-6) or round (Figure 5-7). Air filters must be properly aligned and closed around the filter to ensure good airflow of clean air.

Many air cleaners in recent years have an **airflow restriction indicator** mounted in the air cleaner housing (refer to Figure 5-8). If the air filter element is not restricted, a window in the side of the restriction indicator shows a green color. When the air filter element is restricted, the window in the airflow restriction indicator is orange and "Change Air Filter" appears. Then the air filter must be replaced (Figure 5-8). After the air filter is replaced, a reset button on top of the airflow restriction indicator must be pressed to reset the indicator so it displays green in the window.

An airflow restriction indicator displays the amount of air cleaner element restriction.

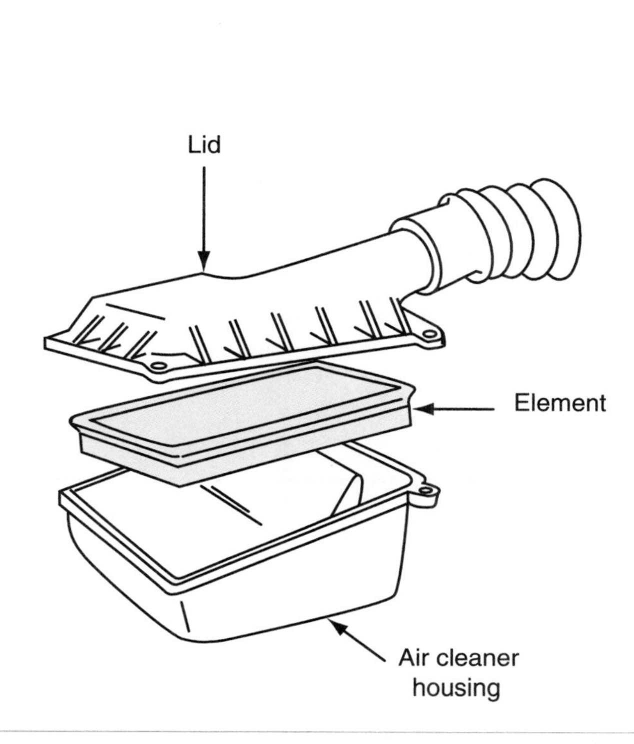


Figure 5-6 Typical flat air cleaner element.

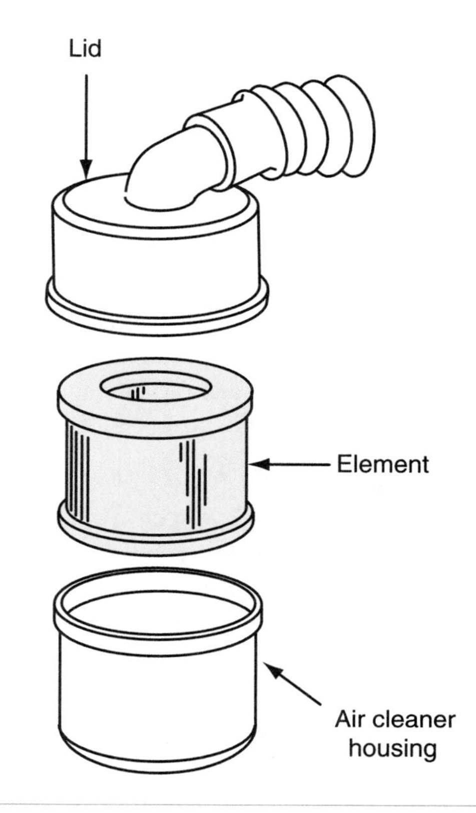


Figure 5-7 Typical round air filter for a late-model vehicle.

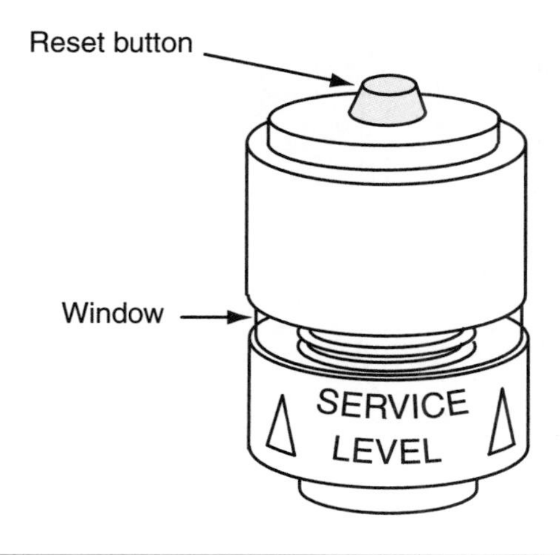


Figure 5-8 Airflow restriction indicator.

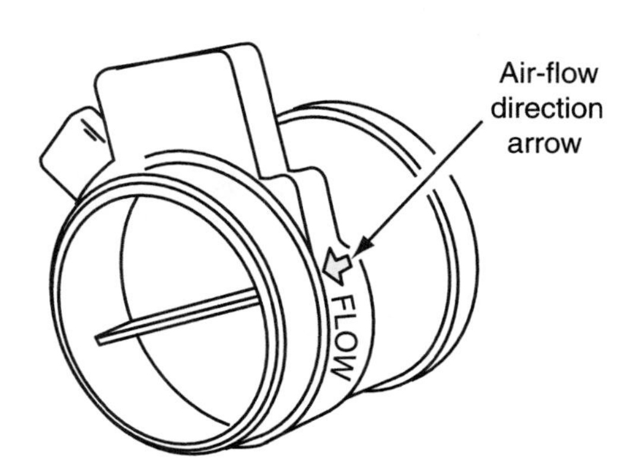


Figure 5-9 The mass airflow (MAF) sensor must be installed in the proper direction.

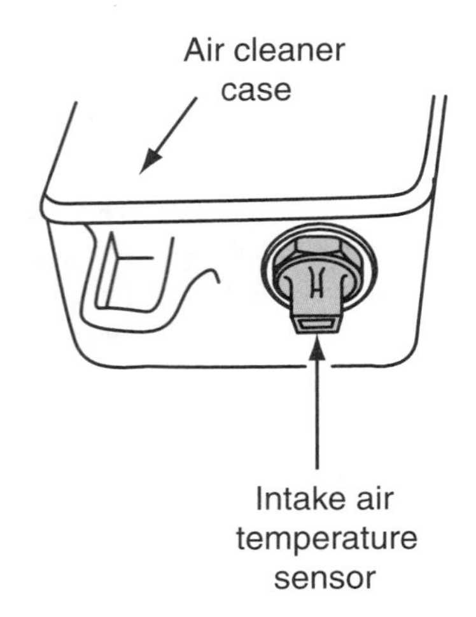


**Figure 5-10** An intake air temperature (IAT) sensor is mounted in some air cleaners.

The **intake air temperature (IAT) sensor** sends an analog voltage signal to the PCM in relation to air intake temperature.

The **powertrain control module (PCM)** is a computer that controls the engine and other functions.

An **intake manifold** is a cast iron, aluminum, or plastic casting with internal passages that conducts air or an air-fuel mixture from the throttle body or carburetor to the cylinder head intake ports.

**Shop Manual**
Chapter 5, page 222

Some air cleaners have a combined MAF sensor and **intake air temperature (IAT) sensor** attached to the air outlet on the air cleaner housing. A duct is connected from the MAF sensor to the throttle body. The MAF sensor must be attached to the air cleaner so air flows through this sensor in the direction of the arrow on the sensor housing (Figure 5-9). If the airflow is reversed through the MAF sensor, the **powertrain control module (PCM)** supplies a rich air/fuel ratio and increased fuel consumption. Other air cleaners contain a separated IAT sensor (Figure 5-10).

# Intake Manifold

The **intake manifold** distributes the clean air or air/fuel mixture as evenly as possible to each cylinder of the engine.

Most older, carbureted engines and engines with throttle-body injection had cast-iron intake manifolds. With this type of engine, the intake manifold delivered air and fuel to the cylinders. Most early intake manifold designs had short runners (Figure 5-11). These manifolds were either wet or dry. **Wet manifolds** had coolant passages cast directly in them. **Dry manifolds** did not have these coolant passages, but some had exhaust passages. Exhaust gases and/or coolant were

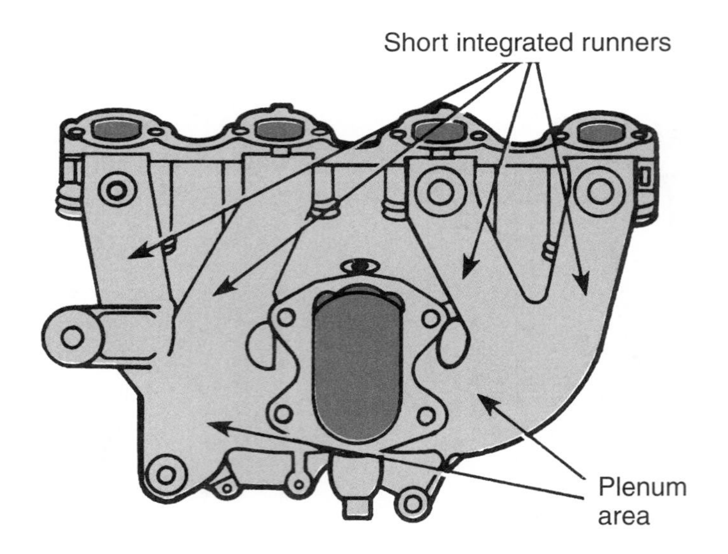


**Figure 5-11** The intake manifold for an in-line 4-cylinder engine.

used to heat up the floor of the manifold. This helped to vaporize the fuel before it arrived in the cylinders. Other dry manifold designs used some sort of electric heater unit or grid to warm up the bottom of the manifold. Heating the floor of the manifold also stopped the fuel from condensing in the manifold's plenum area. Good fuel vaporization and the prevention of condensation allowed for delivery of a more uniform air/fuel mixture to the individual cylinders.

**Wet manifolds** have coolant passages cast directly into them.

**Dry manifolds** do not have coolant passages, but some have exhaust passages.

Modern intake manifolds for engines with port fuel injection are typically made of die-cast aluminum or plastic. These materials are used to reduce engine weight. A plastic manifold transfers less heat to the intake air and this results in a denser air/fuel mixture. Because intake manifolds for port-injected engines only deliver air to the cylinders, fuel vaporization and condensation are not design considerations. These intake manifolds deliver air to the intake ports where it is mixed with the fuel delivered by the injectors (Figure 5-12). The primary consideration of these manifolds is the delivery of equal amounts of air to each cylinder.

Modern intake manifolds also serve as the mounting point for many intake-related accessories and sensors (Figure 5-13). Some include a provision for mounting the thermostat and thermostat housing. In addition, connections to the intake manifold provide a vacuum source for the exhaust gas recirculation (EGR) system, automatic transmission vacuum modulators, power brakes, and/or heater and air-conditioning airflow control doors. Other devices located on or connected to the intake manifold include the manifold absolute pressure (MAP) sensor, knock sensor, various temperature sensors, and EGR passages.

Most engines cannot produce the amount of power they should at high speeds because they do not receive enough air. This is the reason why many race cars have hood scoops. With today's body styles, hood scoops are not desirable because they increase air drag. However, to get high performance out of high-performance engines, more air must be delivered to the cylinders at high engine speeds. There are a number of ways to do this; increasing the air delivered by the intake manifold is one of them. This can be a little tricky though. Too much airflow at low engine speeds hurts the engine's efficiency. Therefore, manufacturers have developed manifolds that deliver more air only at high engine speeds.

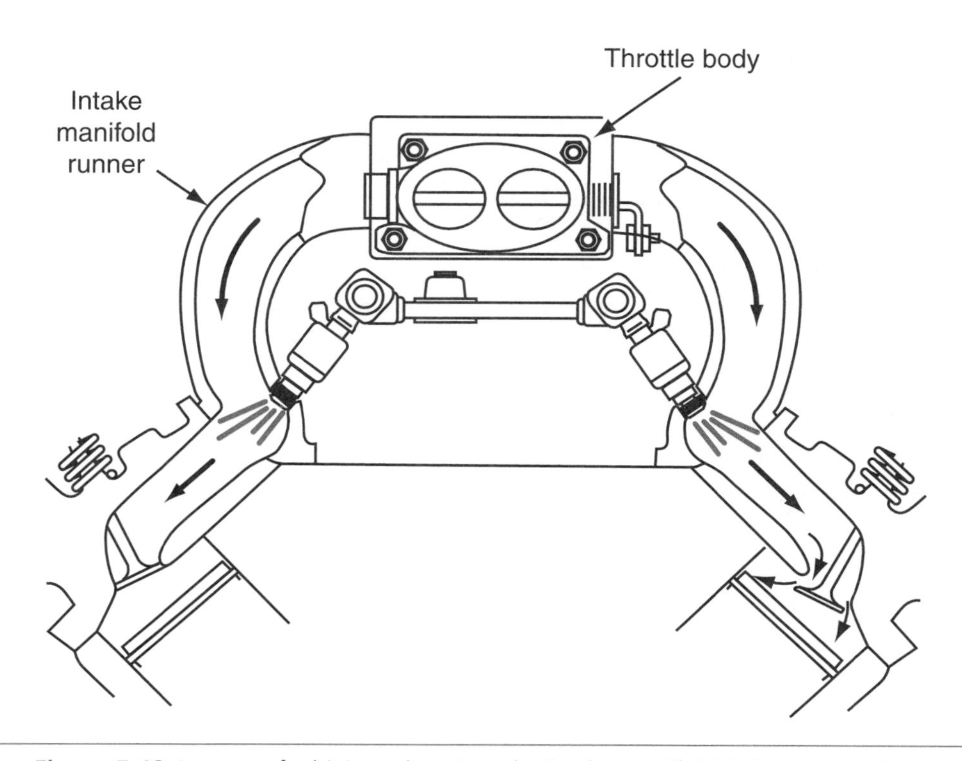


**Figure 5-12** In a port fuel-injected engine, the intake manifold delivers air to the intake ports.

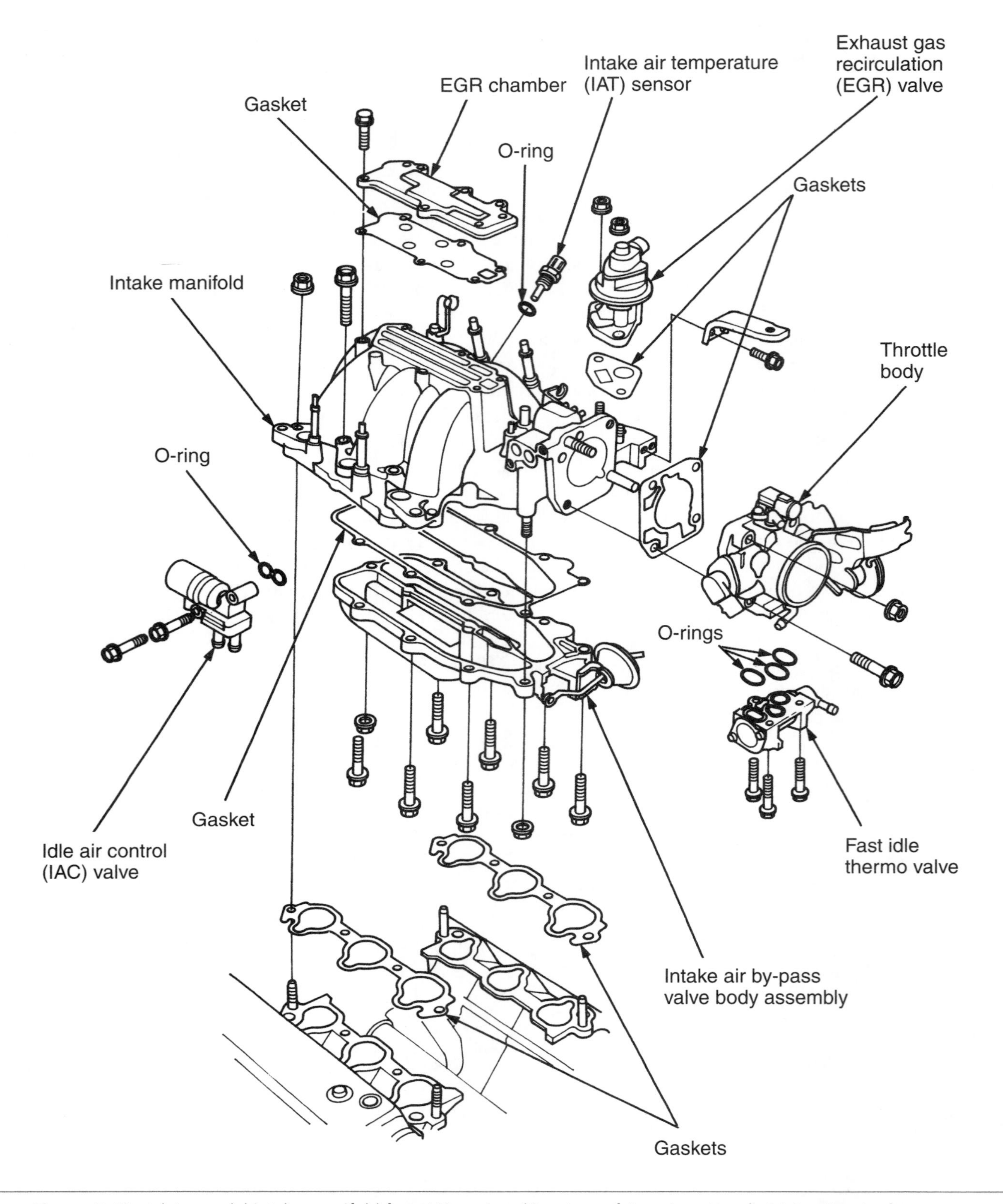


**Figure 5-13** A late-model intake manifold for a V6 engine (Courtesy of American Honda Motor Co., Inc.)

One such system (Figure 5-14) uses a **variable-length intake manifold.** At low speeds the air travels through only part of the manifold on its way to the cylinders. When the engine speed reaches about 3,700 rpm, two butterfly valves open, forcing the air to take a longer route to the intake port. This increases the speed of the airflow, as well as increasing the amount of air

Figure 5-14 A variable-length intake manifold. (Courtesy of Mercedes-Benz of N.A., Inc.)

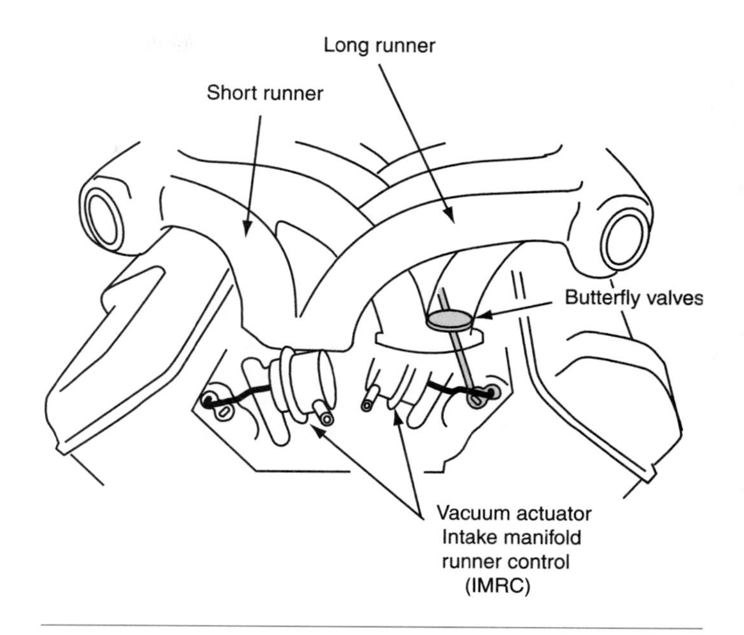


Figure 5-15 A V6 intake manifold with dual runners for each cylinder.

available for the cylinders. As a result, more power is available at high speeds without decreasing low-speed torque and fuel economy and without increased exhaust emissions.

Another approach is the use of two large intake manifold runners for each cylinder (Figure 5-15). Separating the intake runners from the intake ports is an assembly that has two bores for each cylinder. There is one bore for each runner. Both bores are open when the engine is running at high speeds. However, a butterfly valve in one set of the runners is closed by the PCM when the engine is operating at lower speeds. This action decreases airflow speed and volume at lower engine speeds and allows for greater airflow at high engine speeds. The butterfly valves in the intake manifold runners may be operated by a vacuum actuator, and the PCM operates an electric/vacuum solenoid that turns the vacuum on and off to the actuator. In other intake manifolds the valves that open and close some of the runners are operated electrically by the PCM much like a solenoid.

The **variable-length intake manifold** has two runners of different lengths connected to each cylinder head intake port.

**Shop Manual**
Chapter 5, page 225

A PCM operated electric solenoid-type valve that opens and closes some of the intake manifold runners may be called an intake manifold tuning valve (IMTV).

## Vacuum System

The vacuum in the intake manifold is used to operate many systems, such as emission controls, brake boosters, parking brake releases, headlight doors, heater/air conditioners, and cruise controls. Vacuum is applied to these systems through a system of hoses and tubes that can become quite elaborate.

## Vacuum Basics

Vacuum refers to any pressure that is lower than the earth's atmospheric pressure at any given altitude. The higher the altitude, the lower the atmospheric pressure. Vacuum is measured in relation to atmospheric pressure. Atmospheric pressure is the pressure exerted on every object on earth and is caused by the weight of the surrounding air. At sea level, the pressure exerted by the atmosphere is 14.7 psi (101.3 kPa). Atmospheric pressure appears as zero on most pressure gauges. This does not mean there is no pressure, rather it means the gauge is designed to read pressures greater than atmospheric pressure. All measurements taken on this type of gauge are given in pounds per square inch and should be referred to as psig (pounds per square inch gauge). Gauges and other measuring devices that include atmospheric pressure in their readings also display their measurements in psi, however, these should be referred to as psia (pounds per square inch absolute). There is a big difference between 12 psia and 12 psig. A reading of 12 psia is less than atmospheric pressure and therefore would represent a vacuum, whereas 12 psig would be approximately 26.7 psia. Because vacuum is defined as any pressure less than atmospheric, vacuum is any pressure less than 0 psig or 14.7 psia. The normal measure of vacuum is in inches of mercury (in. Hg) instead of psi. Other units of measurement for vacuum are kilopascals and bars. Normal atmospheric pressure at sea level is about 1 bar or 100 kilopascals.

Vacuum in any four-stroke engine is created by the downward movement of the piston during its intake stroke. With the intake valve open and the piston moving downward, a vacuum is created within the cylinder and intake manifold. The air passing the intake valve does not move fast enough to fill the cylinder, thereby causing the lower pressure. This vacuum is continuous in a multicylinder engine, because at least one cylinder is always at some stage of its intake stroke.

The amount of low pressure produced by the piston during its intake stroke depends on a number of things. Basically it depends on the cylinder's ability to form a vacuum and the intake system's ability to fill the cylinder. When there is high vacuum (15 to 22 inches [381 to 559 mm. Hg]), we know the cylinder is well sealed and not enough air is entering the cylinder to fill it. At idle, the throttle plate is almost closed and nearly all airflow to the cylinders is stopped. This is why vacuum is high during idle. Because there is a correlation between throttle position and engine load, it can be said that load directly affects engine manifold vacuum. Therefore, vacuum will be high whenever there is no, or low, load on the engine.

## Vacuum Controls

Engine manifold vacuum is used to operate and/or control several devices on an engine. Prior to the mid-70s, vacuum was only used to operate the windshield wipers and/or a distributor vacuum advance unit. Since then the use of vacuum has become extensive. Today vacuum is typically used to control the following systems:

1. *Fuel Induction System.* Certain vacuum-operated devices are added to carburetors and some fuel-injection throttle bodies to ease engine startup, warm-and-cold engine driveaway, and to compensate for air-conditioner load on the engine.
2. *Emission Control System.* While some emission control output devices are solenoid or linkage controlled, many operate on a vacuum. This vacuum is usually controlled by solenoids that are opened or closed, depending on electrical signals received from the PCM. Other systems use switches that are controlled by engine coolant temperature, such as a **ported vacuum switch (PVS),** or by ambient air, such as a **temperature vacuum switch (TVS).**
3. *Accessory Controls.* Engine vacuum is used to control operation of certain accessories, such as air conditioner/heater systems, power brake boosters, speed-control components, automatic transmission vacuum modulators, and so on.

A **ported vacuum switch (PVS)** is operated by coolant temperature.

A **temperature vacuum switch (TVS)** is controlled by temperature.

## Exhaust System Components

The various components of the typical exhaust system include the following:

- Exhaust manifold
- Exhaust pipe and seal

**Shop Manual**
Chapter 5, page 227

- Catalytic converter
- Muffler
- Resonator
- Tailpipe
- Heat shields
- Clamps, brackets, and hangers
- Exhaust gas oxygen sensors

All the parts of the system are designed to conform to the available space of the vehicle's undercarriage and yet be a safe distance above the road.

## Exhaust Manifold

The **exhaust manifold** (Figure 5-16) collects the burnt gases as they are expelled from the cylinders and directs them to the exhaust pipe. Exhaust manifolds for most vehicles are made of cast or nodular iron. Many newer vehicles have stamped, heavy-gauge sheet metal or stainless steel units.

In-line engines have one exhaust manifold. V-type engines have an exhaust manifold on each side of the engine. An exhaust manifold will have either three, four, or six passages, depending on the type of engine. These passages blend into a single passage at the other end, which connects to an exhaust pipe. From that point, the flow of exhaust gases continues to the catalytic converter, muffler, and tail pipe, then exits at the rear of the car.

The **exhaust manifold** conducts exhaust gases from the cylinder head exhaust ports to the exhaust pipe.

V-type engines may be equipped with a dual exhaust system that consists of two almost identical, but individual, systems in the same vehicle.

Exhaust systems are designed for particular engine–chassis combinations. Exhaust system length, pipe size, and silencer size are used to tune the flow of gases within the exhaust system. Proper tuning of the exhaust manifold tubes can actually create a partial vacuum that helps draw exhaust gases out of the cylinder, improving volumetric efficiency. Separate, tuned exhaust headers (Figure 5-17) can also improve efficiency by preventing the exhaust flow of one cylinder from interfering with the exhaust flow of another cylinder. Cylinders next to one another may release exhaust gas at about the same time. When this happens, the pressure of the exhaust gas from one cylinder can interfere with the flow from the other cylinder. With separate headers, the cylinders are isolated from one another, interference is eliminated, and the engine breathes better. The problem of interference is especially common with V8 engines. However, exhaust headers tend to improve the performance of all engines.

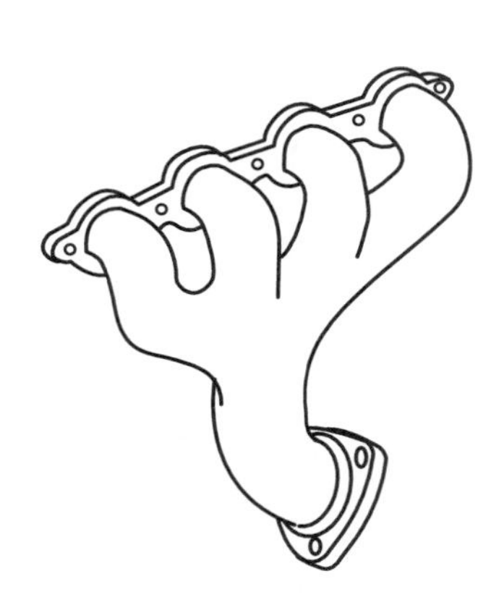

**Figure 5-16** Exhaust manifold.

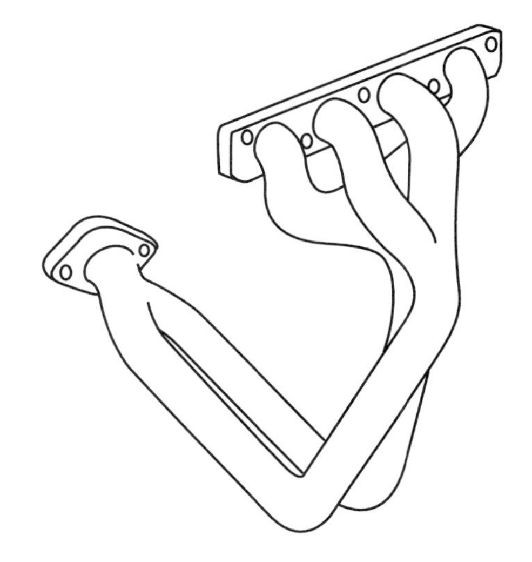

**Figure 5-17** Engine efficiency can be improved with tuned exhaust headers.

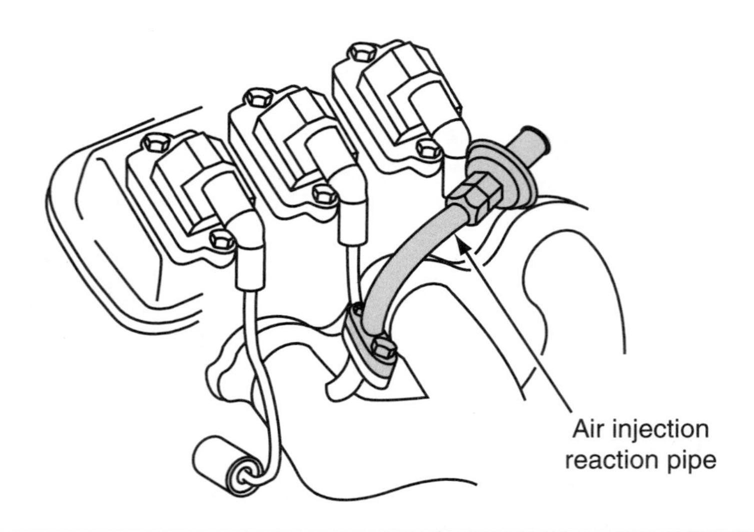


Figure 5-18 AIR pipe mounting on an exhaust manifold.

Exhaust manifolds may also be the attaching point for the air injection reaction (AIR) pipe (Figure 5-18). This pipe introduces cool air from the AIR system into the exhaust stream. Some exhaust manifolds have provisions for the exhaust gas recirculation (EGR) pipe. This pipe takes a sample of the exhaust gases and delivers it to the EGR valve. Also, some exhaust manifolds have a tapped bore that retains the oxygen sensor (Figure 5-19).

## Exhaust Pipe and Seal

The exhaust pipe is metal pipe, either aluminized steel, stainless steel, or zinc-plated heavy-gauge steel, that runs under the vehicle between the exhaust manifold and the catalytic converter (Figure 5-20).

## Catalytic Converters

A **catalytic converter** reduces tailpipe emissions of carbon monoxide, unburned hydrocarbons, and nitrogen oxides.

A **catalytic converter** (Figure 5-21) is part of the exhaust system and a very important part of the emission control system. Because it is part of both systems, it has a role in both. As an emission control device, it is responsible for converting undesirable exhaust gases into harmless gases. As part of the exhaust system, it helps reduce the noise level of the exhaust. A catalytic converter contains a ceramic element coated with a catalyst. A catalyst is a substance that causes a chemical reaction in other elements without actually becoming part of the chemical change and without being used up or consumed in the process.

Catalytic converters may be pellet-type or monolithic-type. A pellet-type converter contains a bed made from hundreds of small beads. Exhaust gases pass over this bed. In a monolithic-type converter, the exhaust gases pass through a honeycomb ceramic block. The converter beads or ceramic block are coated with a thin coating of cerium, platinum, palladium, and/or rhodium, and are held in a stainless steel container. Modern vehicles are equipped with three-way catalytic converters, which means the converter reduces the three major exhaust emissions, hydrocarbons (HC), carbon monoxide (CO), and oxides of nitrogen ($NO_X$). The converter oxidizes HC and CO into water vapor and carbon dioxide ($CO_2$) and reduces $NO_X$ to oxygen and nitrogen.

Many vehicles are equipped with a mini-catalytic converter that is either built into the exhaust manifold or is located next to it (Figure 5-22). These converters are used to clean the exhaust during engine warmup and are commonly called warmup converters. Many catalytic

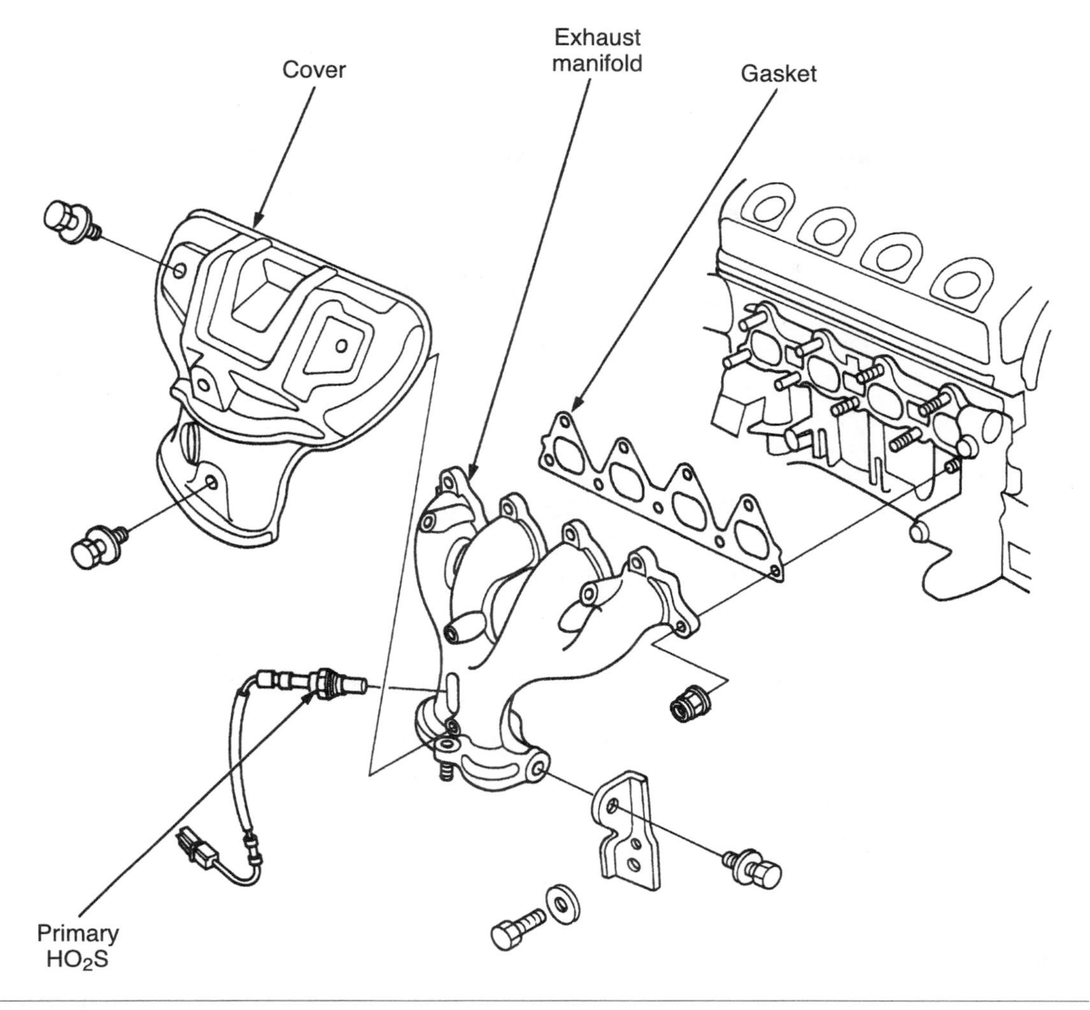


Figure 5-19 An exhaust manifold fitted with a heated oxygen sensor ($HO_2S$). (Courtesy of American Honda Motor Co., Inc.)

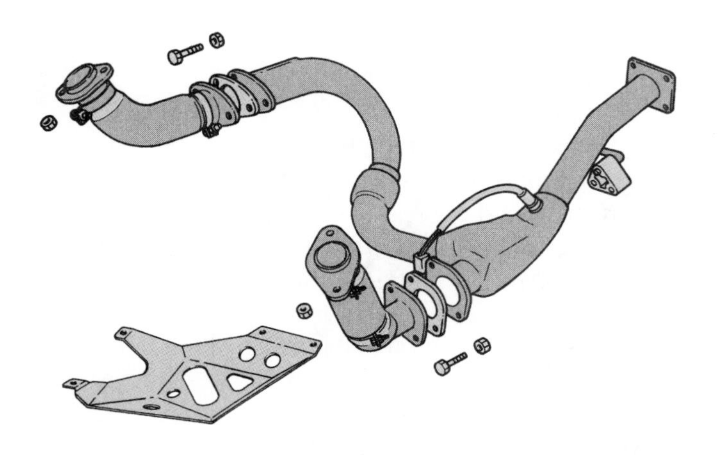

Figure 5-20 The front exhaust pipe for a V6 engine. (Reprinted with permission by American Isuzu Motors, Inc.)

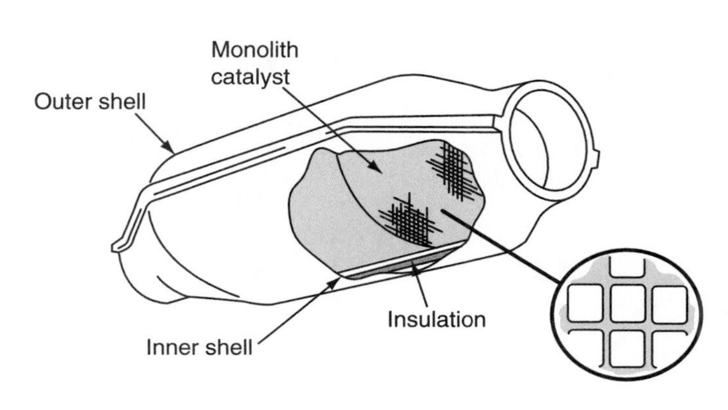


Figure 5-21 Monolithic-type catalytic converter.

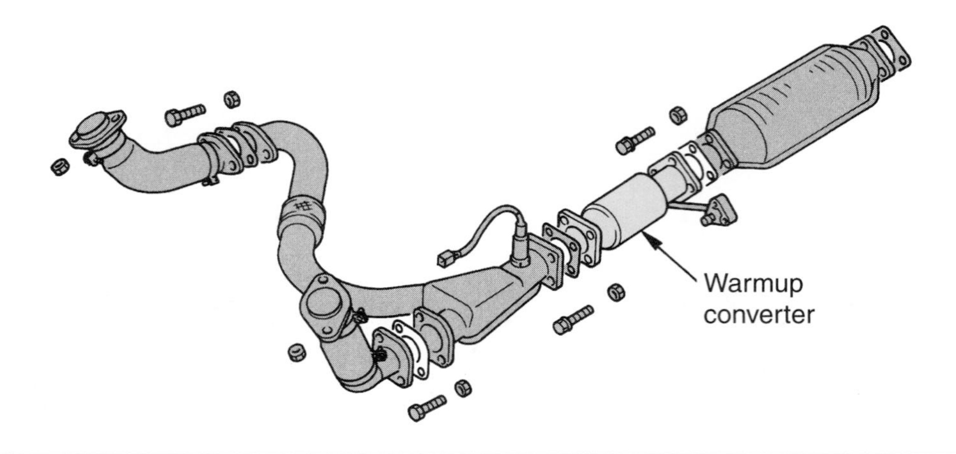


Figure 5-22 A warmup converter located between the front exhaust pipe and the standard converter. (Reprinted with permission by American Isuzu Motors, Inc.)

converters have an air hose connected from the AIR system to the oxidizing catalyst. This air helps the converter work by making extra oxygen available. The air from the AIR system is not always forced into the converter, rather it is controlled by the vehicle's PCM. Fresh air added to the exhaust at the wrong time could overheat the converter and produce $NO_X$, something the converter is trying to destroy.

OBD-II regulations call for a way to inform the driver that the vehicle's converter has a problem and may be ineffective. The PCM monitors the activity of the converter by comparing the signals of an $HO_2S$ located at the front of the converter with the signals from an $HO_2S$ located at the rear (Figure 5-23). If the sensor outputs are the same, the converter is not working properly and the malfunction indicator lamp (MIL) on the dash will light.

### Converter Problems

The converter is normally a trouble-free emission control device, but two things can damage it. One is leaded gasoline. Lead coats the catalyst and renders it useless. The difficulty of obtaining leaded gasoline has reduced this problem. The other is overheating. If raw fuel enters the exhaust because of a fouled spark plug or other problem, the temperature of the converter quickly increases. The heat can melt the ceramic honeycomb or pellets inside, causing a major restriction to the flow of exhaust A plugged converter or any exhaust restriction can cause damage to the exhaust valves due to excess heat, loss of power at high speeds, stalling after starting (if totally blocked), a drop in engine vacuum as engine rpm increases, or sometimes popping or backfiring at the carburetor.

The **muffler** is a device mounted in the exhaust system behind the catalytic converter that reduces engine noise.

## Mufflers

The **muffler** is a cylindrical or oval-shaped component, generally about 2 feet (0.6 meters) long, mounted in the exhaust system about midway or toward the rear of the car. Inside the muffler is a series of baffles, chambers, tubes, and holes to break up, cancel out, or silence the pressure pulsations that occur each time an exhaust valve opens.

Two types of mufflers are commonly used on passenger vehicles (Figure 5-24). Reverse-flow mufflers change the direction of the exhaust gas flow through the inside of the unit. This is the most common type of automotive muffler. Straight-through mufflers permit exhaust gases to pass through a single tube. The tube has perforations that tend to break up pressure pulsations. They are not as quiet as the reverse-flow type.

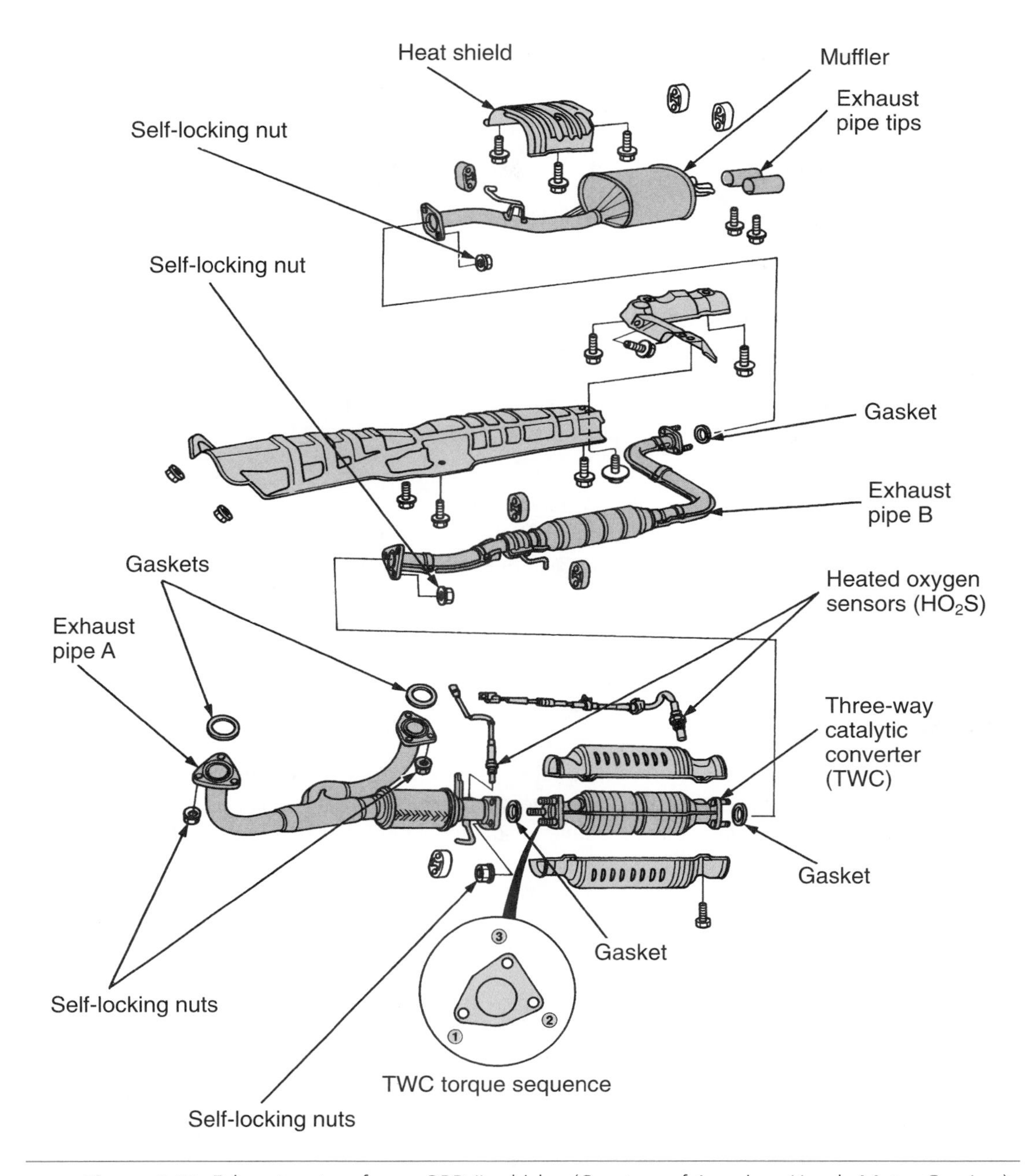


**Figure 5-23** Exhaust system for an OBD-II vehicle. (Courtesy of American Honda Motor Co., Inc.)

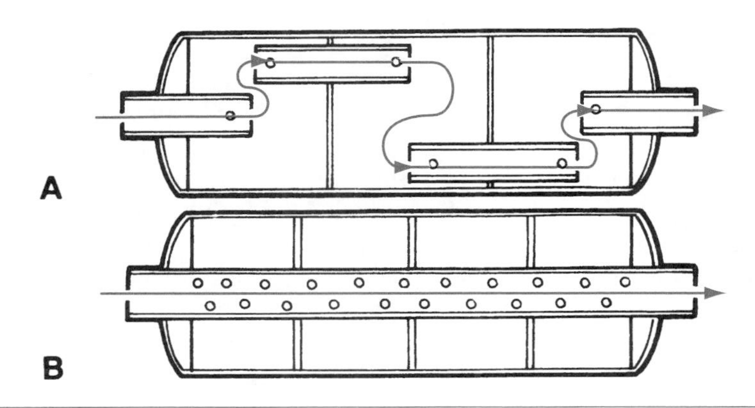


**Figure 5-24** (A) reverse flow muffler; (B) straight-through muffler.

There have been several important changes in recent years in the design of mufflers. Most of these changes have been centered at reducing weight and emissions, improving fuel economy, and simplifying assembly. These changes include the following:

1. *New materials.* More and more mufflers are being made of aluminized and stainless steel. Using these materials reduces the weight of the units as well as extending their lives.
2. *Double-wall design.* Retarded engine ignition timing that is used on many small cars tends to make the exhaust pulses sharper. Many cars use a double-wall exhaust pipe to better contain the sound and reduce pipe ring.
3. *Rear-mounted muffler.* More and more often, the only space left under the car for the muffler is at the very rear. This means the muffler runs cooler than before and is more easily damaged by condensation in the exhaust system. This moisture, combined with nitrogen and sulfur oxides in the exhaust gas, forms acids that rot the muffler from the inside out. Many mufflers are being produced with drain holes drilled into them.
4. *Back pressure.* Even a well-designed muffler will produce some **back pressure** in the system. Back pressure reduces an engine's volumetric efficiency, or ability to breathe. Excessive back pressure caused by defects in a muffler or other exhaust system part can slow or stop the engine. However, a small amount of back pressure can be used intentionally to allow a slower passage of exhaust gases through the catalytic converter. This slower passage results in more complete conversion to less harmful gases. Also, no back pressure may allow intake gases to enter the exhaust.

**Back pressure** reduces an engine's volumetric efficiency.

## Resonator

On some older vehicles, there is an additional muffler, known as a **resonator** or silencer. This unit is designed to further reduce or change the sound level of the exhaust. It is located toward the end of the system and generally looks like a smaller, rounder version of a muffler.

A **resonator** is a device mounted on the tailpipe to reduce exhaust noise further.

## Tailpipe

The **tailpipe** is the last pipe in the exhaust system. It releases the exhaust fumes into the atmosphere beyond the back end of the car.

The **tailpipe** conducts exhaust gases from the muffler to the rear of the vehicle.

## Heat Shields

Heat shields are used to protect other parts from the heat of the exhaust system and the catalytic converter (Figure 5-25). They are usually made of pressed or perforated sheet metal. Heat shields trap the heat in the exhaust system, which has a direct effect on maintaining exhaust gas velocity.

## Clamps, Brackets, and Hangers

Clamps, brackets, and hangers are used to properly join and support the various parts of the exhaust system. These parts also help to isolate exhaust noise by preventing its transfer through the frame (Figure 5-26) or body to the passenger compartment. Clamps help to secure exhaust system parts to one another. The pipes are formed in such a way that one slips inside the other. This design makes a close fit. A U-type clamp usually holds this connection tight (Figure 5-27). Another important job of clamps and brackets is to hold pipes to the bottom of the vehicle. Clamps and brackets must be designed to allow the exhaust system to vibrate without transferring the vibrations through the car.

There are many different types of flexible hangers available. Each is designed for a particular application. Some exhaust systems are supported by doughnut-shaped rubber rings between hooks on the exhaust component and on the frame or car body. Others are supported at the

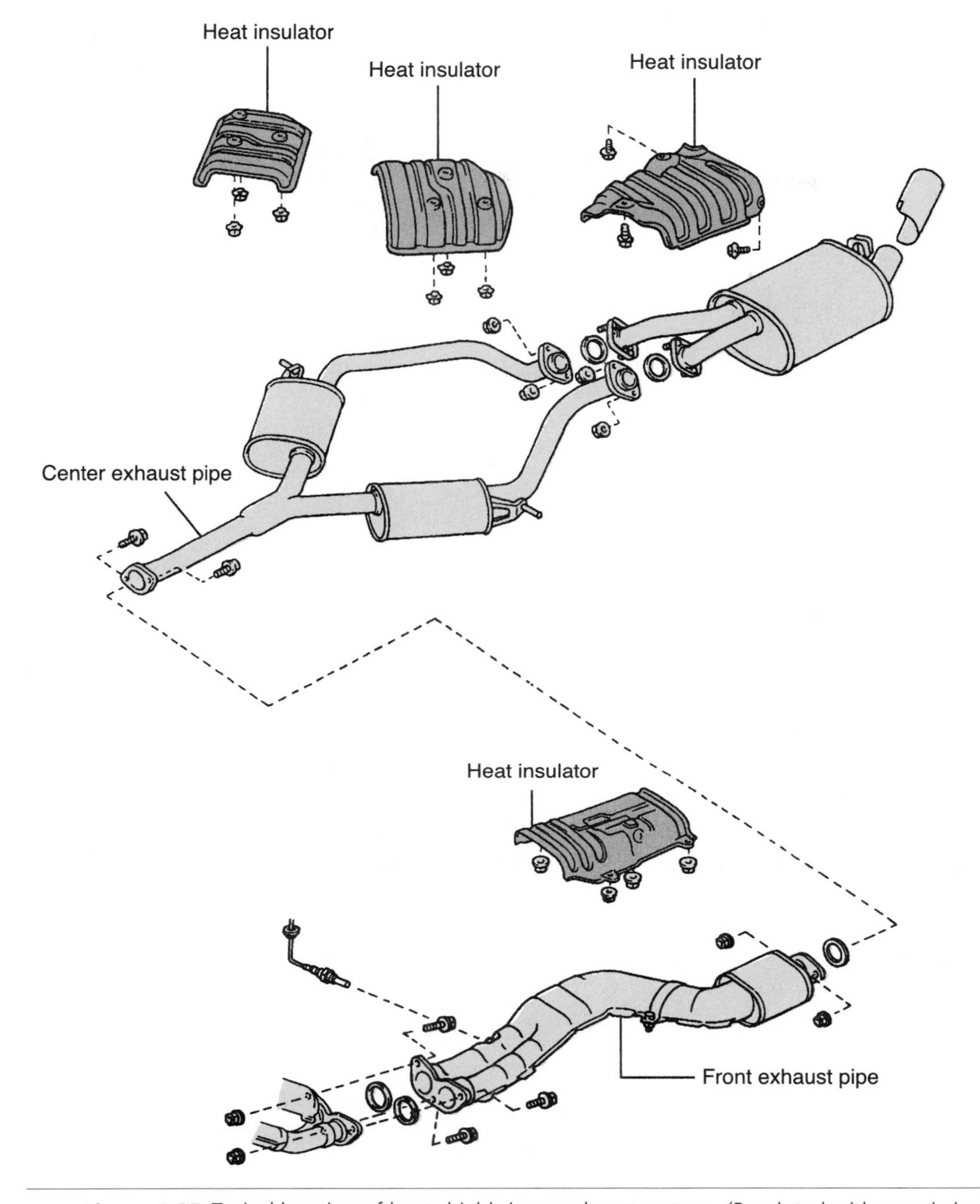


**Figure 5-25** Typical location of heat shields in an exhaust system. (Reprinted with permission)

exhaust pipe and tailpipe connections by a combination of metal and reinforced fabric hanger. Both the doughnuts and the reinforced fabric allow the exhaust system to vibrate without breakage that could be caused by direct physical connection to the vehicle's frame.

Some exhaust systems are a single unit in which the pieces are welded together by the factory. By welding instead of clamping the assembly together, car makers save the weight of overlapping joints as well as that of clamps.

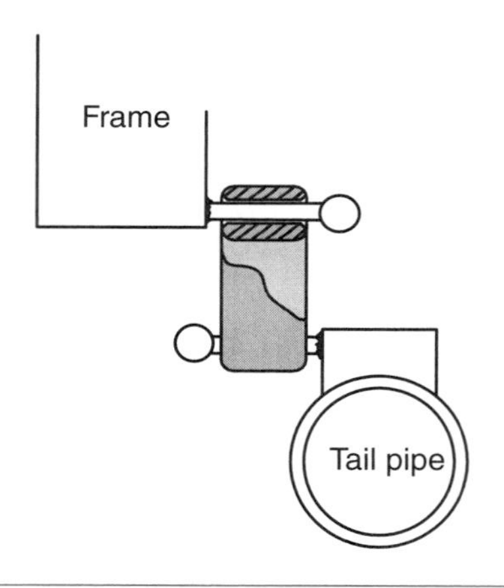


Figure 5-26 Rubber hangers are used to keep the exhaust system in place without allowing it to contact the frame.

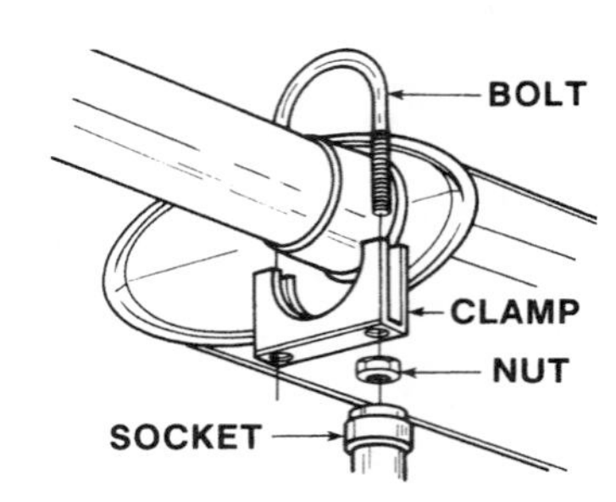


Figure 5-27 A U-clamp is often used to secure two pipes that slip together.

**Shop Manual**
Chapter 5, page 229

**AUTHOR'S NOTE:** During my experience in the automotive service industry, I encountered several cases where restricted exhaust or intake systems were misdiagnosed and confused with ignition system or fuel system defects. Defective fuel system or ignition system components may cause a loss of engine power and reduced maximum speed, but these symptoms are accompanied by cylinder misfiring and/or engine surging. When the exhaust or intake system is restricted, the maximum speed is reduced but the engine does not misfire or surge.

## A BIT OF HISTORY

In the 1930s Charles Nelson Pogue developed the Pogue carburetor. This carburetor used exhaust heat to vaporize the fuel before it was mixed with the air entering the engine. Charles Pogue was issued several patents on this carburetor, and claimed greatly increased fuel mileage. Mr. Pogue never did sell his invention to any of the car manufacturers, but rumors were repeated for years about the fantastic fuel economy supplied by this carburetor.

**Terms to Know**

Air filter
Air flow restriction indicator
Back pressure
Catalytic converter
Dry manifolds
Exhaust manifold
Fresh air tube
Intake air temperature (IAT) sensor
Intake manifold
Mass airflow (MAF) sensor
Muffler
Ported vacuum switch (PVS)
Powertrain control module (PCM)

# Summary

- The air induction system allows a controlled amount of clean, filtered air to enter the engine. Cool air is drawn in through a fresh air tube. It passes through an air cleaner before entering the carburetor or throttle body.
- The air intake ductwork conducts airflow from the remote air cleaner to the throttle body mounted on the intake manifold.
- The air cleaner/filter removes dirt particles from the air flowing into the intake manifold to prevent these particles from causing engine damage.

- The intake manifold distributes the air or air/fuel mixture as evenly as possible to each cylinder, helps to prevent condensation, and assists in the vaporization of the air/fuel mixture. Intake manifolds are made of cast iron, plastic, or die-cast aluminum.
- The vacuum in the intake manifold operates many systems such as emission controls, brake boosters, heater/air conditioners, cruise controls, and more. Vacuum is applied through an elaborate system of hoses, tubes, and relays. A diagram of emission system vacuum hose routing is located on the underhood decal. Loss of vacuum can create many driveability problems.
- A vehicle's exhaust system carries away gases from the passenger compartment, cleans the exhaust emissions, and muffles the sound of the engine. Its components include the exhaust manifold, exhaust pipe, catalytic converter, muffler, resonator, tailpipe, heat shields, clamps, brackets, and hangers.
- The exhaust manifold is a bank of pipes that collects the burned gases as they are expelled from the cylinders and directs them to the exhaust pipe. Engines with all the cylinders in a row have one exhaust manifold. V-type engines have an exhaust manifold on each side of the engine. The exhaust pipe runs between the exhaust manifold and the catalytic converter.
- The catalytic converter reduces HC, CO, and $NO_X$ emissions.
- The muffler consists of a series of baffles, chambers, tubes, and holes to break up, cancel out, and silence pressure pulsations. Two types commonly used are the reverse-flow and the straight-through mufflers.
- The tailpipe is the end of the pipeline carrying exhaust fumes to the atmosphere beyond the back end of the car. Heat shields protect vehicle parts from exhaust system heat. Clamps, brackets, and hangers join and support exhaust system components.
- Exhaust system components are subject to both physical and chemical damage. The exhaust can be checked by listening for leaks and by visual inspection. Most exhaust system servicing involves the replacement of parts.

**Terms to Know**

Resonator

Tailpipe

Temperature vacuum switch (TVS)

Variable-length intake manifold

Wet manifolds

# Review Questions

## Short Answer Essays

1. Explain the operation of an airflow restriction indicator.
2. Describe the result of installing a mass airflow (MAF) sensor backwards.
3. Explain the purposes of the intake manifold.
4. Describe the difference between a wet and dry intake manifold.
5. Explain the advantages of plastic intake manifolds.
6. Explain why fuel vaporization and condensation are not intake manifold design considerations on port-fuel injected engines.
7. Explain the operation of an intake manifold with dual runners at low and high engine speeds.
8. Describe the advantages of tuned exhaust manifolds compared to conventional exhaust manifolds.
9. Explain two catalytic converter operating problems.
10. Describe two different types of mufflers.

## Fill-in-the-Blanks

1. Without proper intake air filtration, contaminants and abrasives in the air will cause severe ________________ damage.
2. If the air filter is restricted, the airflow restriction indicator window appears ________________ in color.
3. In some engines the MAF sensor is mounted on the ______________ ________________ .
4. In a port fuel-injected engine the intake manifold delivers ________________ to the intake ports.
5. A wet intake manifold has ______________ ________________ cast into the manifold.
6. When an intake manifold has dual runners, both runners are open at ____________ engine speeds.
7. In place of cast or nodular iron exhaust manifolds, newer vehicles have manifolds manufactured from stamped, heavy-gauge sheet metal or _____________ ____________.
8. A tuned exhaust manifold prevents exhaust flow from one cylinder from interfering with the ____________ ____________ from another cylinder.
9. A catalytic converter may be overheated by a ____________ air/fuel ratio.
10. The exhaust pipe between the engine and the catalytic converter may be a _______________ _____________ design.

## Multiple Choice

1. A mini-converter is used
   **A.** On small engines where a normal converter will not fit properly.
   **B.** On engines that used leaded fuels.
   **C.** In conjunction with EGR systems to supply clean exhaust for the cylinders.
   **D.** To reduce emissions during engine warm-up.

2. *Technician A* says the positive crankcase ventilation (PCV) system relieves the crankcase of unwanted pressure.
   *Technician B* says the PCV system replaces blowby gases in the crankcase with clean air.
   Who is correct?
   **A.** A only **C.** Both A and B
   **B.** B only **D.** Neither A nor B

3. *Technician A* says a vacuum leak results in less air entering the engine, which causes a richer air/fuel mixture.
   *Technician B* says a vacuum leak anywhere in the system can cause the engine to run poorly.
   Who is correct?
   **A.** A only **C.** Both A and B
   **B.** B only **D.** Neither A nor B

4. *Technician A* says a vacuum leak will cause an engine to run richer than normal.
   *Technician B* says a vacuum leak can cause an engine to detonate.
   Who is correct?
   **A.** A only **C.** Both A and B
   **B.** B only **D.** Neither A nor B

5. Before replacing any exhaust system component, *Technician A* soaks all old connections with a penetrating oil.
   *Technician B* checks the old system's routing for critical clearance points.
   Who is correct?
   **A.** A only **C.** Both A and B
   **B.** B only **D.** Neither A nor B

6. A restricted exhaust system can cause
   **A.** Stalling. **C.** Loss of power.
   **B.** Backfiring. **D.** Acceleration stumbles.

7. *Technician A* says a low vacuum reading can be caused by incorrect ignition timing.
*Technician B* says an engine with low compression will have a low vacuum reading.
Who is correct?
**A.** A only
**B.** B only
**C.** Both A and B
**D.** Neither A nor B

8. *Technician A* says a catalytic converter breaks down HC and CO to relatively harmless byproducts.
*Technician B* says using leaded gasoline or allowing the converter to overheat can destroy its usefulness.
Who is correct?
**A.** A only
**B.** B only
**C.** Both A and B
**D.** Neither A nor B

9. When an airflow restriction indicator window appears orange it is necessary to:
**A.** Replace the airflow restriction indicator.
**B.** Replace the air filter and press the reset button on the airflow restriction indicator.
**C.** Replace the air filter and the airflow restriction indictor.
**D.** Clear the diagnostic trouble codes (DTCs) from the PCM memory.

10. All of these statements about an intake manifold dual runner system are true except:
**A.** Both runners to each cylinder are open at 1,400 engine rpm.
**B.** The butterfly valves in one set of runners may be operated by a vacuum actuator.
**C.** The butterfly valves may be operated electrically by the PCM.
**D.** One runner to each cylinder is open at 900 engine rpm.

CHAPTER 6

# Cylinder Heads

Upon completion and review of this chapter, you should be able to:

- ❒ Explain valve design, including an explanation of the terms, stem, head, face, seat, margin, and fillet.
- ❒ Explain the reasons for burned valves.
- ❒ Describe the causes of valve channeling.
- ❒ Describe two different types of valve seats.
- ❒ Explain the reasons for adding Stellite to valve seats.
- ❒ Explain valve seat recession and explain the causes of this problem.
- ❒ Describe the purposes of valve springs.
- ❒ Explain the purposes of valve stem seals, and list three different types of these seals.
- ❒ Explain the results of worn valve stems and guides.
- ❒ Describe eight different combustion chamber designs.
- ❒ Explain briefly the combustion process.
- ❒ Explain how combustion chamber design can reduce exhaust emissions.

## Introduction

Proper combustion requires the engine to "breathe" properly. The air intake system directs air to the throttle body. From the throttle body, the intake air is directed through the intake manifold, cylinder head, and past the intake valves into the combustion chambers. Once in the combustion chambers, the air/fuel mixture is ignited, and the resulting power output is dependent on chamber design. Finally, the spent gases are expelled from the combustion chambers past the exhaust valves and out through the exhaust manifold and exhaust system.

The intake and exhaust manifolds, cylinder head, valves, and rocker arms must all work together to move the air/fuel mixture into the combustion chambers and expel the exhaust gases. In this chapter, we discuss cylinder heads and related components such as valves, valve seats, valve guides, and rocker arms. Discussion of the common types of failures and their causes is also included. Today's technician must be able to identify a failed component, and determine the cause of the failure. Although lack of lubrication is one of the most common causes of component failure, all engines will exhibit some form of normal wear between moving parts. A technician must be able to classify component wear as normal or abnormal. If abnormal wear is determined, the cause must be identified and corrected before the engine is reassembled.

## Cylinder Heads

On most engines, the **cylinder head** contains the valves, valve seats, valve guides, valve springs, and the upper portion of the combustion chamber (Figure 6-1). In addition, the cylinder head has passages to allow coolant and oil flow through the head.

The **cylinder head** is bolted on top of the engine block and contains the combustion chamber and spark plug opening plus some valve chain components.

The cylinder head is attached to the block above the cylinder bores. This mating surface must be perfectly smooth and flat. A gasket is installed at the mating surface to aid in the sealing of the two parts.

The cylinder head is made of cast iron or aluminum. It is common for engine manufacturers to use aluminum alloys to cast the head and mate it to a cast-iron block. The difference in thermal expansion between the cylinder head and block creates a scrubbing stress that must be withstood by the head gasket.

The aluminum head is also vulnerable to electrolysis corrosion within the cooling system. To combat this problem, the coolant used must contain the correct corrosion inhibitors and be changed on a regular basis.

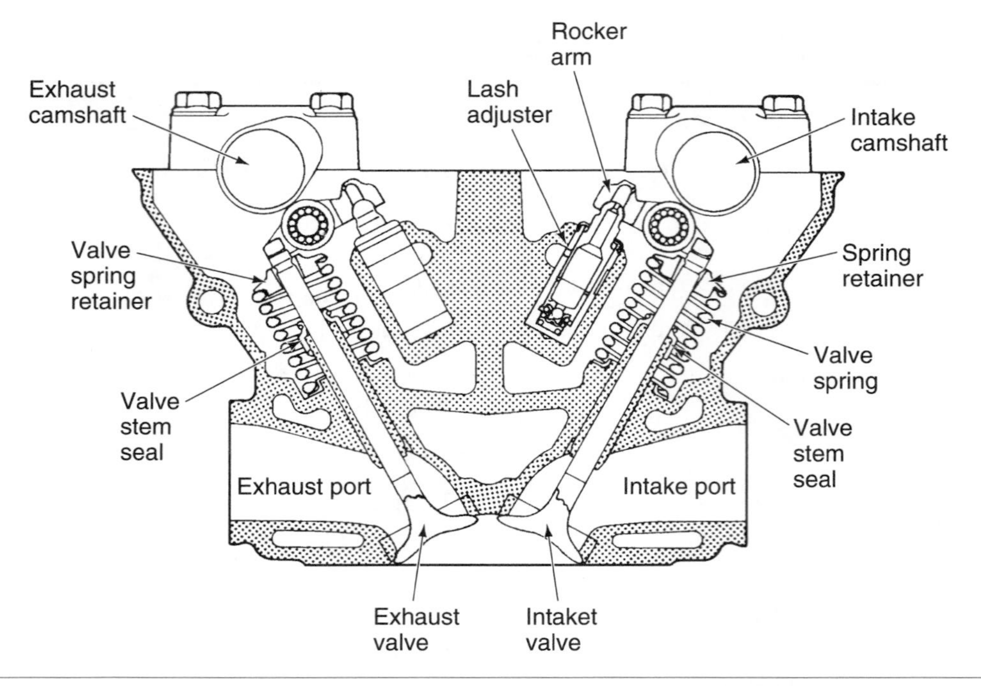


Figure 6-1 Typical cylinder head components. (Courtesy of Hyundai Motor America)

**Shop Manual**
Chapter 6, page 238

**Poppet valves** control the opening or passage by linear movement.

The **head** is the enlarged part of the valve.

The **face** is the tapered part of the section of the valve head that makes contact with the valve seat.

The **valve seat** is the point at which the seal is made by the valve face.

The **margin** is the material between the valve face and the valve head.

The **valve stem** guides the valve through its linear movement.

## Valves

Modern automotive engines use **poppet valves** (Figure 6-2). The poppet valve is opened by applying a force that pushes against its stem. Closing of the valve is accomplished through spring pressure.

The valve as a whole has many parts (Figure 6-3). The large diameter end of the valve is called the **head.** The angled outer edge of the head is called the **face** and provides the contact point to seal the port. The seal is made by the valve face contacting the **valve seat** in the cylinder head. To provide a positive seal, the valve face is cut at an angle. The angle is usually 30 or 45 degrees. The angle causes the seal to wedge tighter when the valve is closed and allow free flow of the air/fuel mixture when the valve is open.

The area between the valve face and the valve head is referred to as the **margin.** The margin allows for machining of the valve face and for dissipation of heat away from the face and stem. The **valve stem** guides the valve through its linear movement. The stem also has valve keeper grooves machined into it close to the top. These grooves are used to retain the valve springs. The stem rides in a valve guide located in the cylinder head.

The head of the intake valve is larger than the exhaust valve's. This is because maximizing intake valve size is considered more important than the exhaust valve. There is only so much area in the combustion chamber to locate the valves, and the larger the intake valve, the smaller the exhaust valve has to be. Smaller exhaust valves are able to work efficiently because the exhaust is being forced out of the chamber by the upward movement of the piston and by the heat generated by combustion, thus they are less sensitive to restrictions, whereas the air/fuel charge is being drawn into the cylinder, and restrictions will reduce the engine's efficiencies. Larger intake valves reduce the restrictions and improve efficiencies.

Most valves are constructed of high-strength steel; however, some manufacturers are experimenting with ceramics. Most valve heads are constructed of 21-2N and 21-4N stainless steel alloys. 21-2N alloy is commonly used in original equipment exhaust valves. 21-4N is a higher grade of stainless steel containing more nickel. One of the primary concerns in valve construction and selection of alloys is the operating temperatures to which they are subjected. Combustion chamber temperatures can range from 1,500 to 4,000°F (82 to 220°C). The types of material used to construct the valve must

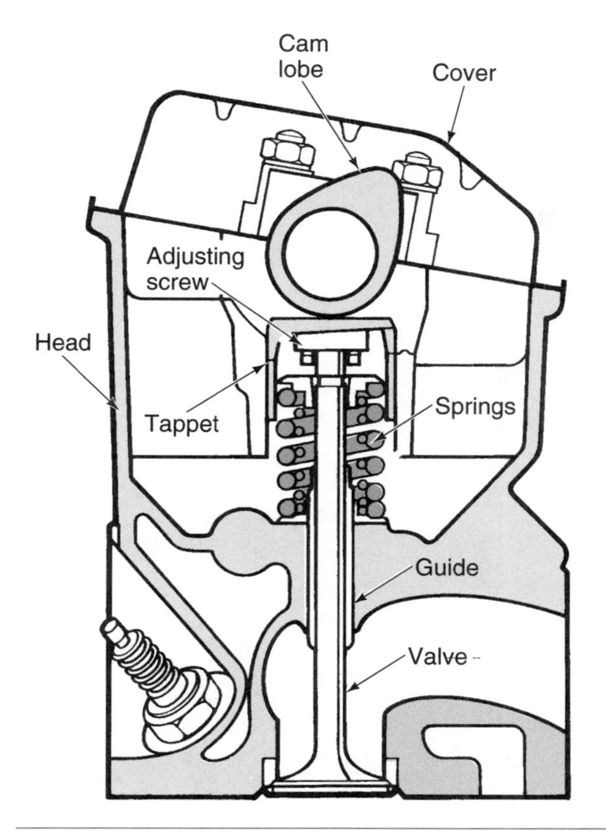


Figure 6-2 Typical poppet valve configuration using an overhead camshaft.

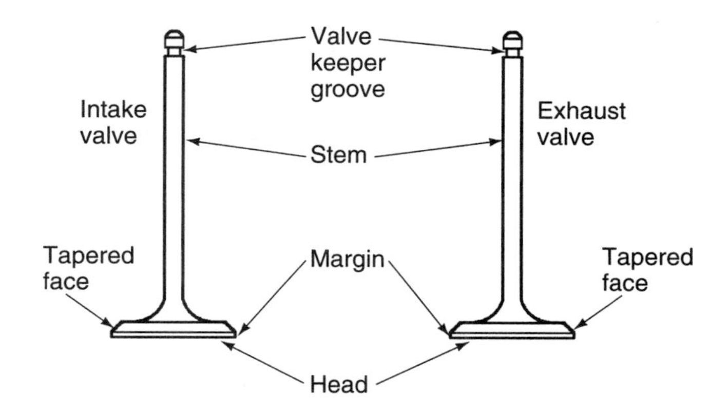


Figure 6-3 The head of the valve contains the face and margin.

be capable of withstanding these temperatures and dissipating the heat rapidly. For every 25°F (14°C) reduction in valve temperature, the valve's burning durability is doubled. There are several alloys available for valve construction that will increase the valves burning durability. An alloy that is being used by many manufacturers is Inconel. Inconel is constructed from a nickel base with 15 to 16 percent chromium and 2.4 to 3.0 percent titanium.

SAE has devised a code method to classify valve alloys. The code uses different letter and number designations corresponding to the types of materials used for example, an NV prefix designates a low-alloy structural steel, while an HNV prefix designates a high-alloy steel. Both of these designations are generally used for intake valves. Exhaust valves must be able to withstand higher temperatures than intake valves. Common prefix designations include EV, which denotes an austenitic steel. HEV denotes a high-strength alloy.

The commonly used 21-4N stainless steel alloy is SAE classified as EV8. This valve contains 21 percent chromium and 3.75 percent nickel. Inconel 751 is classified as an HEV3 alloy.

Ideally, when replacing valves, match the original equipment alloy. Sometimes the only assurance you have concerning the grade of alloy used in a valve is the reputation of the valve manufacturer. A magnet can be used to separate stainless steel from carbon steels. Stainless steel is nonmagnetic, while carbon is magnetic.

Valve construction design is also an important aspect for controlling heat. A valve can be constructed as one piece or two piece. One-piece valves run cooler since the weld of a two-piece valve inhibits heat flow up the stem. Two-piece valves allow the manufacturer to use different metals for the valve head and stem.

In addition to alloy selection, manufacturers design the valve and cylinder head to dissipate heat away from the valve (Figure 6-4). Most of the heat is dissipated through the contact surface of the face and seat. The heat is then transferred to the cylinder head. About 76 percent of the heat is transferred in this manner. Most of the remaining heat is dissipated through the stem to

**Shop Manual**
Chapter 6, page 241

SAE stands for the Society of Automotive Engineers.

Austenitic steel is a corrosive-resistant steel constructed with carbide or carbon alloys.

One-piece valves do not weld the head to the stem like two-piece valves. The advantage of two-piece valves is two different metals can be used to construct the head and the stem.

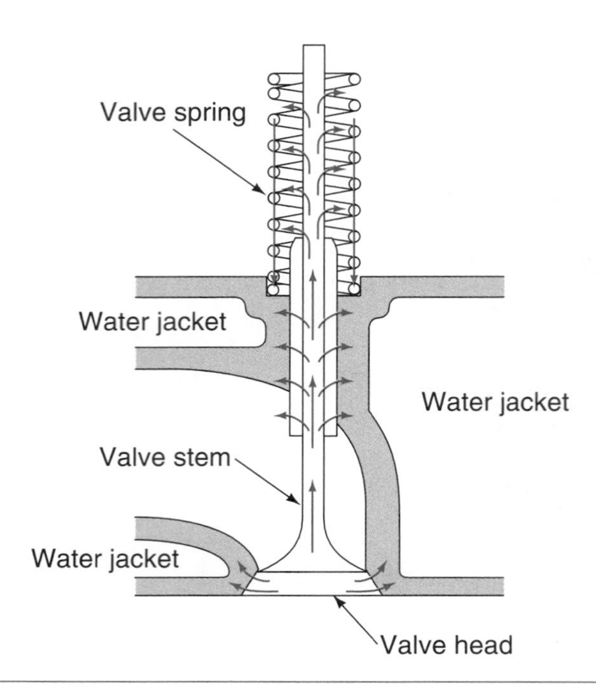


**Figure 6-4** Heat must be removed from the valve.

the valve spring and on to the cylinder head. To increase the transfer of heat through the stem, some stems are filled with sodium.

Many manufacturers use chrome-plated stems to provide additional protection against wear resulting from initial engine starts when oil is not present on the valve stem. In addition, chrome works to protect against galling when cast-iron guides are used. If the valve stem is reground, the chrome plating will be removed. In this case, either the valve will have to be replated or a bronze guide must be used.

The **fillet** is the curved area between the stem and the head.

The **fillet** is the curved area between the stem and the head. It provides structural strength to the valve. The fillet shape also affects the flow of the air/fuel mixture around the valve. A valve with a steeply raked fillet (called a tulip valve) has better flow than a valve with a flatbacked fillet. Fillet shape is only a factor while the valve is opening; after the valve is completely open, the fillet shape has very little effect on airflow.

Although not normally considered a wear area, the fillet must be thoroughly cleaned. Carbon will build up in this area, causing the flow to be disrupted. Excessive carbon in this location can be an indication of worn valve stem seals or guides. A wire brush is usually required to remove carbon buildup on the fillet.

**Aluminized Valves.** Aluminized valves provide increased valve life by reducing the effects of corrosion. Small particles of aluminum are fused to the valve head, causing a reaction that makes the surface corrosion resistant. Aluminized valves should not be reconditioned because this process removes the coating from the head.

## Determining Valve Malfunctions

It is not enough to be able to recognize a failed valve. It is very important to determine the cause of the malfunction to prevent a repeat failure. Different types of failures are usually caused by identifiable conditions (Figure 6-5).

**Deposits on Valves.** Deposits on valve stems (Figure 6-6) are typically the result of excessive valve stem temperatures and the use of improper oil. Lack of heat flow from the valve stem to the guide and into the cylinder head also results in carbon formation.

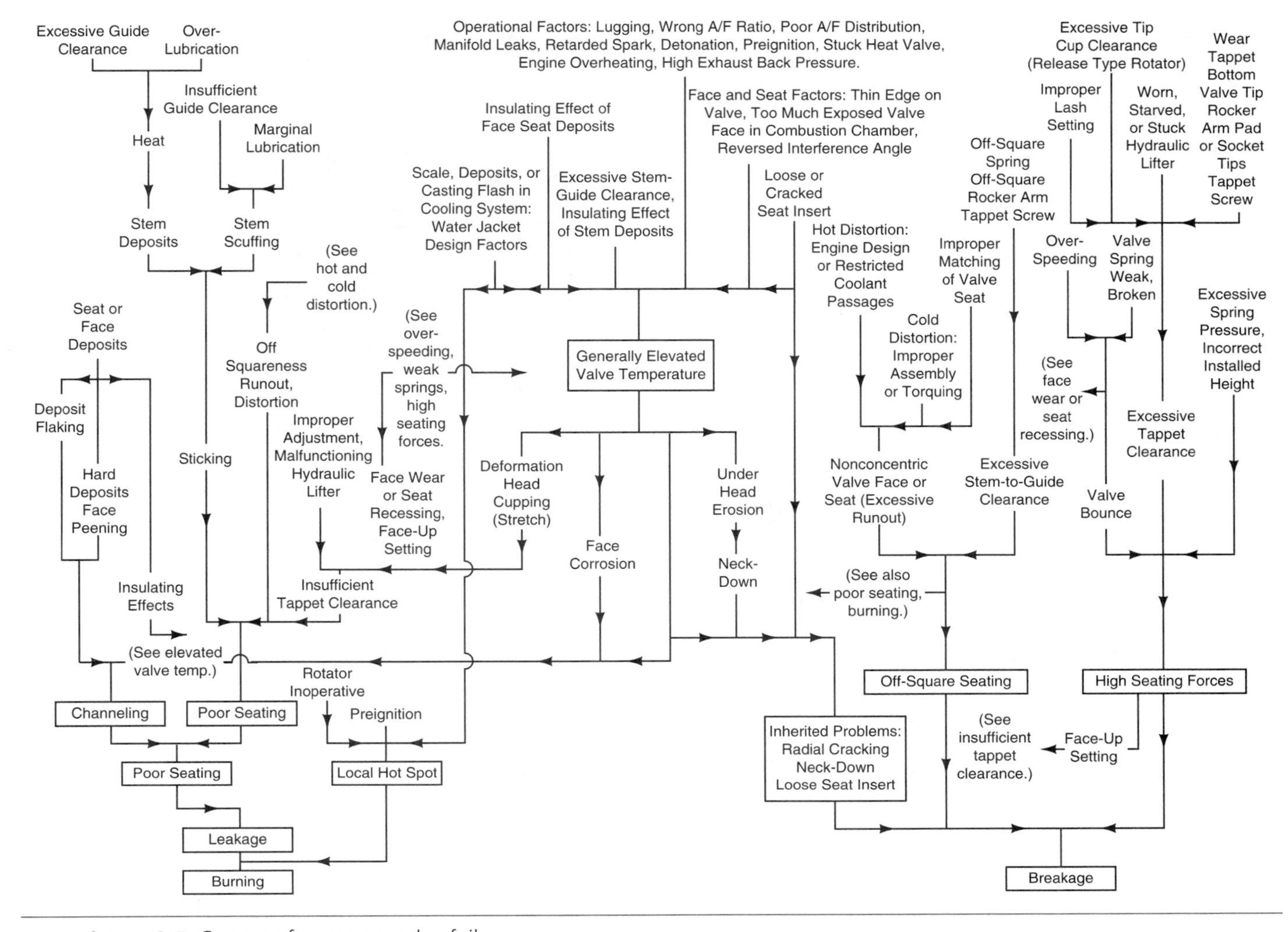


**Figure 6-5** Causes of common valve failures.

**Figure 6-6** Carbon buildup on the neck of the valve interferes with breathing.

**Burned valves** are actually valves that have warped and melted, leaving a groove across the valve head.

**Burned Valves.** A **burned valve** is identifiable by a notched edge beginning at the margin and running toward the center of the valve head (Figure 6-7). The most common cause of a burned valve is a leak in the sealing between the valve face and seat. At the location of the leak, the valve will not cool efficiently. This is because the heat from the valve head cannot be transferred

**Figure 6-7** Burned valves are characterized by a groove from the face toward the fillet.

Exhaust leaks allow cool air to flow past the valve. This cools the valve too fast, changes its structure, and reduces its burning durability.

Valve cupping is a deformation of the valve head caused by heat and combustion pressures.

to the valve seat and cylinder head. As the area increases in temperature, it begins to warp. The warpage results in less contact at the valve seat and compounds the problem of heat transfer. The resultant heat buildup melts the edge of the valve.

Since heat transfer is dependent upon full contact of the face to the seat, anything preventing this contact can result in a burned valve. This includes weak valve springs, carbon buildup, worn valve guides, and valve clearance adjusted too tight (Figure 6-8). Additional causes of burned valves include preignition, exhaust system leaks, and failure of the cooling system.

Not all overheating conditions result in burned valves. The metal of the valve head may become soft and form into a cup shape (Figure 6-9). The force applied by the springs to seat the

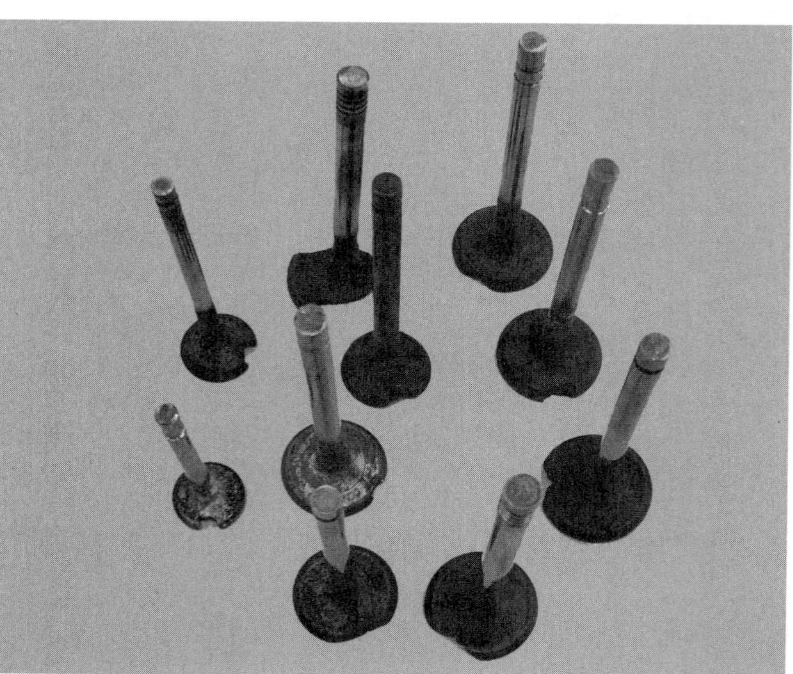

**Figure 6-8** Burned and damaged valves. (Courtesy of JTG Associates)

Figure 6-9 Excessive heating of the valve can lead to cupping.

valve, combined with combustion pressures, deforms the valve head. Because of these overheating malfunctions, it is important to measure the size of the head with an outside micrometer and compare the measurements to specifications. Measure a minimum of two directions to check concentricity. Any warpage not evident to the eye will be indicated by the micrometer readings.

**Channeling.** Under normal conditions, the fillet area of the valve should be the hottest area (Figure 6-10). The temperature around the valve edge should be the coolest. If the valve does not seat properly, the temperature is not transferred to the cylinder head. The area not contacting the seat becomes the hottest area of the valve head (Figure 6-11). The area with the hottest temperature will melt away, developing a channel. **Channeling** usually occurs just prior to valve burning. Channeling is generally due to four main causes:

1. Flaking off of deposits from the valve
2. Extensive valve face peening due to foreign material lodged between the valve face and seat

**Channeling** is referred to as local leakage caused by extreme temperatures developing at isolated locations on the valve face and head.

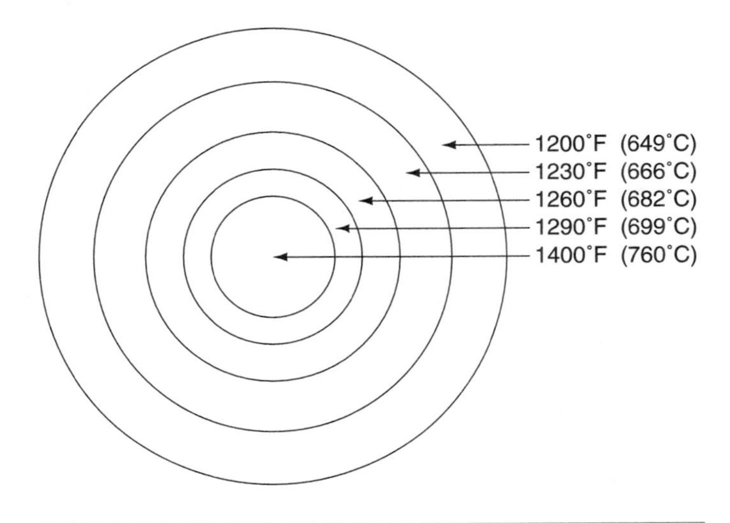


Figure 6-10 Normal vale temperatures as heat is dissipated through the face and seat.

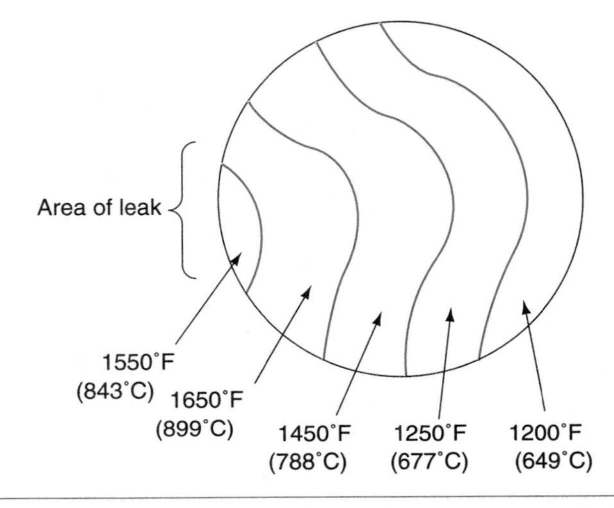


Figure 6-11 Improper face and seat contact results in improper heat dissipation.

3. Corroded valve face
4. Thermal stress of the cylinder head, resulting in radial cracking

**Metal Erosion.** Metal erosion is identified by small pits in the valve face that usually cannot be removed by valve grinding (Figure 6-12). This erosion is generally caused by coolant entering the combustion chamber. The chemical reaction etches away the face material. This condition can also occur if some chemicals enter the combustion chamber. These chemicals can come in the form of coolant system conditioners, oil additives, and fuel additives.

In a **free-wheeling engine,** valve lift and angle prevent valve-to-piston contact if the timing belt or chain breaks. In an **interference engine,** if the timing belt or chain breaks, or is out of phase, the valves will contact the pistons.

**Breakage.** Breakage of the valve can occur due to valve and piston contact. In a **free-wheeling engine,** valve lift and angle prevent valve-to-piston contact if the timing belt or chain breaks. In an **interference engine,** this can be the result of timing chain or belt breakage. Interference designs are commonly used because they offer improved fuel efficiency, increased power output, and lower emissions. However, if the valve timing components fail, the valve can contact the top of the piston or another valve. This contact may cause the valve to break or be drawn into the combustion chamber. If this occurs, major engine damage results. If the engine you are working on has experienced this type of failure, inspect the head and block carefully for cracks and other damage. The head must usually be replaced with this type of failure.

Other than valve-to-piston contact, valves may also experience fatigue and impact breaks (Figure 6-13). A fatigue break is a gradual breakdown of the valve due to excessive heat and pressure. As shown in the top of Figure 6-13, a fatigue break usually shows lines of progression.

**Stem necking** is a condition of the valve stem in which the stem narrows near the bottom.

Impact breakage can result from the valve face being forced into the valve seat with excessive force. As seen in the bottom of Figure 6-13, the impact break shows a crow's feet pattern. Impact breakage can be the result of excessive valve clearance, valve springs with excessive pressure, loss of valve keepers, or any condition that can result in **stem necking.**

Necking is identified by a thinning of the valve stem just above the fillet (Figure 6-14). The head pulls away from the stem, causing the stem to stretch and thin. This condition is caused by overheating or excessive valve spring pressures. In addition, necking can be caused by exhaust gases circling around the valve stem. The corrosive nature of these gases and the action resulting from the swirl eat away the metal of the valve stem. Necking resulting from this action is caused by anything resulting in increased exhaust temperatures, including:

**Off-square** seating or tipping refers to the seat and valve stem not being properly aligned. This condition causes the valve stem to flex as the valve face is forced into the seat by spring and combustion pressures.

- Improper ignition timing
- Improper air/fuel mixtures
- Overloading
- Improper gear ratio in the final drive

An additional cause of valve breakage is referred to as **off-square** seating or tipping. If the valve does not seat properly, it may flex as it attempts to set into the valve seat. This flexing weakens the stem and eventually may lead to its breakage (Figure 6-15). Off-square seating can be caused by worn valve guides, distortion of the valve seat, improper rocker arm alignment, and weak valve springs.

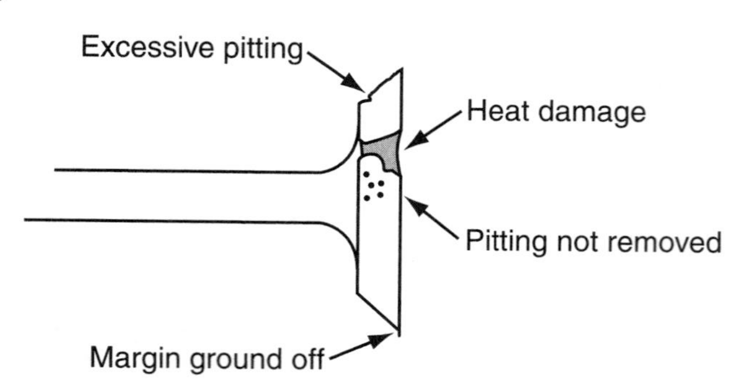


Figure 6-12 Metal erosion of the valve.

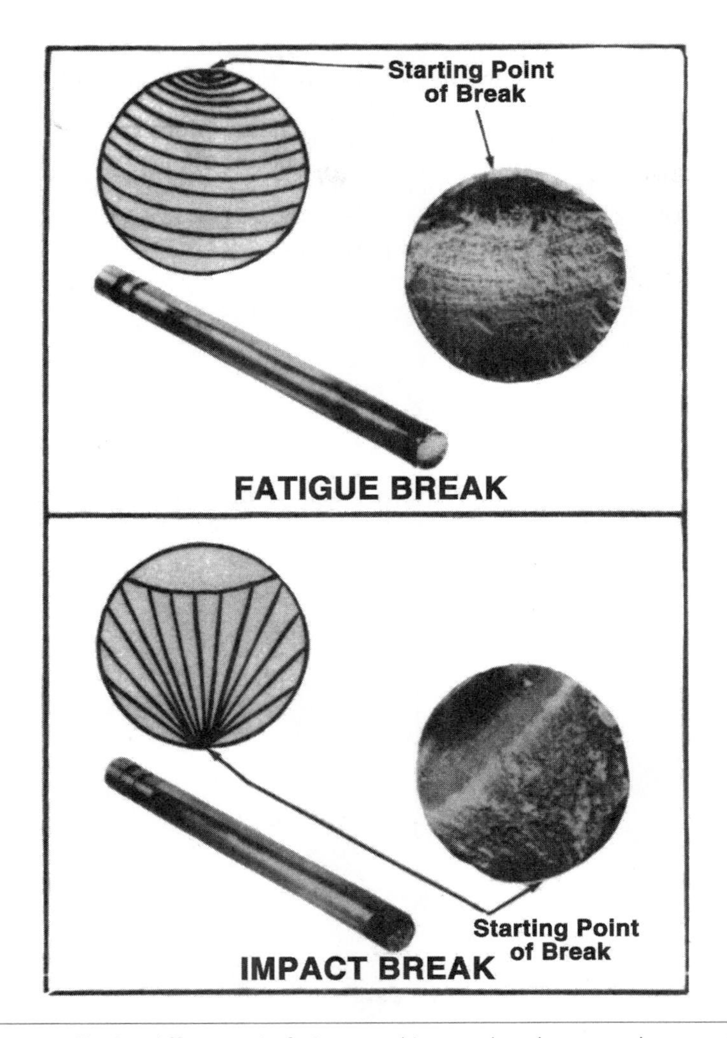


Figure 6-13 Usually the difference in fatigue and impact breakage can be seen in the lines of the break. (Photos courtesy of John Deere)

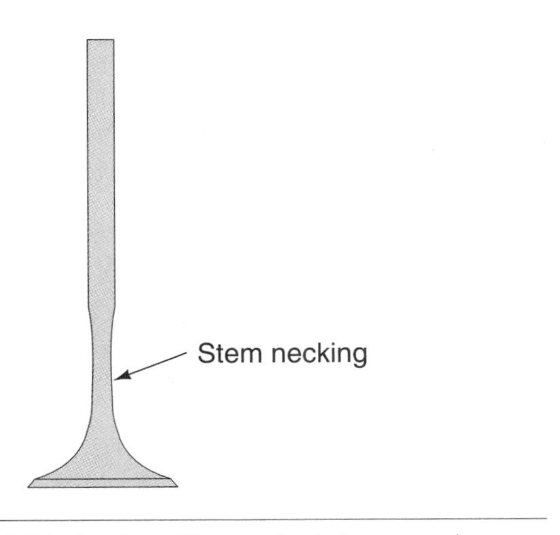


Figure 6-14 A valve with a necked stem must be replaced.

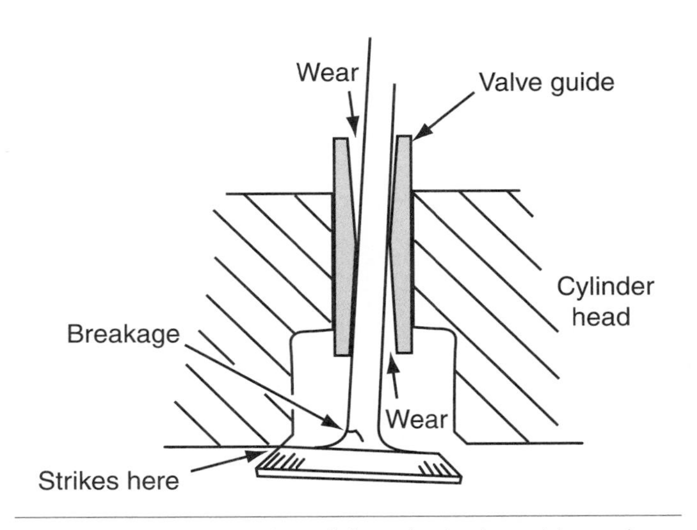


Figure 6-15 Tipping of the valve in the guide can keep the valve from seating properly, resulting in breakage near the fillet.

**Shop Manual**
Chapter 6, page 246

The **valve seat** provides the mating surface for the valve face. **Integral valve seats** are part of the cylinder head. **Valve seat inserts** are pressed into a machined recess in the cylinder head.

**Valve Stem Wear.** When visually inspecting the valve stem, look for galling and scoring (Figure 6-16). Either of these can be caused by worn valve guides or carbon buildup or off-square seating. Also look for indications of stress. A stress crack near the fillet of the valve indicates excessive valve spring pressures or uneven seat pressure. Stress cracks near the top of the stem can be caused by excessive valve lash, worn valve guides, or weak valve springs.

Inspect the tip of the valve stem for wear and flattening. These conditions can be caused by improper rocker arm alignment, worn rocker arms, or excessive valve lash. Separation of the tip from the stem can be caused by these same conditions.

## Valve Seats

When the valve closes, it must form a seal to prevent loss of compression and combustion pressures. The valve face rests against the valve seat to accomplish this task (Figure 6-17). In addition

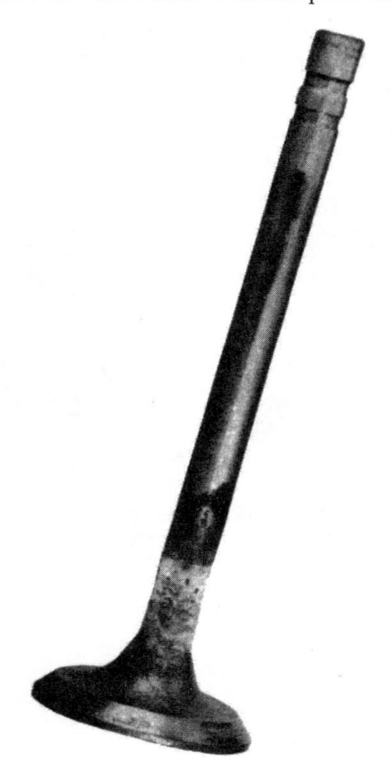

Figure 6-16 Scored valve stem. This wear indicates the guide is worn, which likely caused the face burning. (Photo courtesy of John Deere)

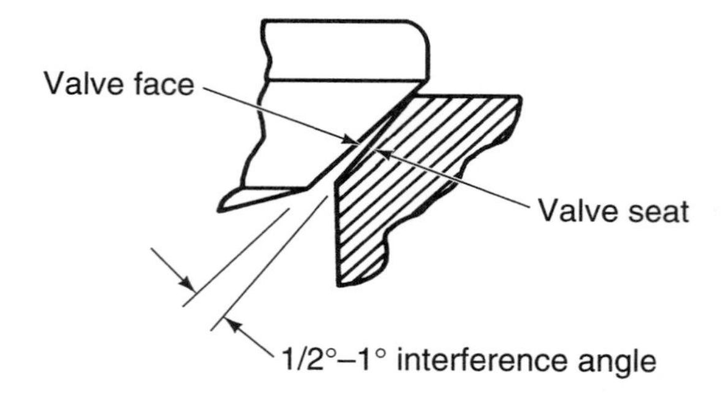


Figure 6-17 The valve face contacts the valve seat to seal the combustion chamber. The face and seat are machined at different angles to create an interference angle which provides a better seal.

to sealing compression pressures, the seat and face provide a path to dissipate heat from the valve head to the cylinder head.

There are two basic types of **valve seats:** integral and removable inserts. **Integral valve seats** are machined into the cylinder head. The seat is induction hardened to provide a durable finish. Most aluminum cylinder heads (and many cast-iron cylinder heads) use **valve seat inserts** that are pressed into a recessed area (Figure 6-18). Common materials used in seat insert construction include cast iron, bronze, and steel alloys. Cast-iron inserts are used to repair cast-iron heads not originally equipped with seat inserts. Bronze inserts are typically used in aluminum cylinder heads. Bronze inserts are claimed to have better heat transfer, thus prolonging valve life. Steel alloys are available in several grades of stainless steel and with several grades of hardness. One of the hardest seats is made of **Stellite.** The Stellite is applied to the valve seat to protect against oxidation and corrosion, and for added wear resistance.

The seat is induction hardened to provide a durable finish. Seat hardening is done by an electromagnet using induction to heat the valve seat to a temperature of about 1,700°F (930°C). This hardens the seat to a depth of about 0.060 in. (1.5 mm). The greatest advantage of integral seats is their ability to transfer heat. Integral seats run about 150°F (83°C) cooler than seat inserts.

Stellite is a hard facing material made from a cobalt-based material with a high chromium content.

Valve seat recession is the loss of metal from the valve seat, causing the seat to recede into the cylinder head.

**Seat Recession.** There are three common causes of **valve seat recession** (Figure 6-19). The first is due to the pressure and heat in the combustion chamber causing local welding of the valve face and valve seat. When the valve is opened again, it forces a breaking of this weld, resulting in the removal of metal. In the past, this action of local welding was controlled by adding lead to the fuel. The lead would be the subject of the welding and breaking instead of the valve face and seat. To counteract the damage to valves and valve seats as a result of the loss of lead, most manufacturers use very hard valve seats.

The second leading cause of valve recession is sustained high engine speeds. The third cause is weak valve springs. Both of these cause valve recession in the same manner. Under these two conditions, the valve rotates faster than normal. This results in excessive grinding between the valve face and seat. If the valve face is a harder material than the seat, the seat will recede.

The solution for seat recession is to install hard valve seats and use high-quality valves. In addition, spring dampeners should be used to reduce chatter.

**Shop Manual**
Chapter 6, page 250

## Valve Springs

There are two basic functions of the **valve spring:** first, it closes the valve against its seat; second, it maintains tension in the valve train when the valve is open to prevent float. The spring must be able to perform these functions while withstanding temperature changes and vibrations. A valve spring is no less a vital part than any other component in the valve train. In today's high-speed,

The valve spring is a coil of specially constructed metal used to force the valve closed. This provides a positive seal between the valve face and seat.

Figure 6-18 Valve seat insert.

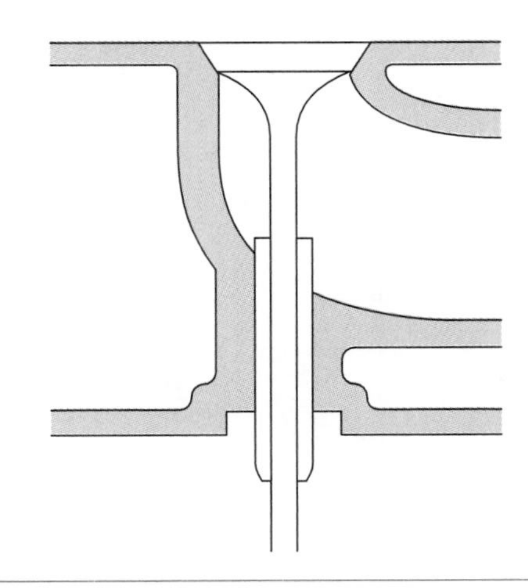

Figure 6-19 Valve seat recession.

high-performance engines, the valve spring becomes a very significant factor in efficiency and power output. The valve spring is strapped with the task of opening and closing the valves fifty times per second, at 6,000 rpm. After miles of use, the spring may lose its tension or break. Of major consideration is the reciprocating valve motion, which produces an inertia effect. Worn or weak valve springs may not have sufficient tension for the lifter to follow the camshaft lobe, resulting in **valve float.**

**Valve float** is a condition that allows the valve to remain open longer than intended. It is the effect of inertia on the valve.

Valve float causes the lifters to leave the camshaft lobe surface. In turn, this results in an abrupt seating of the valve and the lifter impacting the camshaft lobe. Valve float can result in excessive wear of the valve seat, lifter, camshaft, and valve. In addition, weak valve springs can prevent good heat transfer, resulting in burnt valves and valve seat damage by bouncing the valve against it. Weak valve springs usually have wear on their ends.

To accomplish its functions, the valve spring can be designed with different characteristics and features. The most common designs include: dual springs, dampers, and variable rate springs.

Dual springs are basically a set of two springs, with the smaller spring set inside the larger spring. The two springs have different diameters and lengths. The differences in size creates two different harmonic vibrations that cancel each other.

Some manufacturers use a damper inside the valve spring to reduce harmonic vibrations (Figure 6-20). The damper is a flat, wound coil designed to rub against the valve spring. The rubbing effect reduces the vibrations.

Variable rate springs are designed with unequally spaced coils (Figure 6-21). This results in a spring that increases its rate of pressure as it is compressed. The more it is compressed, the more pressure it exerts for the distance compressed; for example, a variable rate spring may exert an additional 40 pounds the first tenth of an inch it is compressed; the next tenth of an inch of compression will result in an additional 50 pounds of pressure; the next tenth of an inch compression will result in an additional 60 pounds of pressure, and so forth. A fixed rate spring, on the other hand, will provide 50 pounds of pressure increase for each tenth of an inch compression.

Along with different shapes and designs, different metals are used to construct valve springs. The steels used to construct a valve spring can vary from the very expensive H-11 tool steel used in Pro Stock drag racing to various grades of stainless steel. The type of steel determines the strength of the valve spring and its ability to transfer heat.

Heat is one of the major contributors to valve spring fatigue. Oil flowing over the valve spring is the only cooling method used. The majority of the heat generated in a valve spring is believed to be the result of friction between the coils of the spring. Some valve spring manufacturers attempt to reduce the amount of friction by coating the valve springs. Common coatings include Teflon® and SDF-1. Teflon® sheds the oil, while SDF-1 retains the oil by use of a wetable matrix. In tests, a coated spring runs about 20°F (11°C) cooler than an uncoated spring.

**Shop Manual**
Chapter 6, page 255

## Valve Guides

**Valve guides** support and guide the valve stem through the cylinder head.

There are two types of **valve guides:** insert and integral. Insert guides are removable liners pressed into the cylinder head (Figure 6-22). This design is common on cylinder heads made of aluminum alloys. The insert is usually constructed of cast iron or bronze.

MECHANICAL VIBRATION DAMPENERS

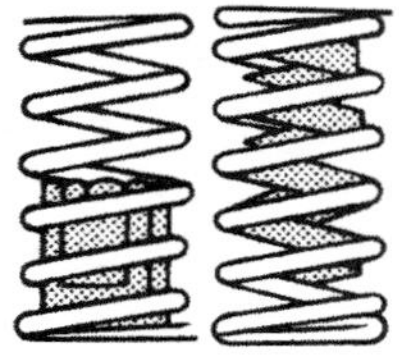

**Figure 6-20** Damper spring.

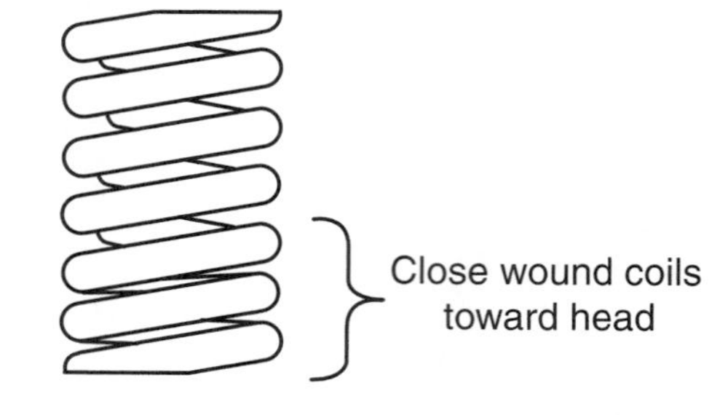


**Figure 6-21** Variable rate valve spring.

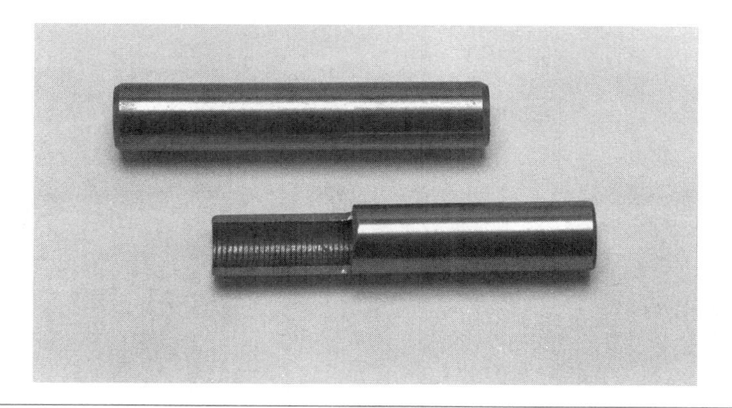

**Figure 6-22** Insert valve guides.

Integral guides are machined into the head and are a part of the cylinder head. Although this design cannot be removed, it can be repaired if it is worn or damaged. Some of the repair methods used include knurling, machining to accept an insert, or inserting a special coil.

## Valve Seals

**Shop Manual**
Chapter 6, page 288

To control the amount of oil allowed to travel down the valve stem to provide lubrication, valve stem seals are used. These seals are constructed of synthetic rubber or plastic and are designed to control the flow of oil, not prevent it. There are three basic types of valve stem seals: O-ring, umbrella, and positive seal.

The O-ring seal is basically a square-cut seal made of rubber (Figure 6-23). The valve stem has a groove machined near the top to accept the O-ring. A shield is also used to deflect oil away from the valve stem.

Umbrella seals fit over the valve guide (Figure 6-24). The seal provides a tight fit around the valve stem and a loose fit around the guide. This design allows the seal to move up and down with the valve.

Positive seals are generally constructed of rubber with a Teflon® insert (Figure 6-25). The design fits the valve guide snugly with either a light spring, spring clip, or press fit. Positive seals are considered by many to be the best type of valve stem seal. It is possible to replace umbrella or O-ring seals with positive seals. If the original guide was not cut for a positive seal, it can be cut to accept the seal.

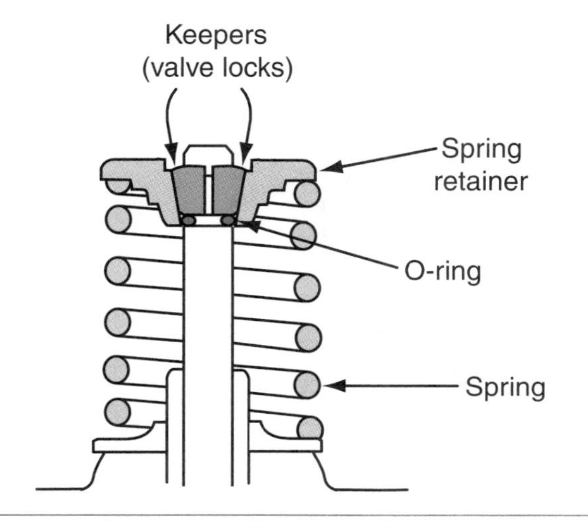


**Figure 6-23** O-ring valve stem seal.

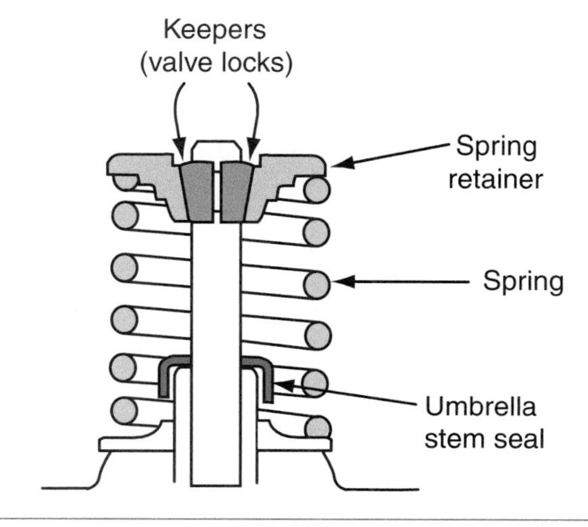


**Figure 6-24** Umbrella seal installed over the valve guide.

Figure 6-25 Positive valve seal designs: Teflon® and rubber; Teflon®; and rubber two ring.

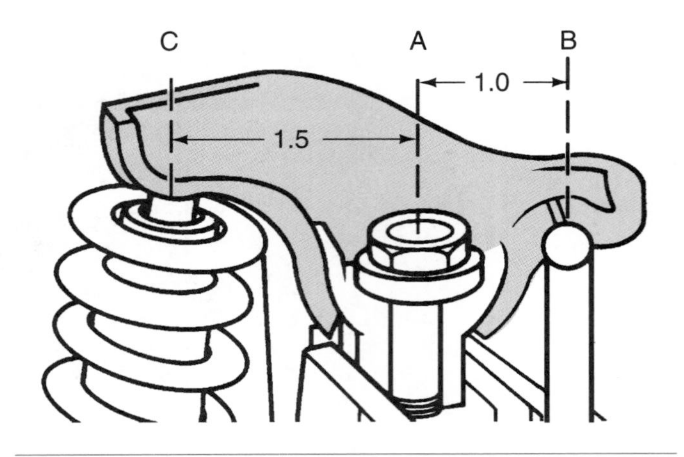


Figure 6-26 Rocker arms are designed to provide the desired amount of valve opening. Mechanical advantage is accomplished by the ratio determined by the distance from A to B and A to C.

**Shop Manual**
Chapter 6, page 285

On overhead cam engines, the lifter is called a follower. Followers are similar to rocker arms except they run directly off of the camshaft.

The **rocker arm** is a pivoting lever used to transfer the motion of the pushrod to the valve stem.

## Rocker Arms

The **rocker arm** is a pivoting lever used to transfer the motion of the pushrod to the valve stem. The design of the rocker arm is such that the side to the valve stem is usually longer than the side to the pushrod (Figure 6-26). This design gives the rocker arms a mechanical advantage. The rocker arm ratio works with the camshaft lobe to provide the desired lift; for example, if the camshaft lobe has a lift of 0.350 in. (8.9 mm) and the rocker arm ratio is 1.5, actual lift is 0.350 × 1.5 = 0.525 in. (assuming zero valve clearance).

# Cylinder Head Component Relationships

All components of the cylinder head associated with valve opening and seat (the valve, spring, guide, and seat) must work together to accomplish their designed task. Wear, damage, or problems in one component will have an effect on the other components. If the valve stem or guide is worn, there will be excessive movement of the valve inside the guide (Figure 6-27). This

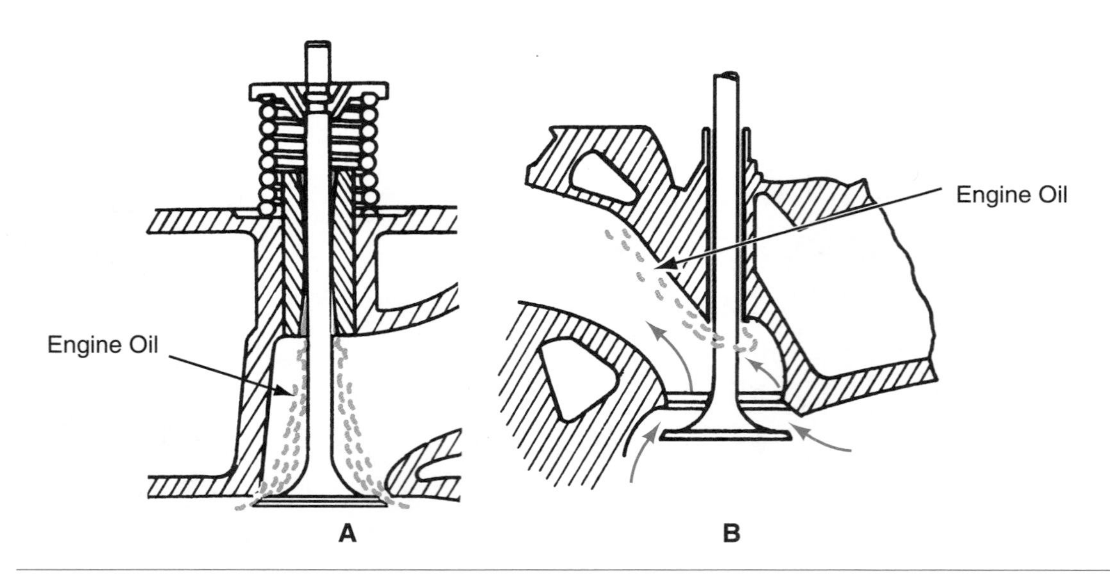


Figure 6-27 Worn valve guides can translate to improper valve sealing and oil consumption.

movement will translate to improper seating and possible burning of the valve. In Figure 6-27, "A" shows oil leaking past a worn intake valve guide. Atmospheric pressure pushes the oil into the air/fuel mixture. When the intake valve opens and the mixture is drawn into the cylinder, the oil is burned with the air/fuel charge. This results in oil consumption and blue smoke from the tailpipe. In Figure 6-27, "B" shows an exhaust valve with a worn guide allowing oil to be sucked by the guide due to venturi action. This too will result in oil consumption and blue smoke.

If a worn valve seat is detected and it is repaired by machining, the valve stem will extend farther through the head by the same amount as removed from the seat. This results in a different angle of the rocker arm and in changes in valve spring height. This condition must be corrected at the time of the valve service.

**AUTHOR'S NOTE:** During the cylinder head reconditioning process, all valve train components must be machined or replaced so they meet the engine manufacturer's specifications. An unsuccessful cylinder head reconditioning job may be defined as one that did not provide a reasonable mileage interval before some valve train component failed. It has been my experience that unsuccessful cylinder head reconditioning jobs are usually caused by failure to make sure that all valve train components are within manufacturer's specifications. For example, the technician may have done an excellent job of reconditioning the valves, seats, and guides, but the condition and measurements on the valve springs were overlooked.

**Wedge chamber** design locates the spark plug between the valves in the widest portion of the wedge in the cylinder head.

**Quenching** is the cooling of gases as a result of compressing them into a thin area. The quench area has a few thousandths of an inch clearance between the piston and combustion chamber. Placing the crown of the piston this close to the cooler cylinder head prevents the gases in this area from igniting prematurely.

The **squish area** of the combustion chamber is the area where the piston is very close to the cylinder head. The air/fuel mixture is rapidly pushed out of this area as the piston approaches TDC, causing a turbulence and forcing the mixture toward the spark plug. The squish area can also double as the quench area.

## Combustion Chamber Designs

The size, shape, and design of the combustion chamber affect the engine's performance, fuel efficiency, and emission levels. Manufacturers can use several different combustion chamber designs to achieve desired engine efficiency results.

### Wedge Chamber

The most common method of creating a **wedge chamber** is to use a flat-topped piston and cast a wedge into the cylinder head (Figure 6-28). Wedge chamber design locates the spark plug between the valves in the widest portion of the wedge. The air/fuel mixture is compressed into the **quench** and **squish areas,** resulting in a turbulence that mixes the air/fuel mixture. Another design method is to use flat chambers in the cylinder head and to offset the cylinder bores. This results in the piston approaching the cylinder head at an angle, and produces the same basic functions of a wedge chamber.

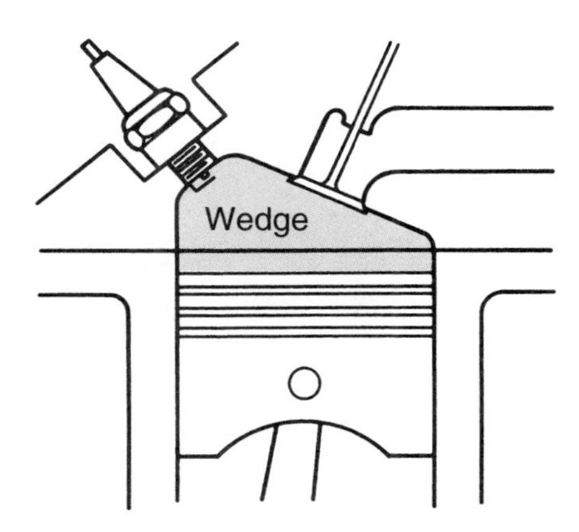


Figure 6-28 Wedge combustion chamber design.

## Hemispherical Chamber

The term "hemi" may be used for an engine with **hemispherical combustion chambers.**

The **hemispherical chamber** is designed in a half-circle with the spark plug located in the center of the dome (Figure 6-29). The valves are located on an incline of 60 to 90 degrees from each other.

The hemi-head design provides an even flame front because the spark plug is centered. The design also allows for easy "breathing" of the engine because the valves are located across from each other. Disadvantages of this design include little turbulence and higher emissions. To counteract this, a domed piston may be used by some manufacturers.

## Swirl Chamber

**Swirl chambers** create an air flow that is in a horizontal direction.

To improve combustion efficiencies, **swirl chambers** are designed with a curved port that causes the air/fuel mixture to swirl in a corkscrew fashion (Figure 6-30). The compression of this

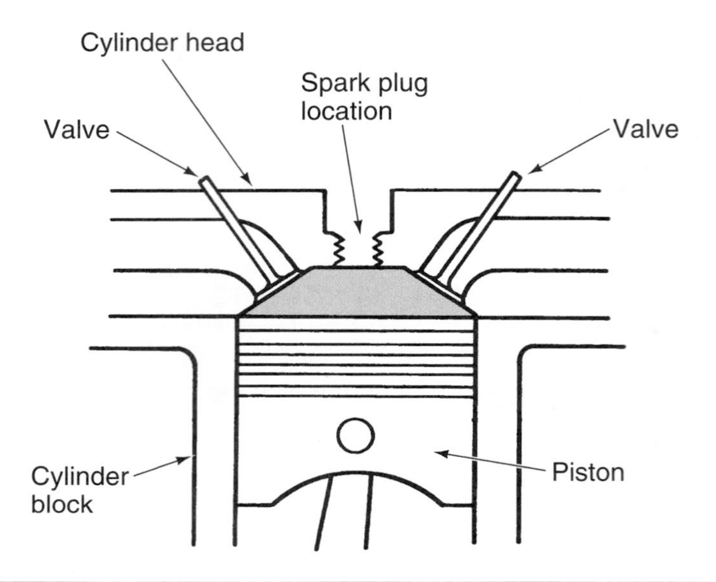


**Figure 6-29** Hemispherical combustion chamber design.

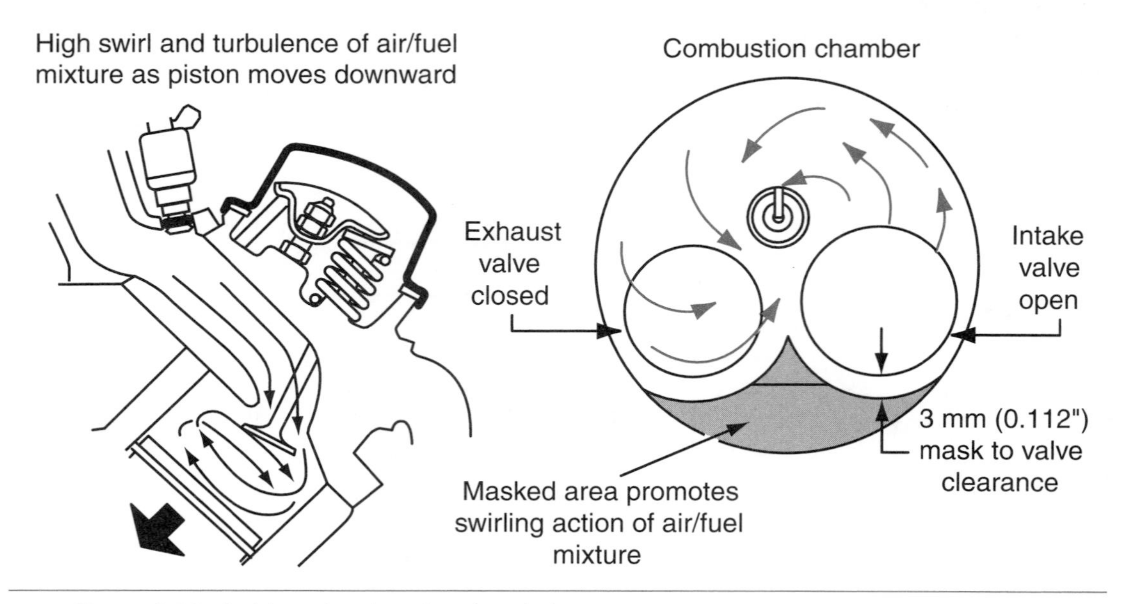


**Figure 6-30** Swirl combustion chamber design.

swirling mixture results in a thorough mixing of the gases, resulting in increased fuel economy and reduced exhaust emissions. Location of the valve and design of the chamber causes the swirl action.

## Chamber-in-Piston

The **chamber-in-piston** design locates the combustion chamber in the piston head instead of the cylinder head (Figure 6-31). The advantage of this chamber design is that the piston remains hot enough to provide proper vaporization of the air/fuel mixture. This design is used on many diesel engines.

A **chamber-in-piston** is a piston with a dish or depression in the top of the piston.

## Pentroof Chamber

**Pentroof combustion chambers** are a common design for multivalve engines using two intake valves with one or two exhaust valves. This style of chamber was first used in production vehicles in the mid 1980s and has grown in popularity. The inverted *V* shape of the chamber provides a good **surface-to-volume ratio,** resulting in lower emissions (Figure 6-32). In addition, this design allows for larger valves, centering the spark plug in the chamber, and good turbulence.

**Pentroof combustion chambers** are a common design for multivalve engines using two intake valves with one or two exhaust valves and provide good surface-to-volume ratio.

**Surface-to-volume ratio** is a mathematical comparison between the surface area of the combustion chamber and the volume of the combustion chamber. The greater the surface area, the more area the mixture can cling to. Mixture clinging to the metal will not burn completely since the metal cools it. Typical surface-to-volume ratio is 7.5:1.

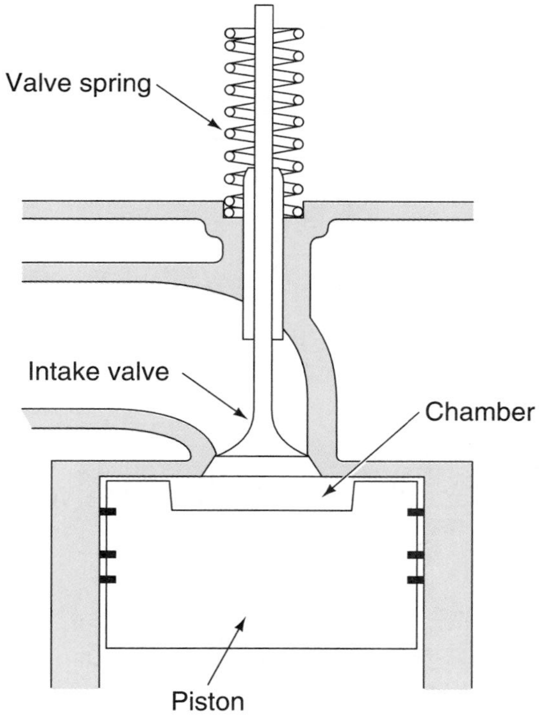

## Fast Burn

A faster combustion can be achieved by directing the incoming air/fuel mixture tangentially through the intake valve (Figure 6-33). This results in a turbulence that mixes the gases. An additional benefit of this design is the compression ratio can be increased without an increase in octane requirements. This is because the fast-burn design has less potential for knock.

Valve spring
Intake valve
Chamber
Piston

Figure 6-31 Chamber-in-piston combustion chamber design.

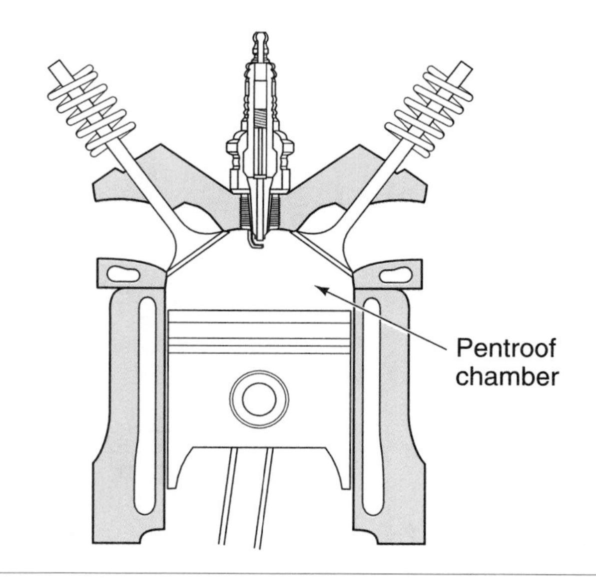


Figure 6-32 Pentroof combustion chamber.

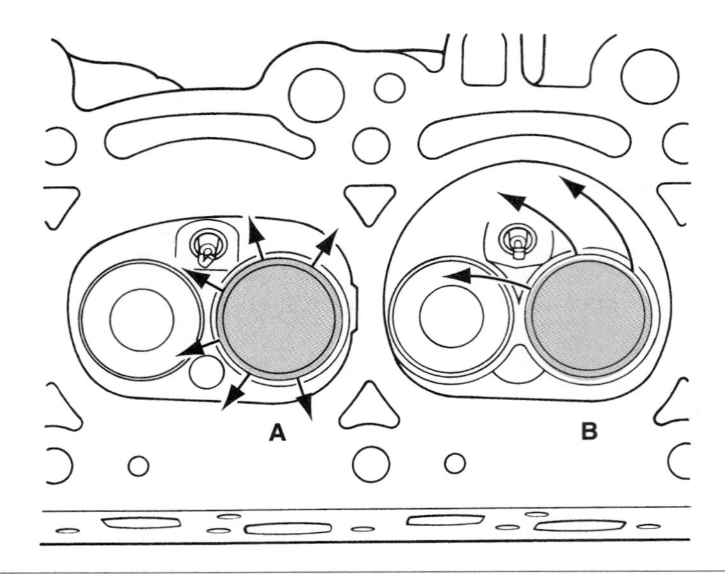


Figure 6-33 (A) Standard uniform flow and (B) fast-burn, tangential-flow combustion chambers.

## Tumble Port

**Tumble port** combustion chambers use a modified intake port design.

An **eddy** is a current that runs against the main current.

**Tumble port** combustion chambers use a modified intake port design, resulting in a tumbling vortex mixture burning technique (Figure 6-34). This design uses the principle of **eddy currents** by using a volume of air that runs against the main current to provide a tumbling condition. Benefits include increased fuel economy and reduced hydrocarbon emissions.

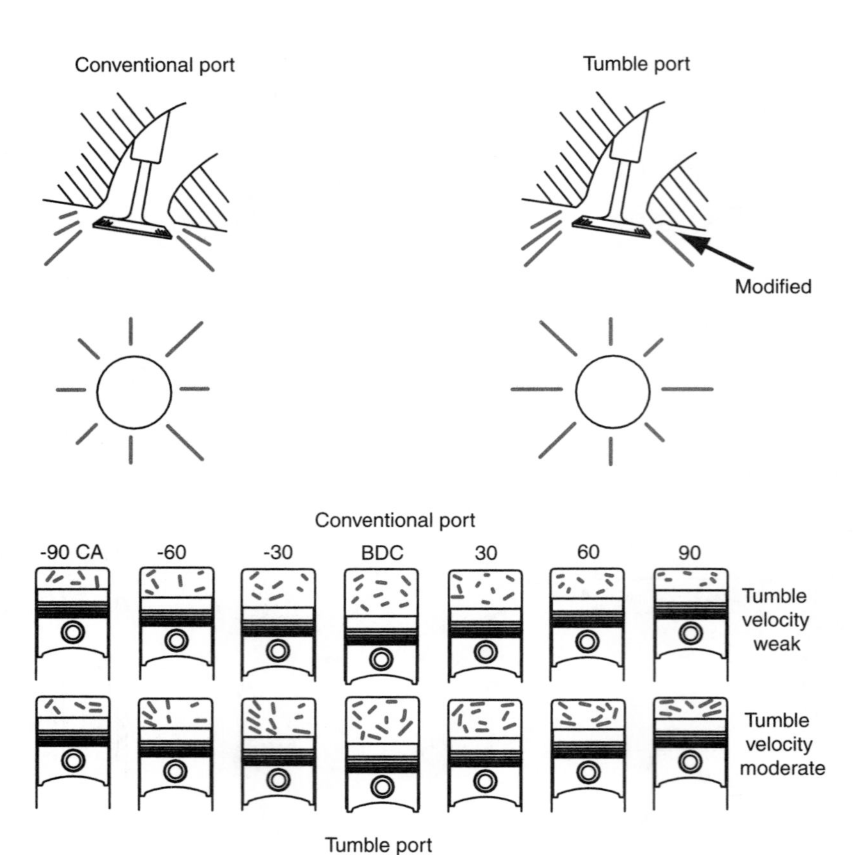


Figure 6-34 Comparison of conventional and tumble port designs.

## Stratified Charge Chambers

The **stratified charge** combustion chamber is actually two chambers contained within the cylinder head (Figure 6-35). The smaller chamber (prechamber) is located above the main chamber and has its own intake valve. A very rich air/fuel mixture is delivered to the prechamber, where the spark plug is located. The rich mixture is very easy to ignite. At the same time, a very lean mixture is delivered to the main chamber. At the completion of the compression stroke, the spark plug fires to ignite the rich mixture in the precombustion chamber. The burning rich mixture will ignite the lean mixture in the main combustion chamber.

A **stratified charge** engine has two combustion chambers. A rich air/fuel mixture is supplied to a small auxiliary chamber and the very lean air/fuel mixture is supplied to the main combustion chamber.

The first manufacturer to mass produce the stratified charge engine was Honda. They used the engine from 1975 to 1987 with the compound vortex controlled combustion (CVCC) design. A three-barrel carburetor was used. Two barrels supplied the lean mixture for the main combustion chamber, while the third barrel supplied the prechamber with the rich mixture.

# The Combustion Process

As was discussed in Chapter Two, the power stroke is accomplished by rapidly expanding the gases resulting from igniting a tightly packed air/fuel mixture. If the flame burns across the combustion area at the proper time and rate, normal combustion has occurred. As the flame burns, heat is produced. The heat causes the burnt gases to expand. The expansion of the gases causes two things to occur: (1) it works against the top of the piston to force it downward; (2) the expansion causes the unburned gases to be packed tighter. Packing the end gases tighter increases their heat to a point very close to self-ignition. To prevent self-ignition, the heat must be removed from the gases. This is done through the top of the piston, the walls of the combustion chamber, and the coolant in the cylinder head.

The heat absorption of the **end gases** can be increased by creating some turbulence within the combustion chamber and by proper design of the squish and quench areas (Figure 6-36). As the gases swirl around, they come into contact with the walls of the combustion chamber or the top of the piston. **Turbulence** prevents the gases from becoming stagnant. Stagnant gases will only dissipate heat from the outer edges, allowing the center to become very hot.

The unburned gases are called **end gases.**

**Turbulence** prevents gases from becoming stagnant within the comubstion chamber.

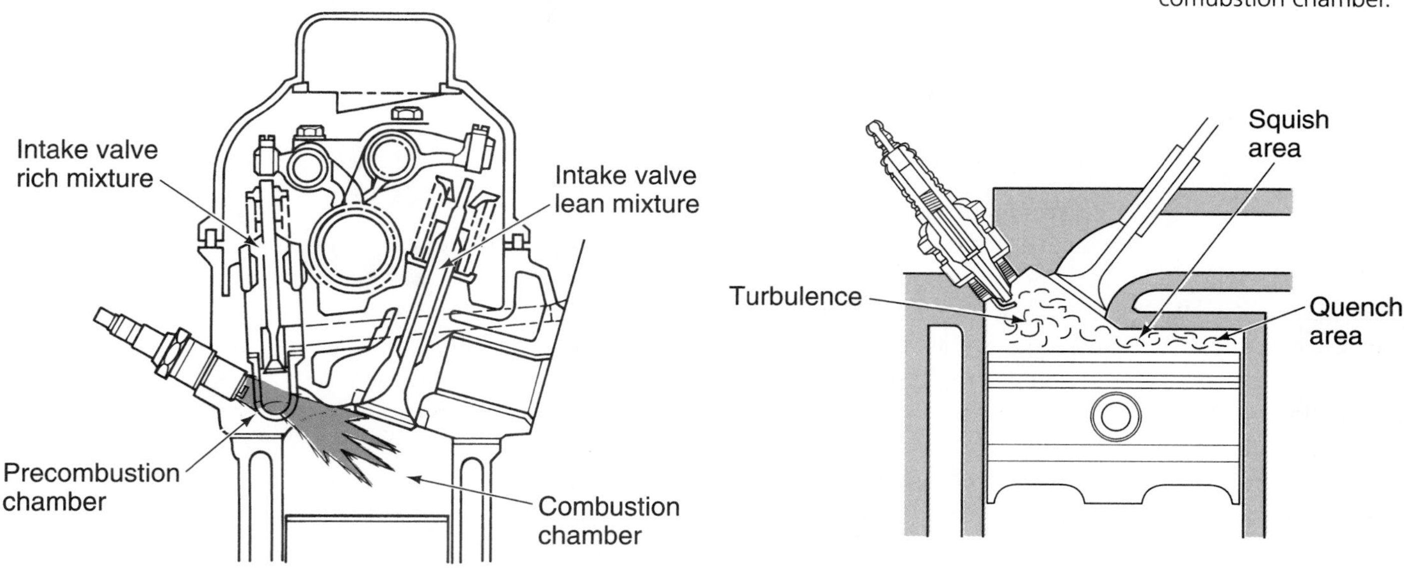


**Figure 6-35** Stratified combustion chamber.

**Figure 6-36** Quenching and turbulence are used to cool the temperatures of the end gases to prevent self-ignition.

Turbulence also causes the flame to burn faster. The turbulence can be promoted by piston top design and the squish area of the chamber. The squish area causes the gases to be squirted out of an area of the chamber as the piston compresses the gases.

As the gases burn, the end gases are packed into the quench area of the chamber. This thins the gases and allows them to dissipate heat faster. The quench area design affects the emission output of the engine. As we discussed, to prevent the end gases from igniting on their own, heat is removed from them. This unburned mixture has a thickness of about 0.002 to 0.010 in. (0.1 to 0.3 mm). This thin layer of gases never burns because the temperature is too low, and it is expelled into the exhaust system as unburned hydrocarbons. The larger the combustion chamber, the more the surface area for the formation of the **skin effect.**

**Skin effect** is a small layer of unburned gases formed around the walls of the combustion chamber.

Manufacturers have reduced hydrocarbon emissions by redesigning the quench area. Reduction of hydrocarbons can be accomplished by increasing the height of the quench area. This increases the temperature of the end gases and reduces the thickness of the skin effect. Another means is by eliminating undesirable quench areas. This is done by removing sharp corners and pockets. Cylinder head gasket redesign and improved fit have eliminated many of these undesired quench areas.

Finally, skin effect has been reduced by lowering the compression ratio of the engine. This is not desirable from a performance point of view, but it was necessary to reduce emissions. Most engine manufacturers reduced the output of their engines during the late 1970s to the mid 1980s. Better understanding of the combustion process and chamber designs has made a return of higher compression engines possible in the last few years.

## A BIT OF HISTORY

With the great amount of moving parts within the engine, it is truly surprising how well it works. Unfortunately, engines requiring carbon-based fuels pollute the air and use a nonreplenishable fuel source. Today, most manufacturers are attempting to design a reliable, cost-efficient electric vehicle. This idea is not new. During the early years of the "horseless carriage," over one-third were driven electrically.

## Preignition

As we discussed in earlier, preignition results in a knocking or chatter sound. Preignition occurs when a flame is ignited in the combustion chamber prior to the spark plug firing. It is usually the result of hot spots in the chamber. This can be caused by excessive combustion chamber pressures. An excessively high compression ratio causes the temperature of the fuel charge to be raised above its flash point. A change in compression ratio can occur anytime there is a decrease in combustion chamber volume. This can be caused by excessive carbon buildup. Other things that change compression ratio include piston design and milling of the cylinder head and/or the block. Another cause of preignition is the use of excessively low **octane** fuel. The octane required to properly operate the engine is based on the compression ratio of the engine. The higher the ratio, the higher the octane required.

**Octane** is the fuel's ability to resist detonation. The lower the octane number, the easier it detonates.

Any area of the combustion chamber that retains heat can cause preignition. These include:

- Thin valve margins
- Gasket edges
- Sharp metal edges

- Spark plugs
- Carbon

The result of preignition is high pressures within the combustion chamber before the piston reaches TDC. This pressure forces the piston to reverse direction too soon, wasting useful combustion energy. It is possible to blow a hole in the top of the piston near the outer edge as a result of preignition (Figure 6-37).

### Detonation

Detonation is another example of improper combustion resulting in engine noise. Detonation occurs when a second flame front is started after the spark plug is fired. When the two flames collide, a knocking sound is generated. Detonation also results in higher-than-normal combustion pressures and temperatures, which can burn the top of the piston.

## Multivalve Engines

Many of today's engines are designed with more than one intake and/or exhaust valve per cylinder (Figure 6-38). This multivalve design allows more air/fuel mixture to enter the cylinder faster, and allows for better expulsion of the burned gases. The most common arrangement is four valves per cylinder. Two valves are intake valves and two are exhaust valves. Three valves per cylinder is also a common arrangement.

Another variation is the use of a jet valve. The jet valve directs air from above the throttle plates into the combustion chamber. At low engine speeds, this air causes a swirling action in the combustion chamber (Figure 6-39). The resultant turbulence increases the efficiency of the burn. Since the air inlet to the jet valve is above the throttle plates, the amount of air entering the valve decreases as engine speed increases.

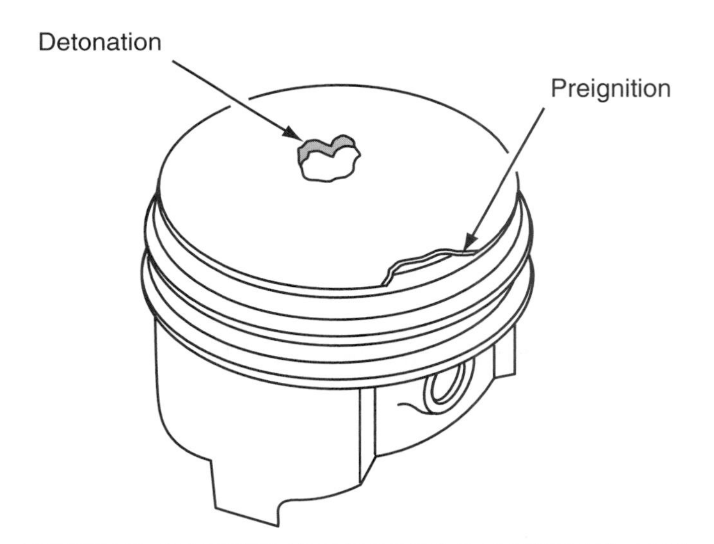


**Figure 6-37** Piston damage caused by abnormal combustion.

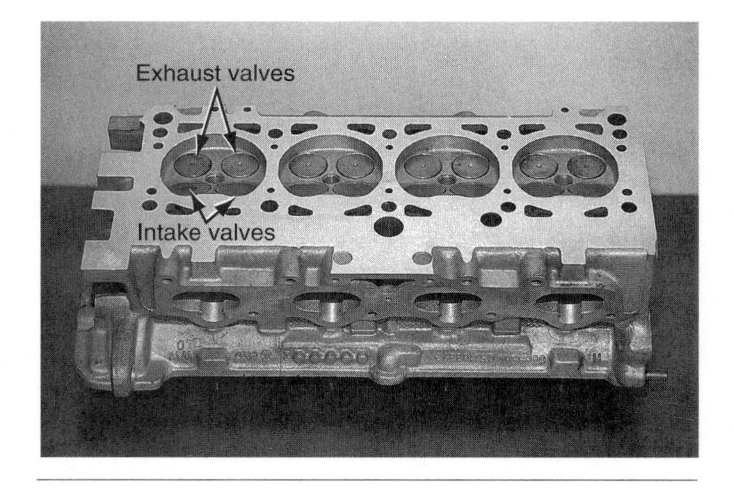


**Figure 6-38** Multivalve cylinder head.

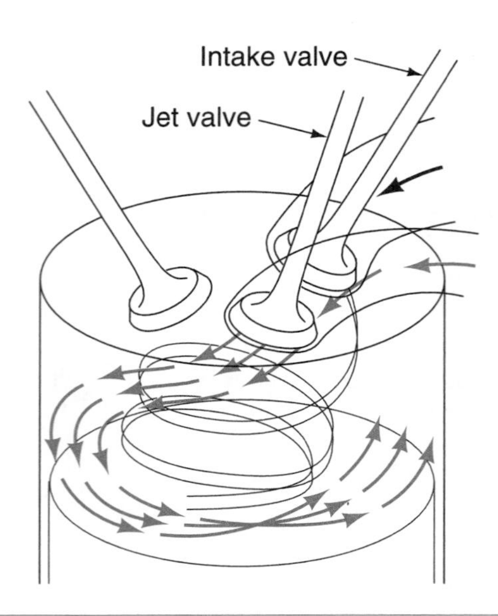


Figure 6-39 Some multivalve engines use a jet valve to increase the burn effectiveness at low engine speeds.

**A BIT OF HISTORY**

The Honda Insight and the Toyota Prius are the first hybrid cars to be introduced in the North American market. As the name suggests, these cars have two power sources to supply torque to the drive wheels. These power sources are a mechanical power source from the engine and an electric power source from batteries and an electric motor. These cars may be referred to as parallel hybrids because power may be supplied to the drive wheels from the engine or from the electric motor. An on-board computer system controls the power source to the drive wheels. When the vehicle is decelerating, the electric motor acts as a generator to help charge the batteries. The batteries do not require charging from an external source.

## Summary

- On most engines, the cylinder head contains the valves, valve seats, valve guides, valve springs, and the upper portion of the combustion chamber.
- The poppet valve is opened by applying a force that pushes against its stem. Closing of the valve is accomplished through spring pressure.
- The valve as a whole has many parts. The large diameter end of the valve is called the head. The angled outer edge of the head is called the face and provides the contact point to seal the port. The area between the valve face and the valve head is referred to as the margin. The stem guides the valve through its linear movement.
- Manufacturers use several methods to dissipate the heat from the valve. First, the contact surface of the face and seat is used to transfer the heat to the cylinder head. The second method is through the stem, valve guide, valve spring, and on to the cylinder head.
- A burned valve is identifiable by a notched edge beginning at the margin and running toward the center of the valve head.

**Terms to Know**

Burned valve
Chamber-in-piston
Channeling
Cylinder head
Eddy currents
End gases
Face
Fillet
Free-wheeling engine
Head
Hemispherical chamber

- ❒ Metal erosion is identified by small pits in the valve face and is generally caused by a chemical reaction in the combustion chamber.
- ❒ Breakage of the valve can occur due to valve and piston contact, valve springs that are too tight, loss of valve keepers, any condition that can result in stem necking, and off-square seating.
- ❒ The continual movement of the valve may result in wear on the portion of the stem traveling in the guide. As the stem wears, the oil clearance increases, resulting in excessive oil being digested into the combustion chamber and the formation of deposits on the valve.
- ❒ The valve seat is located in the cylinder head and provides a seal when the valve face contacts it. It also provides a path to dissipate heat from the valve head to the cylinder head.
- ❒ There are two basic types of valve seats: integral and removable inserts. Integral valve seats are machined into the cylinder head. Most aluminum cylinder heads (and many cast-iron cylinder heads) use valve seat inserts that are pressed into a recessed area.
- ❒ Valve guides support and guide the valve stem through the cylinder head. There are two types of valve guides used: insert and integral. Insert guides are removable tubes that are pressed into the cylinder head; integral guides are machined into the head.
- ❒ There are two basic functions of the valve spring: first, it closes the valve against its seat; second, it maintains tension in the valve train when the valve is open to prevent float.
- ❒ To accomplish its functions, the valve spring can be designed with different characteristics and features. The most common designs include dual springs, dampers, and variable rate springs.
- ❒ Valve seals control the amount of oil allowed to travel down the valve stem to provide lubrication.
- ❒ The rocker arm is a pivoting lever used to transfer the motion of the pushrod to the valve stem.
- ❒ The size, shape, and design of the combustion chamber affect the engine's performance, fuel efficiency, and emission levels.

**Terms to Know**

Integral valve seats
Interference engine
Margin
Octane
Off-square
Pentroof cumbustion chamber
Poppet valves
Quench
Rocker arm
Skin effect
Squish area
Stellite
Stem necking
Stratified charge
Surface-to-volume ratio
Swirl chambers
Tumble port
Turbulence
Valve float
Valve guides
Valve seat
Valve seat insert
Valve seat recession
Valve spring
Valve stem
Wedge chamber

# Review Questions

## Short Answer Essays

1. List the purpose of the following components:
   Cylinder head — Exhaust valve
   Intake valve — Valve seat
2. List and describe the common types of combustion chambers.
3. List the common causes for the following valve failures: burned valves, channeling, metal erosion, and breakage.
4. What faults can weak valve springs cause?
5. List the causes of valve seat recession.
6. What are the differences between interference and free-wheeling engines?
7. List and describe the functions of the parts of a valve.
8. Describe the functions of the valve spring.
9. List and describe the three common types of valve stem seals.
10. Describe the purpose of the valve guides and the difference between integral and insert types.

## Fill-in-the-Blanks

1. The area between the valve head and the valve face is called the valve ________.
2. ________________ control the flow of gases into and out of the engine cylinder.
3. Valve stem ________________ control the amount of oil allowed to travel down the valve stem and provide lubrication.
4. When the valve closes, it must form a seal to prevent loss of compression. The ________________ ________________ rests against the valve seat to accomplish this task.
5. A ________________ ________________ has two basic functions: first, it closes the valve against its seat; second, it maintains tension in the valve train when the valve is open to prevent valve float.
6. ________________ guides are removable tubes that are pressed into the cylinder head. ________________ guides are machined into the head and are a part of the cylinder head.
7. Most of the heat is dissipated from the valve through the contact area between the valve ________________ and ________________.
8. The most common cause of a burned valve is a leak in the sealing between the ________________ ________________ and ________________ .
9. Metal erosion is generally caused by ________________ entering the combustion chamber.
10. Necking is identified by a ________________ of the valve stem just above the ________________ .

## Multiple Choice

1. Valve faces are typically ground to an angle of:
   **A.** 15 degrees or 20 degrees.
   **B.** 20 degrees or 30 degrees.
   **C.** 30 degrees or 45 degrees.
   **D.** 45 degrees or 60 degrees.

2. Valve stem seals are being dicussed.
   *Technician A* says worn valve stem seals may cause rapid valve stem and guide wear.
   *Technician B* says worn valve stem seals may cause excessive oil consumption.
   Who is correct?
   **A.** A only
   **B.** B only
   **C.** Both A and B
   **D.** Neither A nor B

3. Valve malfunctions are being discussed.
   *Technician A* says the most common cause of a burned valve is coolant entering the combustion chamber.
   *Technician B* says valve breakage can be caused by valve springs with excessive pressures.
   Who is correct?
   **A.** A only
   **B.** B only
   **C.** Both A and B
   **D.** Neither A nor B

4. While inspecting a valve stem, it is noticed the stem is necked.
   *Technician A* says this is identified by a thinning of the valve stem just above the fillet.
   *Technician B* says this may have been caused by overheating or excessive valve spring pressures.
   Who is correct?
   **A.** A only
   **B.** B only
   **C.** Both A and B
   **D.** Neither A nor B

5. *Technician A* says the face of the valve is cut at an angle to improve its sealing capabilities.
   *Technician B* says the fillet allows the valve face to be resurfaced.
   Who is correct?
   **A.** A only
   **B.** B only
   **C.** Both A and B
   **D.** Neither A nor B

6. *Technician A* says the exhaust valve runs cooler since it is only exposed to spent gases.
*Technician B* says the intake valve usually has a smaller head diameter than the exhaust valve.
Who is correct?
**A.** A only
**B.** B only
**C.** Both A and B
**D.** Neither A nor B

7. *Technician A* says seat recession can be caused by lack of lead in the fuel.
*Technician B* says seat recession can be caused by weak valve springs.
Who is correct?
**A.** A only
**B.** B only
**C.** Both A and B
**D.** Neither A nor B

8. *Technician A* says valve springs are used to close the valve against its seat.
*Technician B* says valve float can occur if the valve springs are too strong.
Who is correct?
**A.** A only
**B.** B only
**C.** Both A and B
**D.** Neither A nor B

9. *Technician A* says valve stem seals prevent oil from entering the valve guide.
*Technician B* says the umbrella seals fit over the valve guide and move up and down with the valve.
Who is correct?
**A.** A only
**B.** B only
**C.** Both A and B
**D.** Neither A nor B

10. Combustion chamber designs are being discussed.
*Technician A* says the wedge chamber design locates the spark plug between the valves in the widest portion of the wedge.
*Technician B* says the chamber-in-piston type is the most common for multivalve engines.
Who is correct?
**A.** A only
**B.** B only
**C.** Both A and B
**D.** Neither A nor B

# Camshafts and Valve Trains

Upon completion and review of this chapter, you should be able to:

- ❑ Explain the function of the valve train.
- ❑ Describe the purpose and operation of common style automatic belt tensioners.
- ❑ List the components of the valve train.
- ❑ Explain the purpose and function of the camshaft.
- ❑ Describe the relationship among the camshaft lobe design and lift, duration, and overlap.
- ❑ Explain the purpose of the lifters.
- ❑ Describe the operation of hydraulic lifters.
- ❑ Explain the purpose of pushrods.
- ❑ List and explain the common methods of mounting the rocker arms.
- ❑ Explain rocker arm geometry.
- ❑ Describe the advantages of variable valve timing.
- ❑ Explain the operation of a variable valve timing system.

## Introduction

Proper breathing of the engine depends on the valves opening and closing at the correct time. This is the function of the **valve train.** All of the components that make up the valve train work together to perform this task. Everything in the valve train is related and dependent upon each other; thus, a failure anywhere in the system will result in poor performance.

The **valve train** is made up of components that open and close valves to allow the engine to breathe correctly.

## Valve Timing

The timing of the valve opening is critical to engine operation. The intake valve must begin to open prior to the piston completing the exhaust stroke (Figure 7-1). Also, the valve must be closed at the proper time to prevent loss of compression. If the exhaust valve is not opened and

**Shop Manual**
Chapter 7, page 307

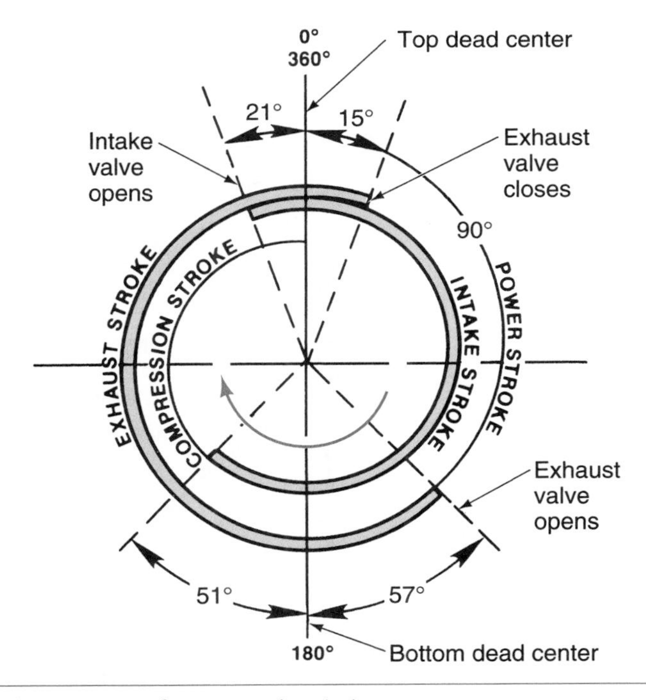


Figure 7-1 The importance of correct valve timing.

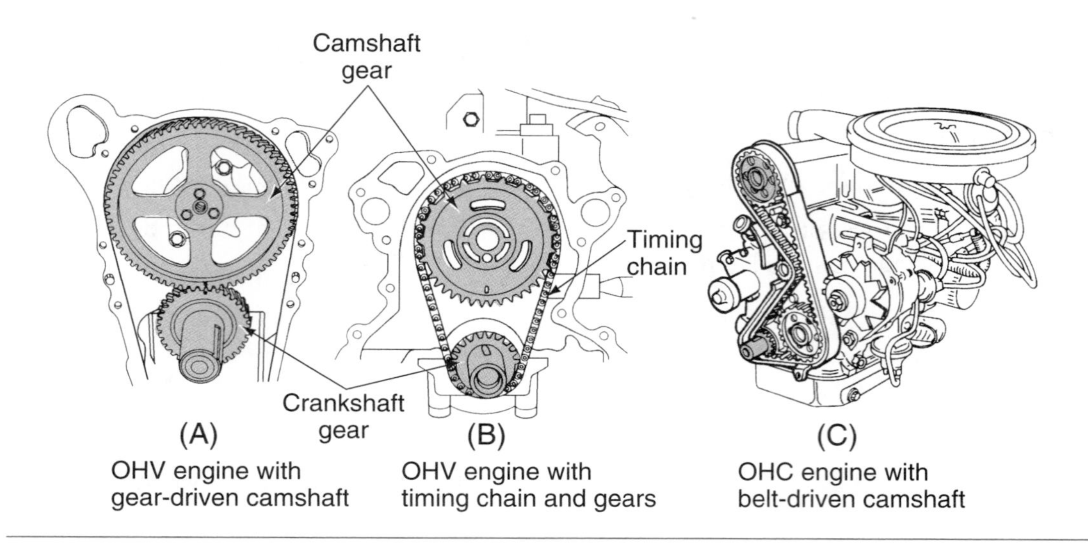


**Figure 7-2** Common valve drive train types include gear drive (A), chain drive (B), and belt drive (C).

closed at the precise time, the engine will not be able to breathe properly and draw in a fresh air/fuel charge.

The **camshaft** controls valve opening, rate of valve opening, and how long it is open.

Valve timing is a function of the **camshaft,** which is driven off of the crankshaft. The camshaft is driven through a valve train consisting of gears, chain, or belt (Figure 7-2). Since the camshaft must rotate at half the crankshaft speed, the gear or sprocket on the camshaft is twice the diameter of the gear or sprocket on the crankshaft. The camshaft sprocket has twice as many teeth as the crankshaft sprocket.

There are several different drive systems for overhead camshafts. In some dual overhead camshaft (DOHC) V6 engines, a single timing belt surrounds the crankshaft and camshaft sprockets and all other pulleys (Figure 7-3). In this type of belt drive, the belt drives the overhead camshafts on both cylinder heads.

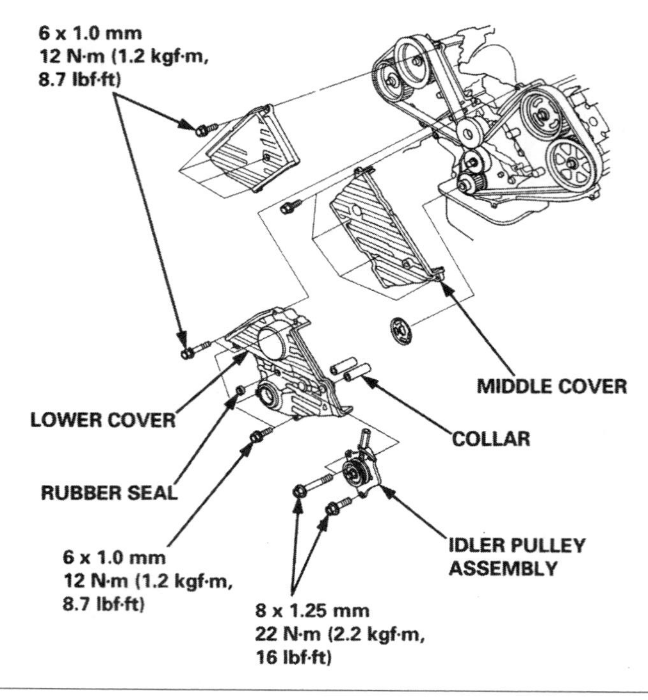


**Figure 7-3** DOHC engine timing belt that drives all the camshafts. (Courtesy of American Honda Motor Co., Inc.)

In other DOHC V6 engines, a timing chain surrounds the crankshaft sprocket and intermediate shaft sprocket, and this chain drives the intermediate shaft (Figure 7-4). A front cover is mounted over the chain drive, and an outer sprocket is mounted on the intermediate shaft on the outer side of the front cover. A timing belt surrounds the outer sprocket on the intermediate shaft and the overhead camshafts on both sides of the engine. The back side of the belt contacts two idler pulleys and a tensioner pulley (Figure 7-5).

In some DOHC V6 engines, a primary timing chain surrounds the crankshaft sprocket, both intake camshaft sprockets, and the balance shaft sprocket. The intake camshafts have dual sprockets, and secondary timing chains are connected from the intake camshafts to the exhaust camshaft sprockets on both cylinder heads (Figure 7-6).

Timing chains are more durable compared to timing belts. Manufacturers of engines with timing belts usually recommend a specific mileage interval when the timing belt must be

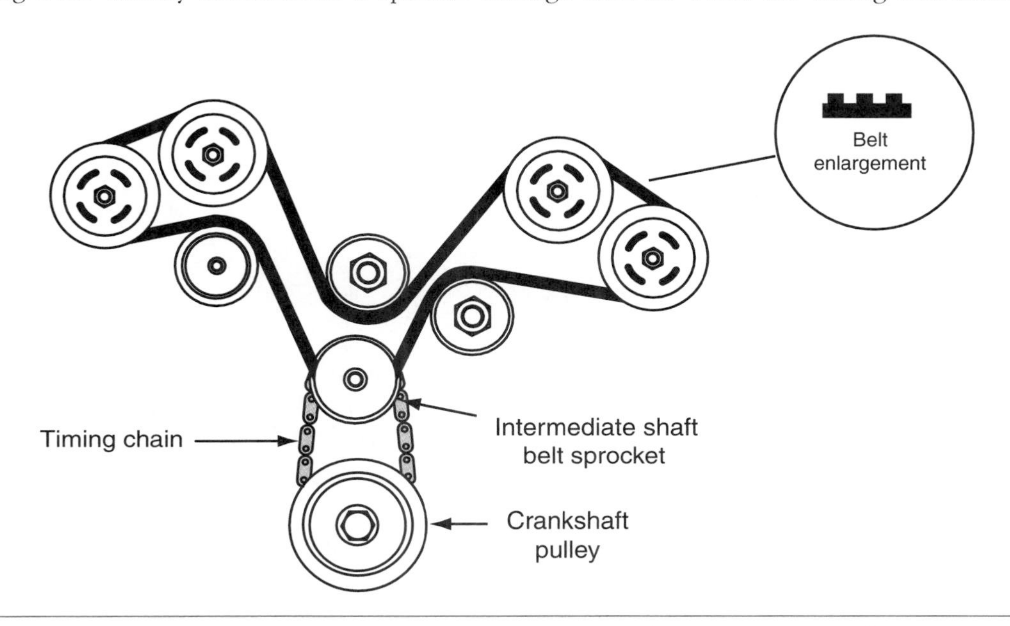


**Figure 7-4** DOHC engine with a timing chain that drives the intermediate shaft.

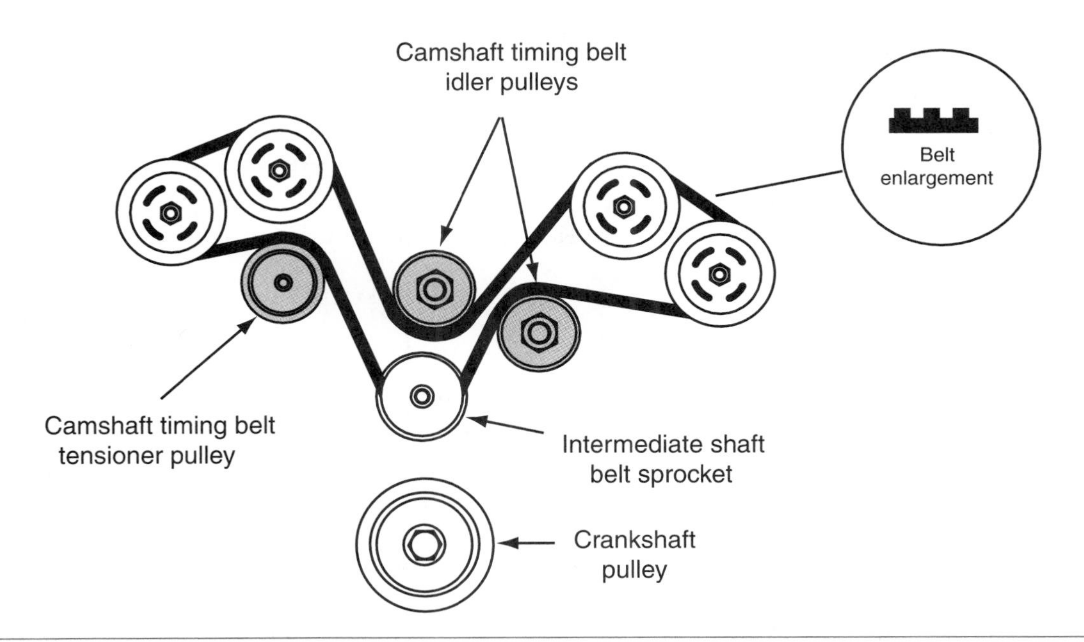


**Figure 7-5** DOHC engine with a timing belt that drives all the camshafts from the intermediate sprocket.

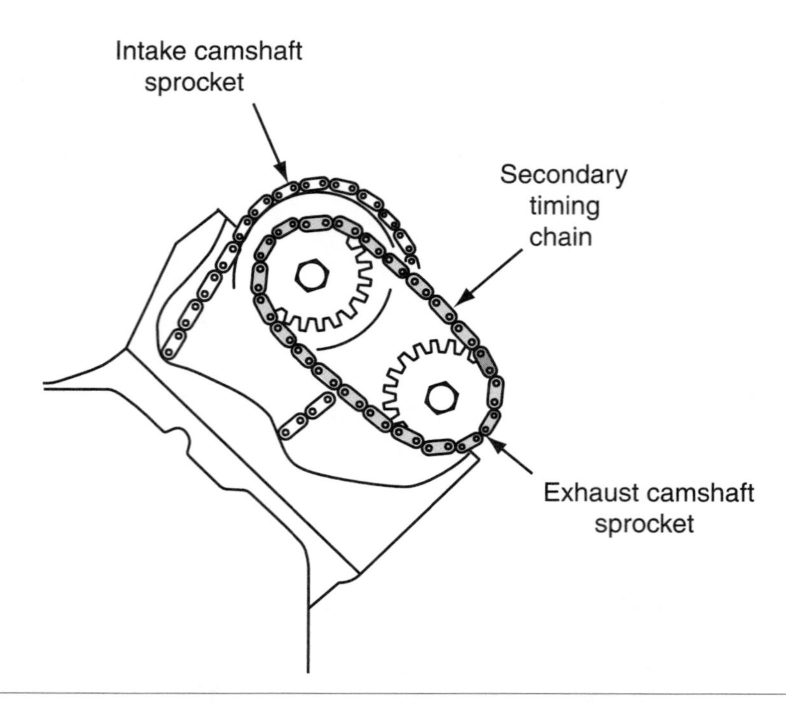


Figure 7-6 DOHC engine with primary and secondary timing chains.

An **interference engine** has valves that strike the top of the pistons if the valve timing is not correct.

A **free-wheeling engine** has pistons that do not strike the valves if the valve timing is not correct.

replaced. An **interference engine** is one in which the pistons will strike the valves if the valve timing is not correct. In this type of engine, if the timing belt breaks with the engine running, the valves will be bent. A **free-wheeling engine** is one in which the pistons do not strike the valves if the valve timing is not correct.

**AUTHOR'S NOTE:** During my experience in the automotive service industry, I have encountered some expensive repairs on interference engines when the timing belt or chain became worn and jumped some teeth on the sprockets. The reason for the expensive repairs was that some of the valves were bent when the pistons hit the valves after the timing belt or chain jumped some sprocket teeth. Many engine manufacturers recommend timing belt replacement at a specific mileage interval to reduce the possibility of these expensive repairs, and these belt replacements are much less expensive than a belt replacement plus removing the cylinder heads and reconditioning or replacing the valves.

For proper valve timing and engine operation, the camshafts and the crankshaft must remain in the same relative position to each other. When installing the timing components, proper initial relationship must be observed. Timing marks are provided to assist in proper installation (Figure 7-7).

**Shop Manual**
Chapter 7, page 317

## Belt Tensioners

Many engines using timing belts have an automatic adjustable tensioner to maintain a constant tension on the belt throughout its life (Figure 7-8). The chambers on both sides of the piston are filled with silicon oil. If the belt tension increases, the piston moves downward. The piston movement causes an increase in the hydraulic pressure in the lower chamber (Figure 7-9). Since the check ball is closed, the oil must seep past the piston and cylinder wall to go into the reservoir chamber. As the oil seeps into the reservoir, the tension piston moves slowly down the cylinder until the load from the belt balances with the spring tension.

Silicon is a nonmetallic element that can be doped to provide good lubrication properties. The melting point of silicon is 2,570°F (1,410°C).

If the belt tension is decreased, spring tension causes the piston to move up the cylinder. The hydraulic pressure below the piston lowers, and increased pressure in the reservoir opens

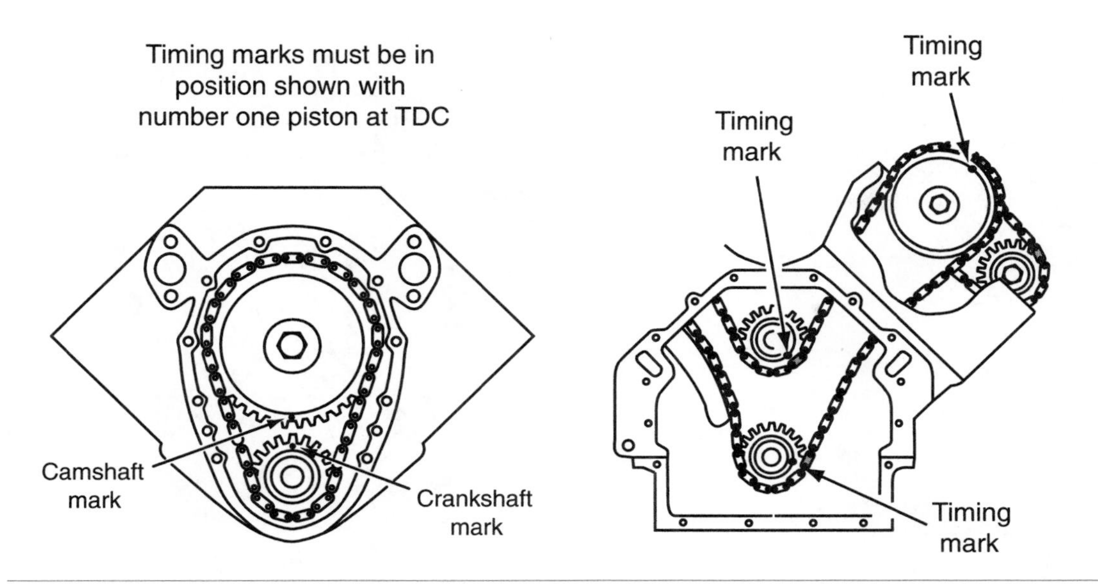


Figure 7-7 During valve train component replacement, it is important to align the marks to assure proper valve timing.

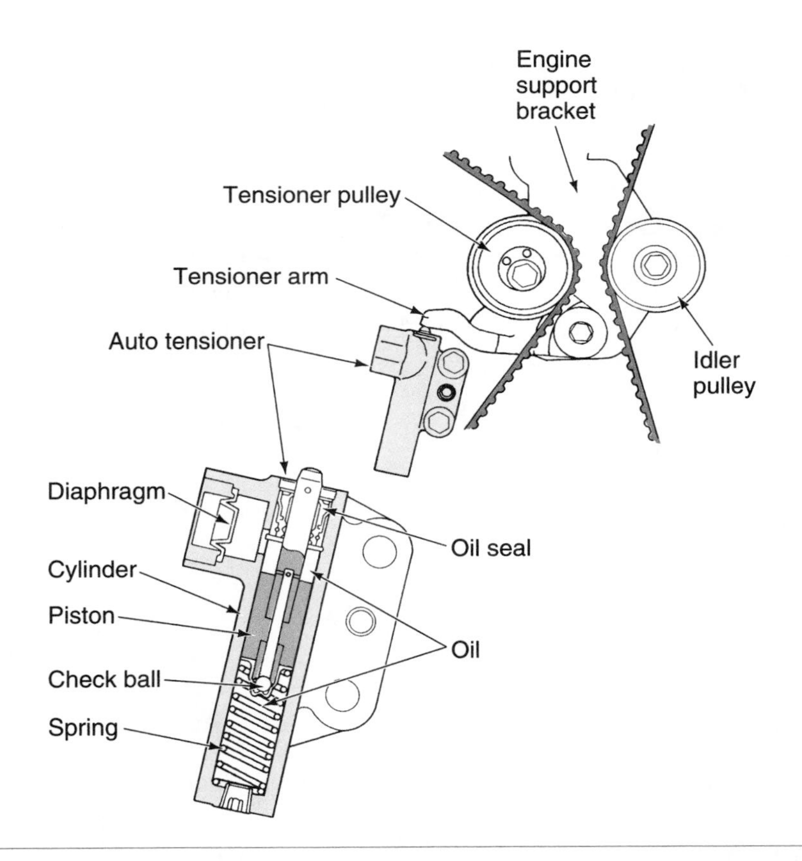


Figure 7-8 An automatic adjusting tensioner may be used to maintain proper belt tension. (Courtesy of Hyundai Motor America)

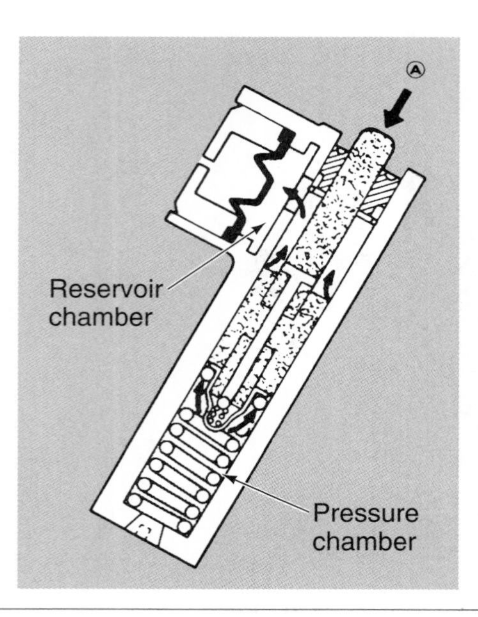


Figure 7-9 Hydraulic pressure flow through the tensioner when belt tension increases. (Courtesy of Hyundai Motor America)

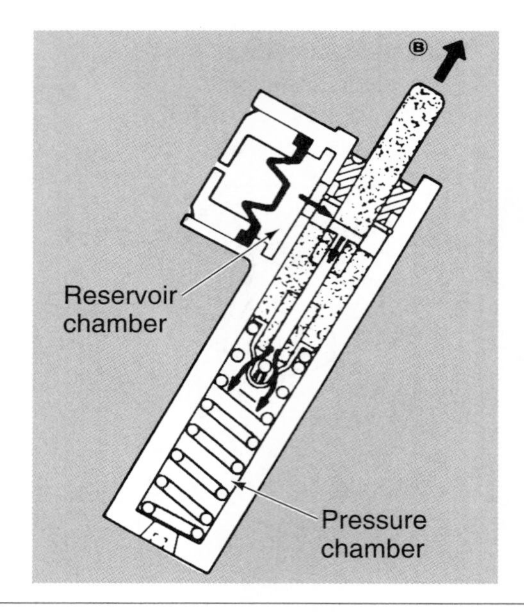


Figure 7-10 Hydraulic pressure flow through the tensioner when belt tension is released. (Courtesy of Hyundai Motor America)

the check valve (Figure 7-10). The piston stops moving when the belt tension balances with the spring tension.

## Valve Train Components

The valve train components work together to open and close the valves.

The valve train components work together to open and close the valves. For proper engine operation, this function must be performed at the precise time without loss of motion. The components of the valve train vary depending on engine design. The common components include:

- Camshaft
- Lifters or lash adjusters
- Pushrods
- Rocker arms or followers
- Valves
- Timing chain or belt
- Camshaft and crankshaft sprockets

### Camshafts

**Shop Manual**
Chapter 7, page 320

Simply stated, the camshaft is used to control valve opening (Figure 7-11). However, its function is actually more complicated. The camshaft also controls the rate of opening and closing, and how long the valve is open. Camshafts are usually constructed of nodular cast iron, although some high-performance engines use cast steel. Since roller lifter camshafts have the highest amount of load per square inch, most of these style camshafts are manufactured from forged steel billets.

Some late-model engines have camshafts made from a steel tube. The lobes are bonded to the steel tube, and a nose piece is friction welded to the tube. Camshaft end play is controlled by a thrust wall at the front of the camshaft. This type of camshaft is more rigid than a camshaft made from cast or forged steel. The increased camshaft rigidity reduces engine vibration.

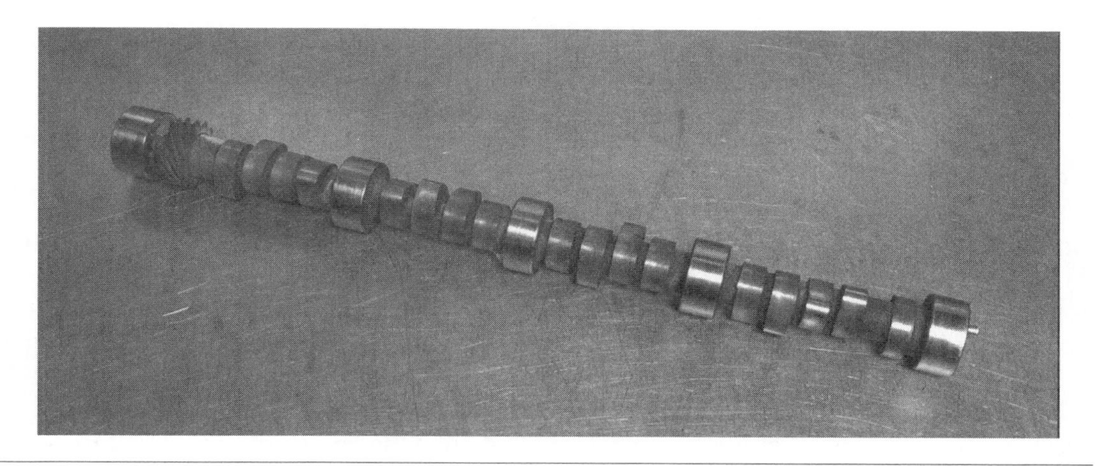

**Figure 7-11** Typical camshaft design.

The **lobes** of the camshaft push the valves open as the camshaft is rotated. There is usually one lobe for each intake and exhaust valve. The lobe design is a factor in determining how far the valve will open and for how long (Figure 7-12). The height of the lobe determines the amount of valve lift or opening.

The amount of **lobe lift** affects both gross valve lift and net valve lift. Additional factors of valve lift include rocker arm ratio, pushrod deflection, and lifter condition. Valve lift is an important factor in determining how well the engine can breathe (ingest the air/fuel mixture and expel it). The valve is a restriction to airflow into the combustion chamber. This is true until the valve is lifted about 0.300 in. (8 mm) off its seat. At this point, the valve offers very little restriction to airflow. However, there is still an advantage to increasing valve lift over 0.300 in. (8 mm).

The longer the valve remains open after reaching the 0.300 in. (8 mm) lift, the better the airflow. Many manufacturers have increased the amount of lift in their camshafts to increase the amount of valve opening. Lift does not adversely affect fuel economy, power, or emissions (as does excessive duration). For this reason, many manufacturers have maintained moderate duration of the camshafts and increased lift in an attempt to increase airflow. One of the limita-

The **lobes** of the camshaft push the valves open as the camshaft is rotated.

**Lobe lift** is the valve lifter movement created by the rotating action of the camshaft lobe. Typical lobe lift of original equipment camshafts is between 0.240 and 0.280 in. (6 and 6.6 mm).

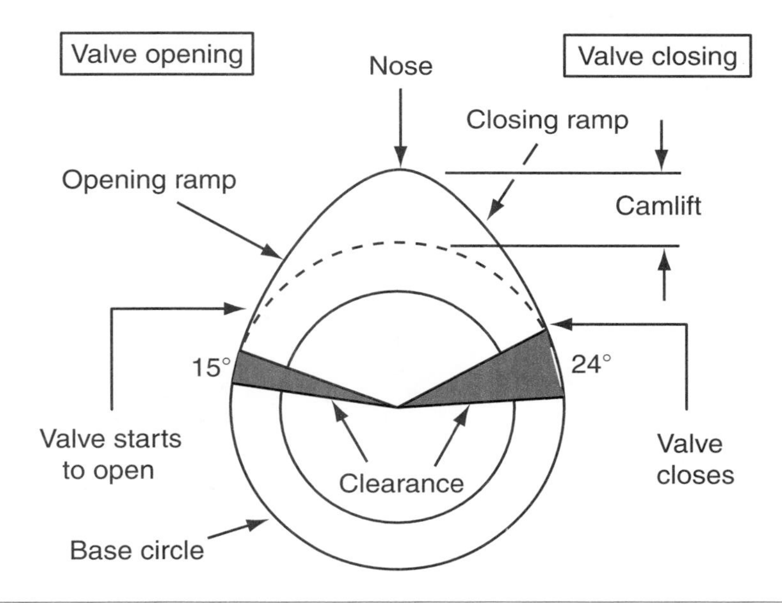


**Figure 7-12** The lobe design determines how far the valve will open and for how long.

**Cam lift** is a measurement of lobe height. **Valve lift** is the maximum distance the valve is lifted from its seat.

**Duration** is the length of time, expressed in degrees of crankshaft rotation, the valve is open.

The area of the mechanical lifter camshaft from base circle to the edge comprises the **clearance ramp.**

**Valve overlap** is the length of time, measured in degrees of crankshaft revolutions, that the intake and exhaust valves of the same combustion chamber are open simultaneously.

**Split overlap** means the intake and exhaust valves are equally open at TDC.

**Single-pattern** lobes are ground the same on both opening and closing ramps. Lobes ground differently on the two ramps are **asymmetrical.**

tions on maximum valve lift is the amount of travel of the valve spring. If the valve spring is required to travel a greater distance, its coils can bind.

Lift can be expressed in two methods. The first is **cam lift,** which is the measured lift of the cam lobe. The second is **valve lift.** This is the cam lift multiplied by the rocker arm ratio to provide a specification for the amount of valve movement.

If a valve opens at top dead center (TDC) and closes at bottom dead center (BDC), the **duration** the valve is open is 180 degrees. This is not practical for real-world engine operation. In actuality, the intake valve must open before TDC and close after BDC. The additional degrees of rotation must be calculated into the total duration degrees. If the intake valve opens 6 degrees before TDC and closes 12 degrees after BDC, total duration is 198 degrees of crankshaft rotation.

Since duration allows more time for the air/fuel mixture to fill the combustion chamber, duration has more effect on engine performance and efficiency than any other valve timing specification. There are instances when an engine may require longer durations to perform a specified function; for example, the high-performance engines used in most racing applications require a camshaft with higher durations. This camshaft allows the engine to breathe at higher rpm. The drawback is that the engine performs poorly at lower engine rpms.

If a replacement camshaft is to be purchased, there are two methods used to measure duration. SAE (Society of Automotive Engineers) specifications are measured after 0.006 in. (0.15 mm) of lift on hydraulic lifter camshafts. Performance camshafts are usually measured for duration at 0.050 in. (1.25 mm) lift. An SAE-measured camshaft will have approximately 40 degrees to 50 degrees more duration than a camshaft measured at 0.050 in. (1.25 mm) lift.

**Clearance ramps** are ground into the lobe to prevent the valve train components from hammering on each other. The clearance ramps are located between the base circle and the opening or closing ramps. The clearance ramps remove clearance in the valve train at a slower rate than the rate of lift (Figure 7-13). Most camshafts used with hydraulic lifters have a clearance ramp between 0.005 and 0.010 in. (0.13 and 0.25 mm). This ramp assures the lifter check valve is closed to provide a solid connection.

Engine performance characteristics are also affected by **valve overlap** (Figure 7-14). Most stock camshafts are ground with a **split overlap.** The camshaft can be ground as **advanced** or **retarded.** In addition, it can be ground as **single-pattern** or **asymmetrical.** When measuring the amount of valve overlap, lobe centers must be considered. The intake lobe center can be found using a degree wheel and a dial indicator. The degree wheel indicates the number of degrees after TDC where the maximum amount of valve lift is achieved for the intake valve. The exhaust lobe centerline is found in the same manner. The degree wheel also indicates the number of degrees before TDC where the valve is fully opened. Lobe spread is the average of the two readings; for example, if the intake lobe center is 110 degrees after TDC and the exhaust center is 120 degrees before TDC, the lobe spread is 115 degrees. If the intake lobe center is less

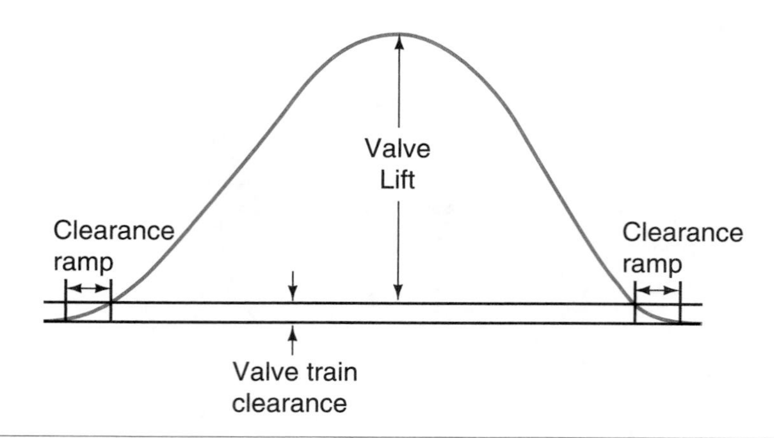


**Figure 7-13** The clearance ramp removes valve train clearance at a slower rate than valve lift.

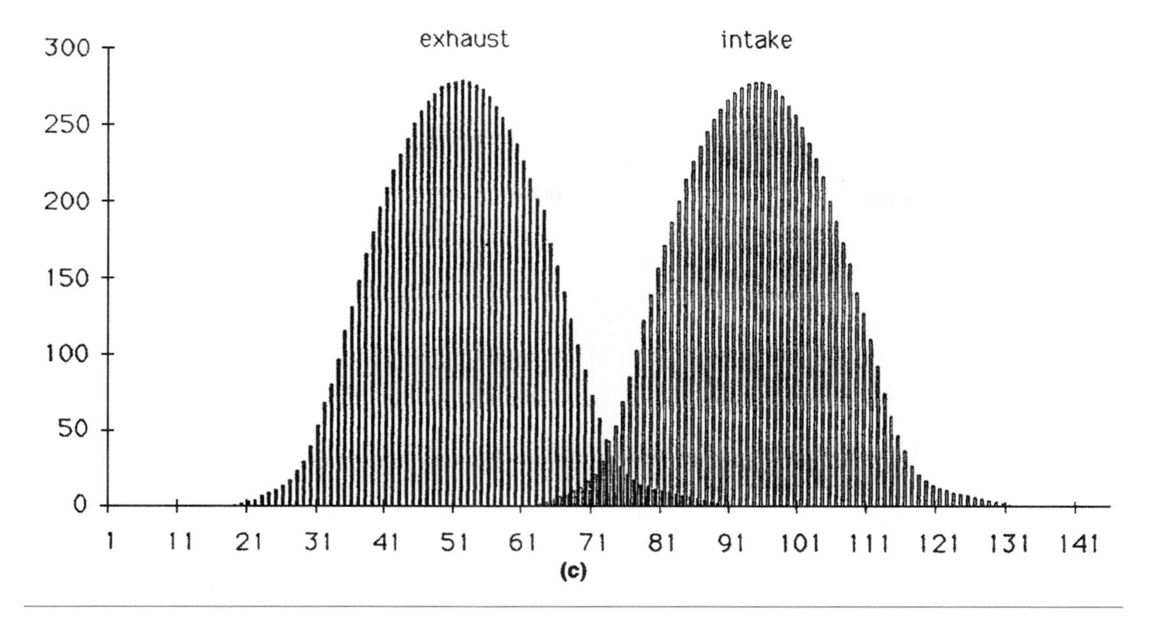


Figure 7-14 Valve overlap. (Courtesy of Elgin Racing Cams)

than the lobe spread, it is advanced. In the preceding example, the advance of the intake center is 5 degrees.

The camshaft lobes generally have a very slight taper of 0.0007 to 0.002 in. (0.0178 to 0.0508 mm) across its face (Figure 7-15). Along with offsetting the centerlines of the **lifter** and the lobe, this taper provides for lifter rotation. The bottom of the lifter assembly is machined with a spherical shape. This prevents early wear, since the lifter's edge is not riding on the edge of the camshaft lobe. The rotation of the lifter and designs of the camshaft and lifter assemblies spread the load of the valve train against more of the lobe face. Rotating the lifter is desirable because it helps to dissipate the nearly 100,000 psi (690 mpa) load the lifter is subjected to.

**Advanced camshafts** have the intake valve opened more than the exhaust valve at TDC. This type of camshaft is used to increase low rpm performance (higher torque output). **Retarded camshafts** are designed to have the exhaust valve open more than the intake valve at TDC.

**Lifters** are mechanical (solid) or hydraulic connections between the camshaft and the valves.

A solid lifter may be called a tappett.

**Camshaft Bearings.** Camshafts that are installed within the engine block are usually fitted with full round bearings (Figure 7-16). These are one-piece bearings that are pressed into the block, and the journal of the camshaft is slid into place. Many overhead camshaft engines use split bearings with camshaft caps. Modern machining and oil improvements have made it possible for some manufacturers to eliminate the use of camshaft bearings in their overhead camshaft engines.

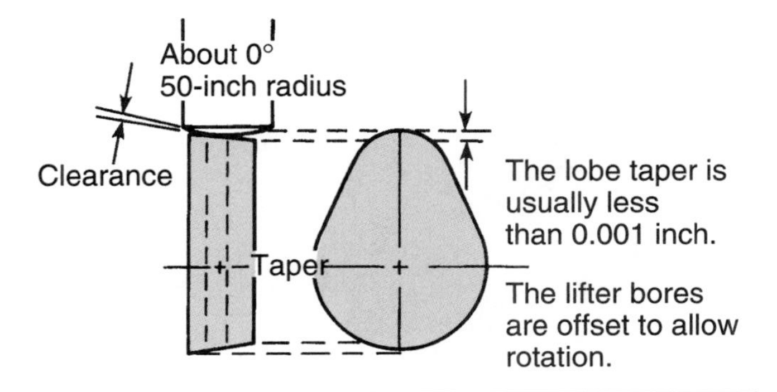


Figure 7-15 The taper of the camshaft lobes will provide for lifter rotation to help prevent early wear.

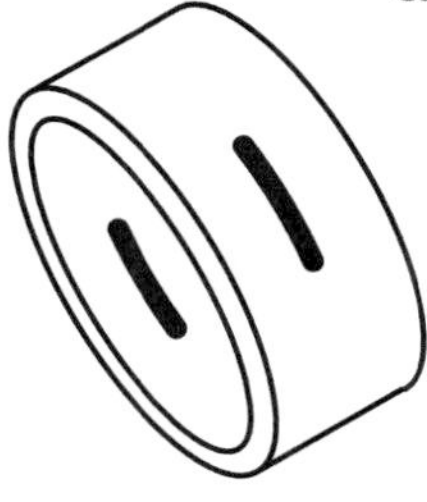


Figure 7-16 Full round camshaft bearings are used in engines with the camshaft in the block and in many overhead camshaft engines.

## Lifters

Lifters are mechanical (solid) or hydraulic connections between the camshaft and the valves. Lifters follow the contour of the cam lobes to lift the valve off its seat. Solid lifters provide a rigid connection, while hydraulic lifters use oil to absorb the resultant shock of valve train operation. Each lifter design has its advantages and disadvantages.

Solid lifters (Figure 7-17) require a specified clearance between the components of the valve train. Periodic adjustment is required to maintain this clearance. If the clearance is too tight, there is no room for expansion due to heat. This can result in loss of power and parts damage. If the clearance is too great, excessive noise may be generated. In addition to noise, a change of 0.001 in. (0.03 mm) in valve lash may result in about a 3-degree change in duration.

**Shop Manual**
Chapter 7, page 331

Hydraulic lifters (Figure 7-18) use oil to automatically compensate for the effects of engine temperature and valve train wear. Some engine manufacturers use a roller-type hydraulic lifter in an effort to reduce friction (Figure 7-19). This type of lifter is fitted with a roller that rides on the camshaft lobe. Friction is reduced since the lifter body is not rubbing against the camshaft lobe. Roller lifters can also be designed as solid lifters for high-performance and heavy-duty diesel applications.

**Zero lash** is the point where there is no clearance or interference between components of the valve trains.

The hydraulic lifter automatically maintains **zero lash.** When the hydraulic lifter is resting against the base of the camshaft lobe, the valve is in its closed position (Figure 7-20). In this posi-

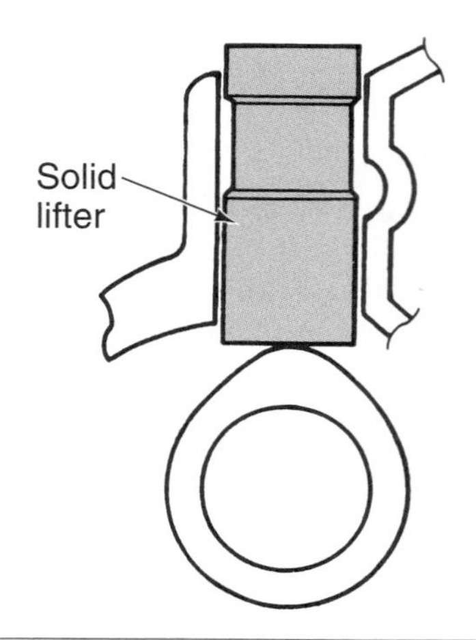


**Figure 7-17** A solid lifter provides a rigid connection between the camshaft and pushrod.

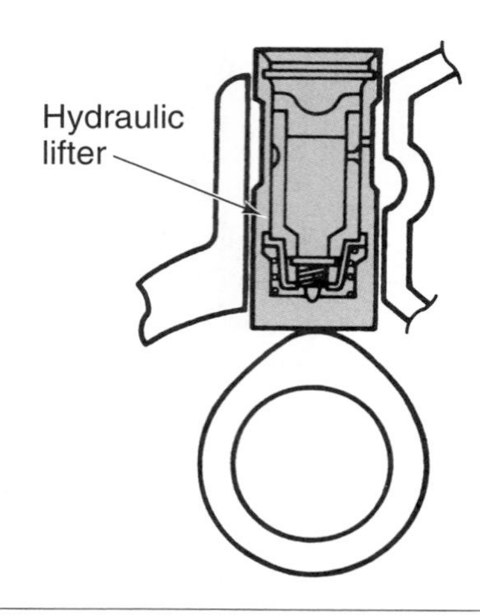


**Figure 7-18** Hydraulic lifter compensates for temperature and wear by automatically adjusting valve lash.

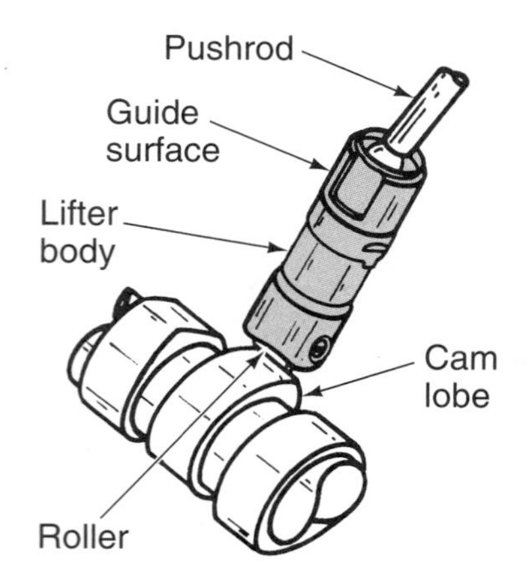


**Figure 7-19** Roller-type valve lifters reduce friction and wear by using a rolling ball.

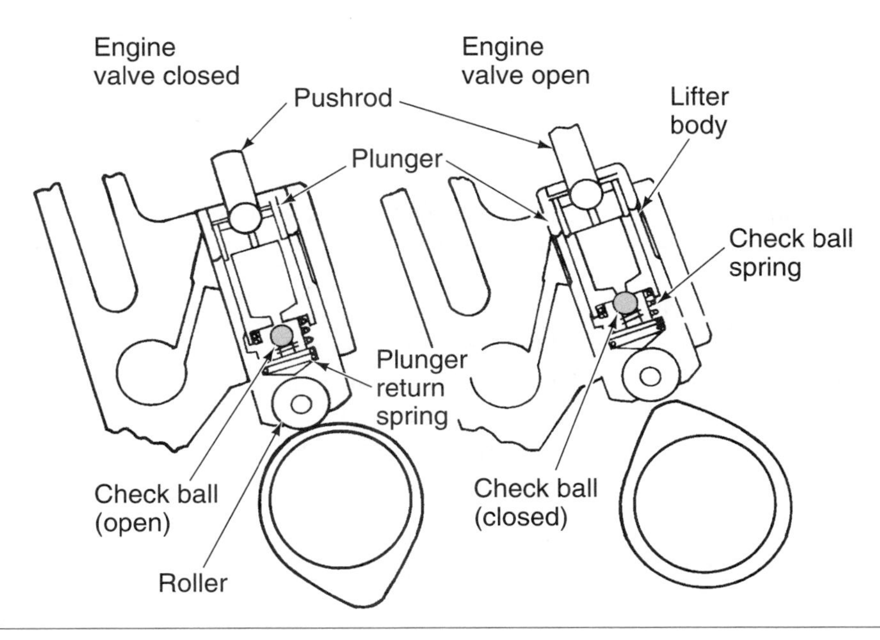


**Figure 7-20** Hydraulic valve lifter operation comparison between valve closed and valve open.

tion, the plunger spring maintains a zero clearance in the valve train (Figure 7-21). Oil supplied by the oil pump is pushed through the oil feed holes of the lifter bore and into the lifter body. Oil is able to enter the lifter body only when the lifter is in this position, since the feed through the engine block is not aligned with the lifter body when it is on top of the lobe. The oil passes through the body and enters the inside of the plunger through an oil passage on its side. The check valve allows oil to flow from the plunger into the body, but not from the body to the plunger. Oil continues to flow into the lifter until it is full. Pressurized oil is trapped in the lifter by the closed check valve. The spring under the plunger pushes the lifter body toward the camshaft lobe and the plunger toward the pushrod, removing any play in the valve train. The position of the check valve is determined by the pressure applied on the pushrod cap. This pressure is transferred from the closed valve through the pushrod to the cap. When the camshaft lobe attempts to raise the lifter body, the pressure on the cap and plunger is increased and the check valve closes, preventing the oil in the area below the plunger from escaping. Since the oil cannot be compressed, a rigid connection is formed, causing the plunger to raise with the lifter body and open the valve.

As the temperature of the engine increases, parts begin to expand, reducing the amount of clearances. This causes the plunger of the lifter to be held lower in the lifter body, causing some of the oil that was trapped below the plunger to leak out from the lifter body. In addition, when the lifter is moved up the block bore by the camshaft lobe, the pressures of attempting to open the valves push the plunger down in the body. This results in leakdown. When the lifter follows the cam lobe back to the base, the oil feed passages align and the lifter is filled with oil

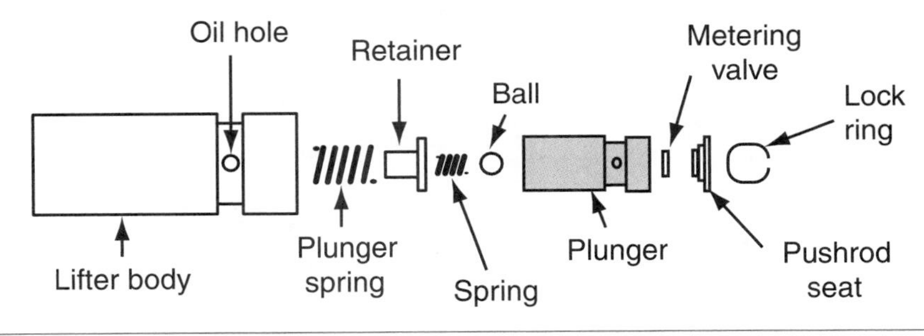


**Figure 7-21** Internal components of a hydraulic valve lifter.

again. This process of filling and leakdown automatically adjusts the hydraulic lifter to compensate for temperature changes and component wear.

Lifter **pump-up** occurs when excessive clearance in the valve train allows a valve to float. The lifter attempts to compensate for the clearance by filling with oil. The valve is unable to close since the lifter does not leak down fast enough.

**Valve float** occurs when the valve continues to open or stays open after the camshaft lobe has passed from the lifter.

Solid lifters are often used in high-performance engines because they cannot **pump up.** Since hydraulic lifters adjust automatically, they can overcompensate at higher engine speeds. At high engine rpm, valve train inertia causes the valves to open farther than what is accomplished by the cam lobe and rocker arm ratio. This **valve float** causes additional clearance to occur in the valve train. The lifter compensates for the additional clearance as it would under normal clearance corrections because the spring pushes the plunger up to take up the additional clearance. At this time, the body cavity fills with oil from the plunger cavity. The filling of the plunger cavity causes the lifter to be slightly longer than normal as it approaches the heel of the camshaft lobe. Again, the lifter compensates for this. On the next valve opening, the cavity is filled again and the cycle starts over. At high rpm, the lifter is unable to return to its normal size when it is approaching the heel of the lobe. The result is that the valve can be held open. When the engine speed is reduced, lifter operation returns to normal.

## OHC Valve Train Designs

Some overhead camshaft (OHC) engines use hydraulic lash adjusters (Figure 7-22). The lash adjuster works in the same manner as the hydraulic lifter. In some V6 DOHC valve trains, the camshaft journals ride in cylinder head bores, and holders and holder plates retain the camshafts to the cylinder heads (Figure 7-23). The rocker arm shafts and rocker arms are mounted below the camshafts in the cylinder heads (Figure 7-24).

In other V6 DOHC engines, the camshafts and valve lifters are mounted in cam carriers mounted on top of each cylinder head (Figure 7-25). A gasket is located between each cam carrier and the cylinder head (Figure 7-26).

## Pushrods

**Pushrods** connect the valve lifters to the rocker arms.

Overhead valve (OHV) engines, with the camshaft located in the block, use **pushrods** to transfer motion from the lifter to the rocker arms (Figure 7-27). Besides being a link to the rocker arms, some pushrods are also used to feed oil from the lifter to the rocker arms (Figure 7-28). The oil is sent up the hollow portion of the pushrod.

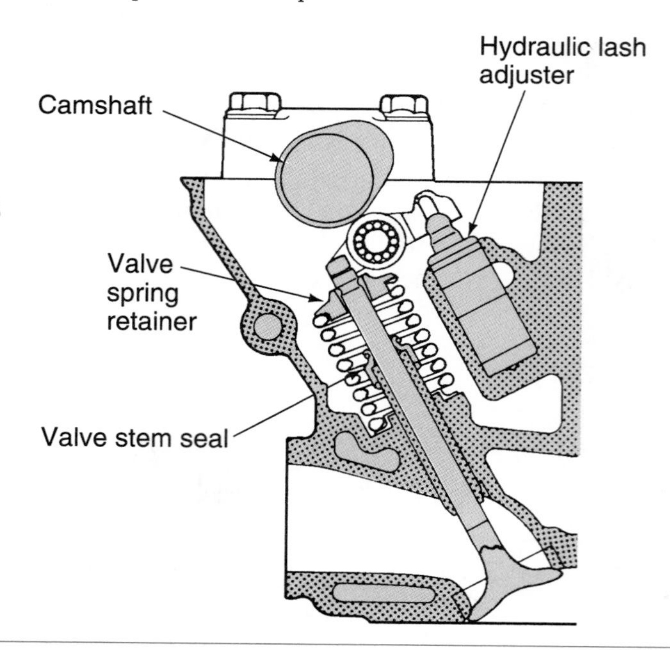


Figure 7-22 Some OHC engines use hydraulic valve lash adjusters. (Courtesy of Hyundai Motor America)

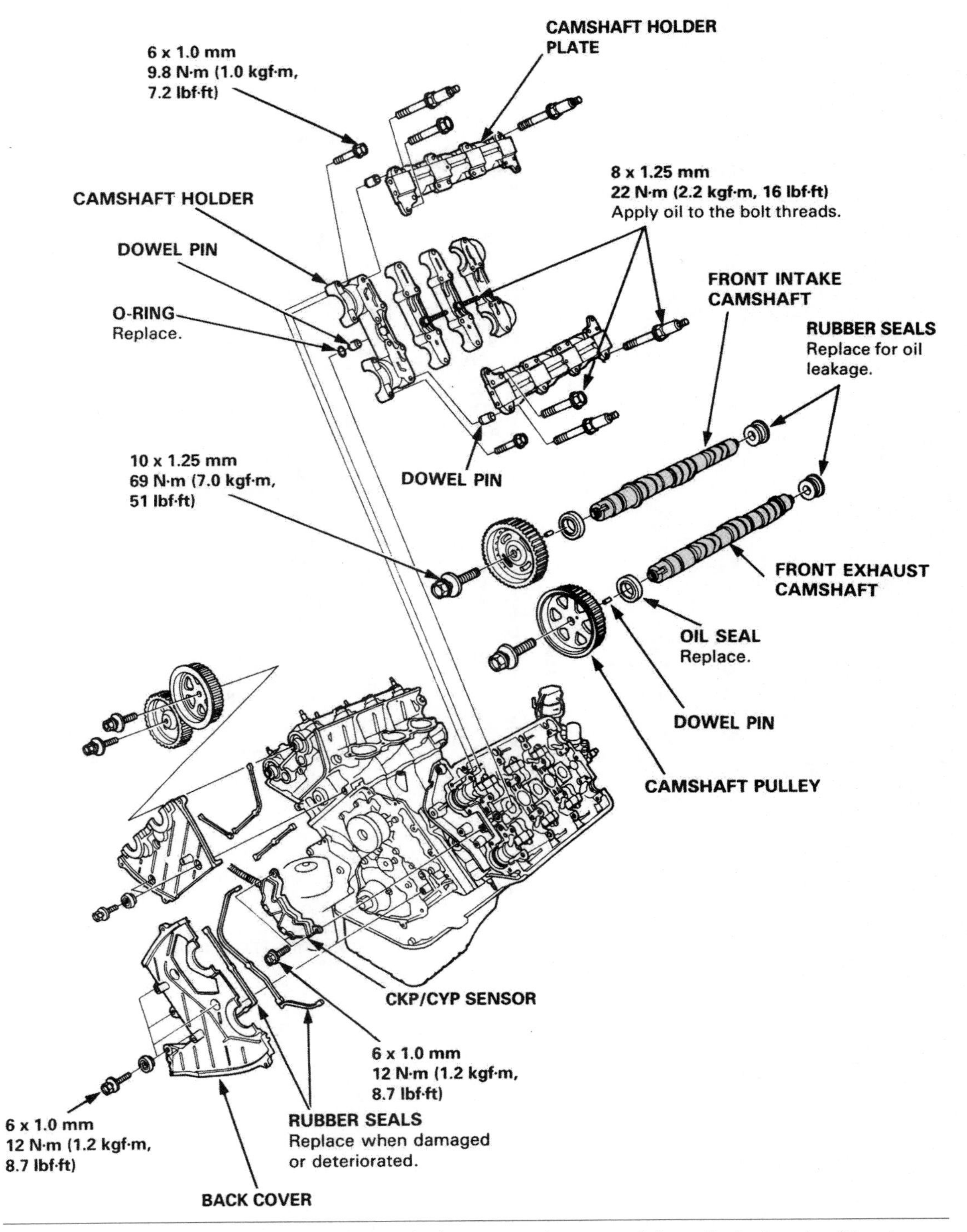


Figure 7-23 DOHC engine with camshafts mounted on top of the cylinder heads. (Courtesy of American Honda Motor Co., Inc.)

Most pushrods are constructed from seamless carbon steel tubing. To decrease wear, some pushrods may have a hardened tip insert at each end. High-performance engines may use pushrods that are made of chromium-molybdenum tubular steel for increased strength. Chrome-moly pushrods may also have a thinner wall thickness to reduce weight and inertia.

The bottom of the pushrod fits into a socket in the lifter. The recesses of the socket are deep enough to prevent the pushrod from falling out during operation. The top of the pushrod fits into a small cup in the rocker arm.

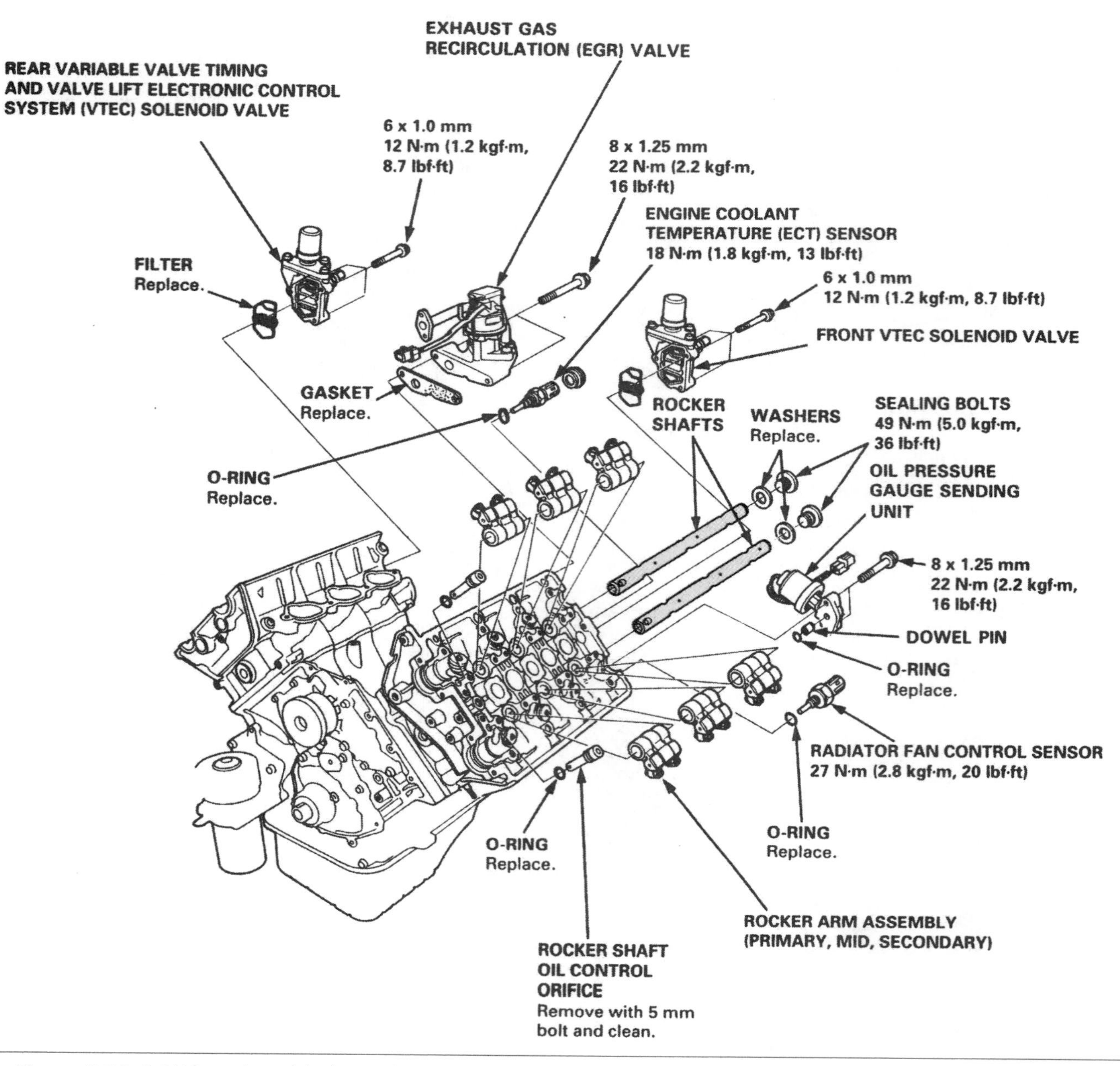


**Figure 7-24** DOHC engine with the pushrods mounted in the cylinder heads under the camshafts. (Courtesy of American Honda Motor Co., Inc.)

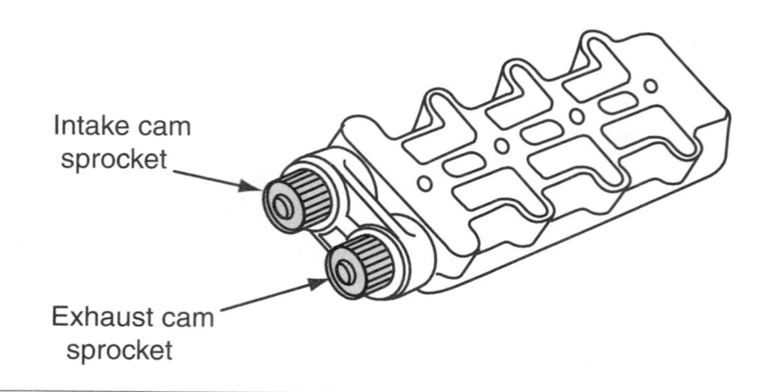


**Figure 7-25** DOHC engine with camshafts and lifters mounted in a cam carrier.

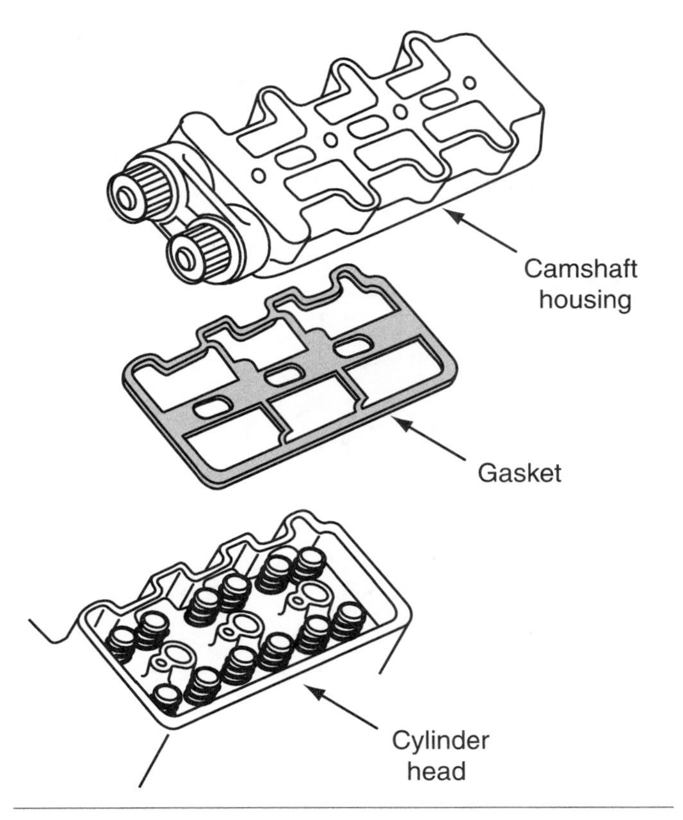


Figure 7-26 Cam carrier mounting to cylinder head.

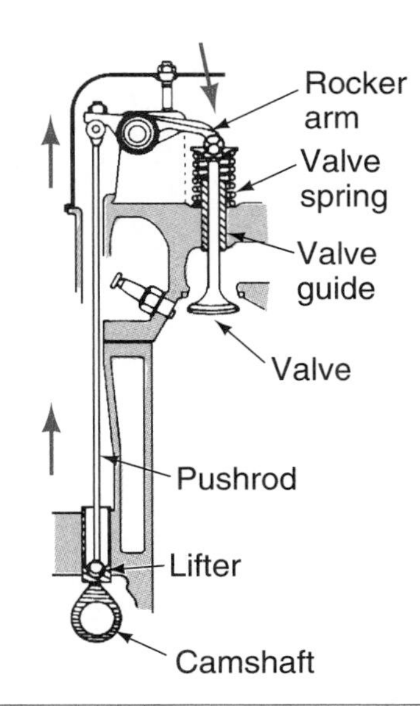


Figure 7-27 The valve lifter transfers force to the pushrod.

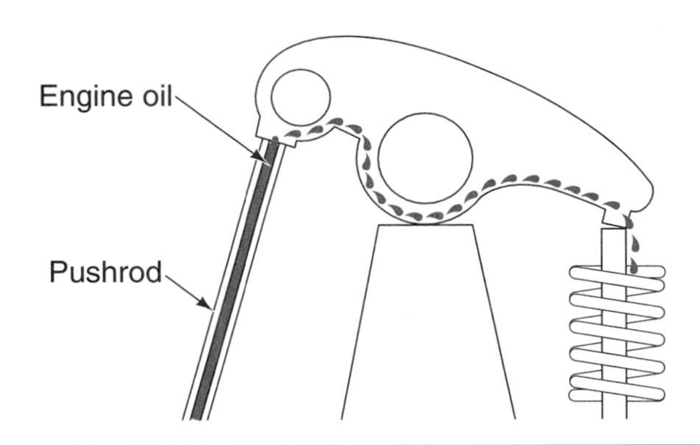


Figure 7-28 The pushrod can be used to deliver oil to the rocker arm.

Tolerances within the valve train generally allow some machining of the cylinder head without the need to correct pushrod length; however, if the deck surface of the cylinder head and/or the engine block is machined or if valve stem length is increased, it may be necessary to change the pushrods to maintain the correct valve train geometry.

## Rocker Arms

The **rocker arm** is a pivoting lever used to transfer the motion of the pushrod to the valve stem. In overhead camshaft engines, the camshaft may operate the rocker arms directly. Rocker arms changes the direction of the cam lift and spring closing forces, and it provides a leverage during valve opening. Most rocker arms are constructed of cast iron, cast aluminum, or stamped steel.

The **rocker arm** is a pivoting lever used to transfer the motion of the pushrod to the valve stem.

The difference in the rocker arm dimensions from center to valve stem and center to pushrod is called the **rocker arm ratio**. It is a mathematical expression of the leverage being applied to the valve stem.

The design of the rocker arm is such that the side to the valve stem is usually longer than the side to the pushrod (Figure 7-29). This design gives the rocker arms a mechanical advantage. The **rocker arm ratio** works with the camshaft lobe to provide the desired lift. For example, if the camshaft lobe has a lift of 0.350 in. (8.9 mm) and the rocker arm ratio is 1.5, actual lift is $0.350 \times 1.5 = 0.525$ in. (13.3 mm).

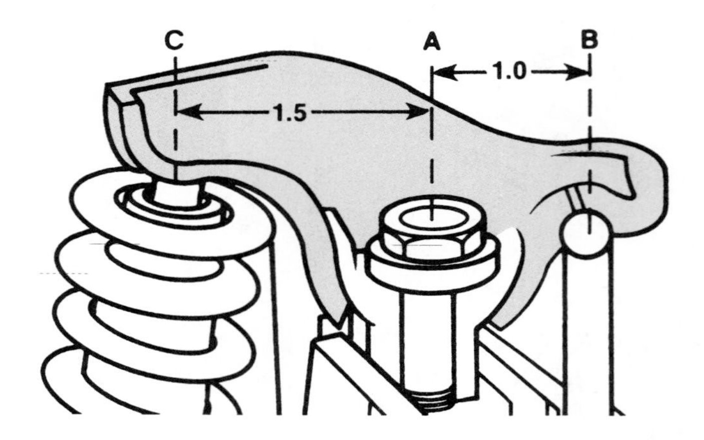


Figure 7-29 Rocker arms are designed to provide a mechanical advantage. The ratio is determined by the distance from A to B and from A to C.

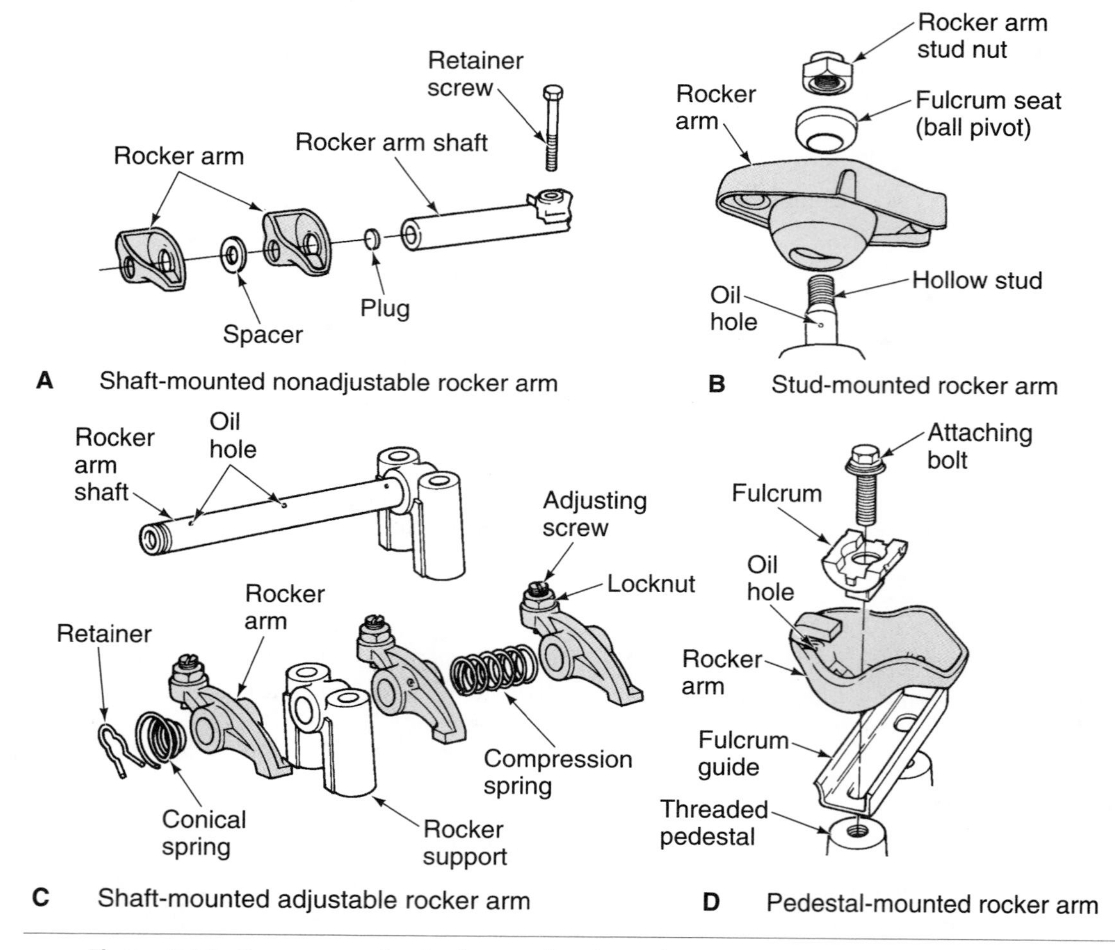


Figure 7-30 Common methods of mounting the rocker arm.

There are three common methods of mounting the rocker arm (Figure 7-30):

- Shaft-mounted
- Stud-mounted
- Pedestal-mounted

**Shaft-mounted rocker arms** are located on a heavy shaft running the length of the cylinder head. The shaft is supported by stands. **Stud-mounted rocker arms** are used on overhead valve engines. Each rocker arm is mounted on a stud. A split ball is used for a pivot point for the rocker. **Pedestal-mounted rocker arms** are similar to the stud-mounted assemblies, except the rocker pivots on a split shaft.

**Shaft-mounted rocker arms** are used in a mounting system that positions all rocker arms on a common shaft mounted above the cylinder head.

A **stud-mounted rocker arm** is mounted on a stud that is pressed or threaded into the cylinder head.

A **pedestal-mounted rocker arm** is mounted on a threaded pedestal that is an integral part of the cylinder head.

**Rocker Arm Geometry.** When the valve is opened 50 percent of its travel, the rocker arm-to-stem contact should be centered on the valve stem (Figure 7-31). If the rocker arm is not centered, it will cause excessive side thrust on the stem. Rocker arm geometry does not usually need to be checked or corrected; however, if a different camshaft with a higher lift than the original camshaft is installed, rocker arm geometry will be affected. The higher lift causes the valve to move farther, causing the rocker arm tip to swing down farther in its arc. As the tip travels a greater distance, the tip moves away from the stem center. Other causes of incorrect rocker arm geometry include:

- Cylinder head resurfacing
- Cylinder block deck resurfacing
- Sinking the valve seat
- Worn or incorrect pushrods
- Worn rocker arms

**Shop Manual**
Chapter 7, page 334

In engines that use pushrods, the angle of the rocker arm in relation to the pushrod is also important. Incorrect rocker arm geometry may result in the pushrod hitting the side of its passage in the cylinder head or the pushrod guide plate.

In addition, on engines that use hollow pushrods to supply oil to the upper engine, incorrect rocker arm geometry may prevent proper lubrication. These engines have a small hole in the recessed cup that accepts the pushrod tip. When the opening in the pushrod aligns with the hole in the rocker arm, oil is sprayed over the rocker arms and valve springs. The holes should align

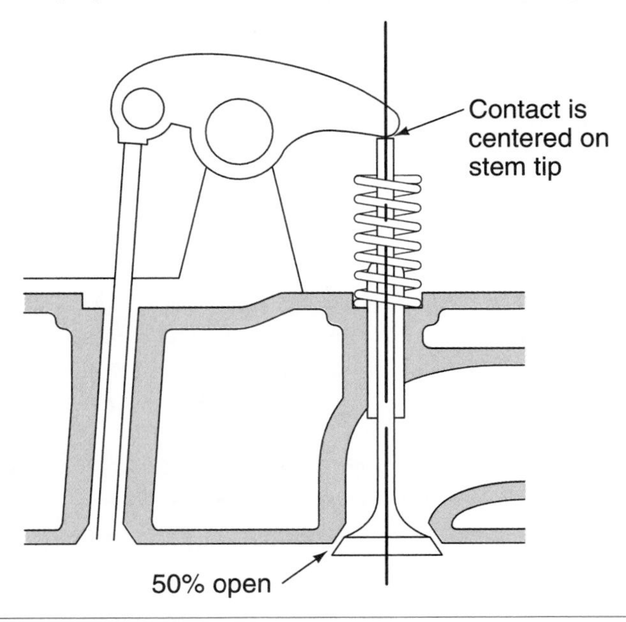


Figure 7-31 Proper rocker arm geometry places the arm center on the stem and reduces side thrust.

when the rocker arm is in the lash position. As the rocker arm is moved to open the valve, the holes are no longer aligned and oil flow stops. Incorrect rocker arm geometry may cause the holes not to align properly and result in too much or too little oil flow.

Any machining operations that alter the distance between the camshaft and the rocker arm may affect rocker arm geometry. Even changes to camshaft lift and duration can change the geometry. Usually the tolerances in the valve train will allow for some changes and machining without problems, but the geometry should be checked while rebuilding the engine.

## Variable Camshaft Timing Principle

Some engines are equipped with variable camshaft timing (VCT). On an engine with fixed valve timing, the valve timing is a compromise among providing a reasonable amount of performance, fuel economy, and emissions. If the valves are open longer at low speed and have more overlap, some unburned fuel is swept into the exhaust stream; however, if the duration of valve opening and valve overlap are reduced, engine performance is decreased. VCT provides increased performance at higher engine rpm, and improved fuel economy and emissions. Under certain engine operating conditions, such as cruising speed, the variable valve timing allows the exhaust valves to remain open longer. When the exhaust valves close later, some exhaust gas recirculates back into the combustion chambers, and this reduces oxides of nitrogen ($NO_x$) emissions. Some engines with variable valve timing, such as the Ford 2.0L Zetec engine, do not have an exhaust gas recirculation (EGR) valve, because the variable valve timing system reduces $NO_x$ emissions.

Some variable valve timing mechanisms contain a helical shaft or gear mounted in an outer housing on the front of the camshaft. The outer housing is attached to the camshaft sprocket. The helical shaft is bolted to the camshaft, and a linear to rotary piston is mounted over the helical shaft. The helical shaft and linear-to-rotary piston have matching helical grooves. The outer surface of the linear-to-rotary piston is mounted in the outer housing, and the cam-shaped outer surface of this piston prevents the piston from rotating in the outer housing (Figure 7-32); however, the linear-to-rotary piston can move horizontally. Horizontal or linear piston movement is controlled by oil pressure supplied to the front of the sealed area at the front of the linear-to-rotary piston.

The powertrain control module (PCM) operates two solenoids that control the supply of oil pressure to the front of the linear-to-rotary piston. One solenoid supplies oil pressure to the linear-to-rotary piston, and the other solenoid allows oil to return from this piston to the oil sump. When the PCM energizes the return solenoid and de-energizes the supply solenoid, oil is drained from the linear-to-rotary piston, and the camshaft timing is unchanged. The PCM receives a number of input signals, including a signal from the camshaft position sensor. If these sensor inputs inform the PCM that advanced camshaft timing is required, the PCM closes the return solenoid and opens the supply solenoid. Under this condition, oil pressure is supplied through the supply solenoid to the area in front of the linear-to-rotary piston. This oil pressure moves the linear-to-rotary piston, and the spiral grooves on this piston rotate the camshaft a maximum of 15 degrees in relation to the sprocket and outer housing. The PCM provides the camshaft advance that supplies optimum engine performance, economy, and emissions. On some DOHC engines, such as the Ford 2.0L Zetec engine, the VCT mechanism is mounted only on the exhaust camshaft.

## Variable Camshaft Timing and Lift System

Some DOHC engines with four valves per cylinder have a VCT system that also varies the valve lift. In these systems, a valve actuating T-shaped lever is mounted between each pair of roller-type rocker arms. The camshaft lobes contact the rocker arm rollers, and the outer ends of the T-shaped lever contacts contact the top of the valve stems. Hydraulic pistons in the rocker arms connect or disconnect the T-shaped lever from the rocker arms. The hydraulic pistons are con-

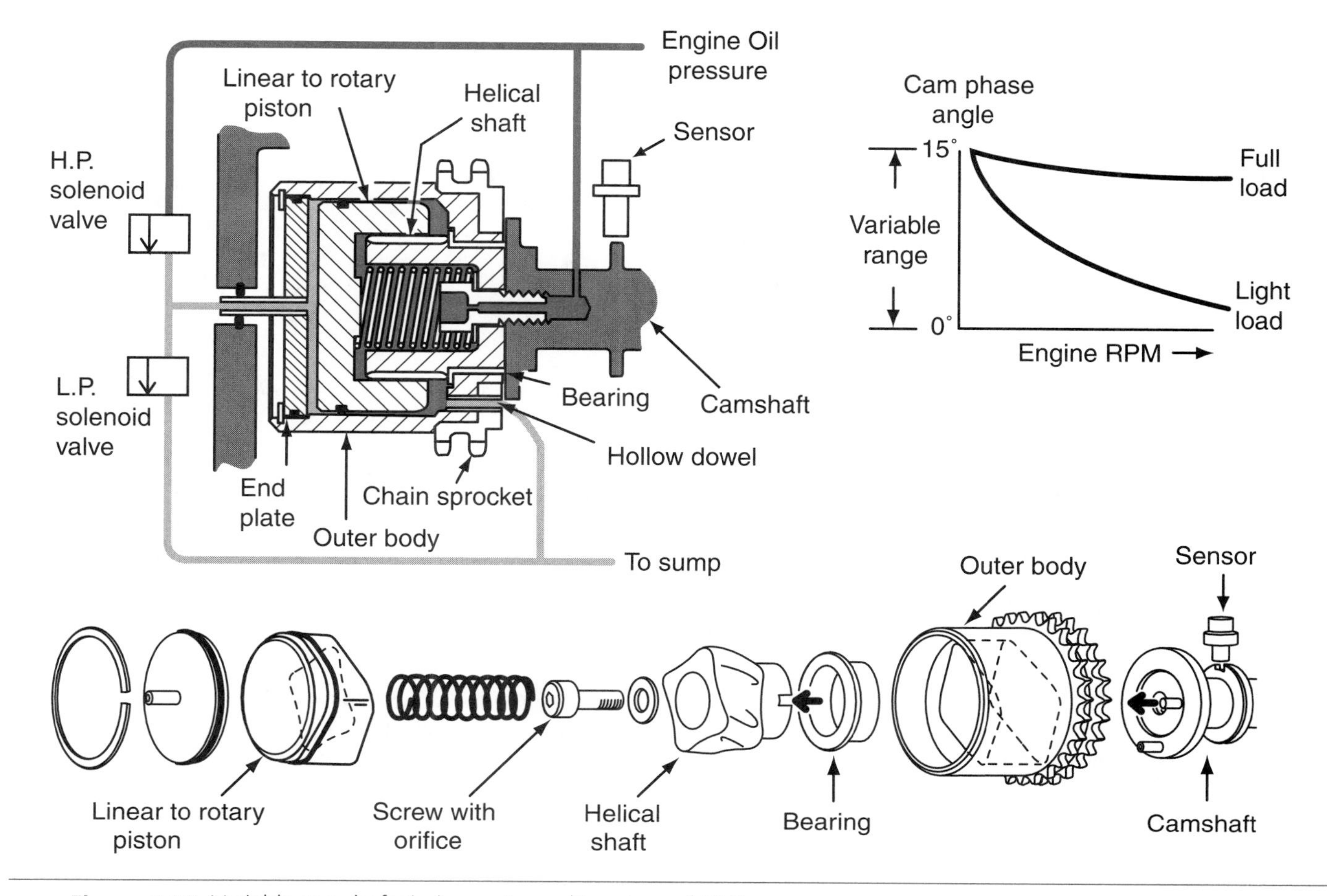


**Figure 7-32** Variable camshaft timing system. (Courtesy of SAE)

trolled by oil pressure supplied through PCM-controlled solenoids. The camshaft has low-speed and high-speed lobes that contact each pair of rocker arms. The PCM controls the hydraulic pistons in the rocker arms to deactivate both valves, activate the low-speed rocker arms, or activate the high-speed rocker arms. Piston H in the high-speed rocker arm is normally pulled into the rocker arm shaft by spring force, while piston L in the low-speed rocker arm is pushed out above the rocker arm shaft by spring force.

When the PCM input signals indicate that the deactivate valve mode is desirable, the PCM supplies this mode on every second cylinder in the firing order. Under this condition, the PCM activates the solenoids to supply oil pressure to the low-speed rocker arms. This action forces the hydraulic pistons in the low-speed rocker arms into the rocker arm shaft so they do not engage the low-speed rocker arms to the T-shaped levers. The spring force on piston H keeps this piston pushed into the rocker arm so it does not connect the high-speed rocker arm to the T-shaped lever. Since neither of the rocker arms is connected to the T-shaped lever, the valves are inoperative (Figure 7-33). Disabling every second cylinder in the firing order provides improved fuel economy at moderate cruising speed and during deceleration.

In the low-speed cam mode, the PCM operates the control solenoids so oil pressure is not supplied to any of the rocker arm pistons. In this mode, the spring force pushes the pistons out of the low-speed rocker arms so these pistons engage the low-speed rocker arms to the T-shaped levers. Under this condition, the low-speed cam lobes operate the low-speed rocker arms and the T-shaped levers to provide valve opening in each cylinder. The low-speed camshaft lobes operate the valves to provide optimum fuel economy and emissions.

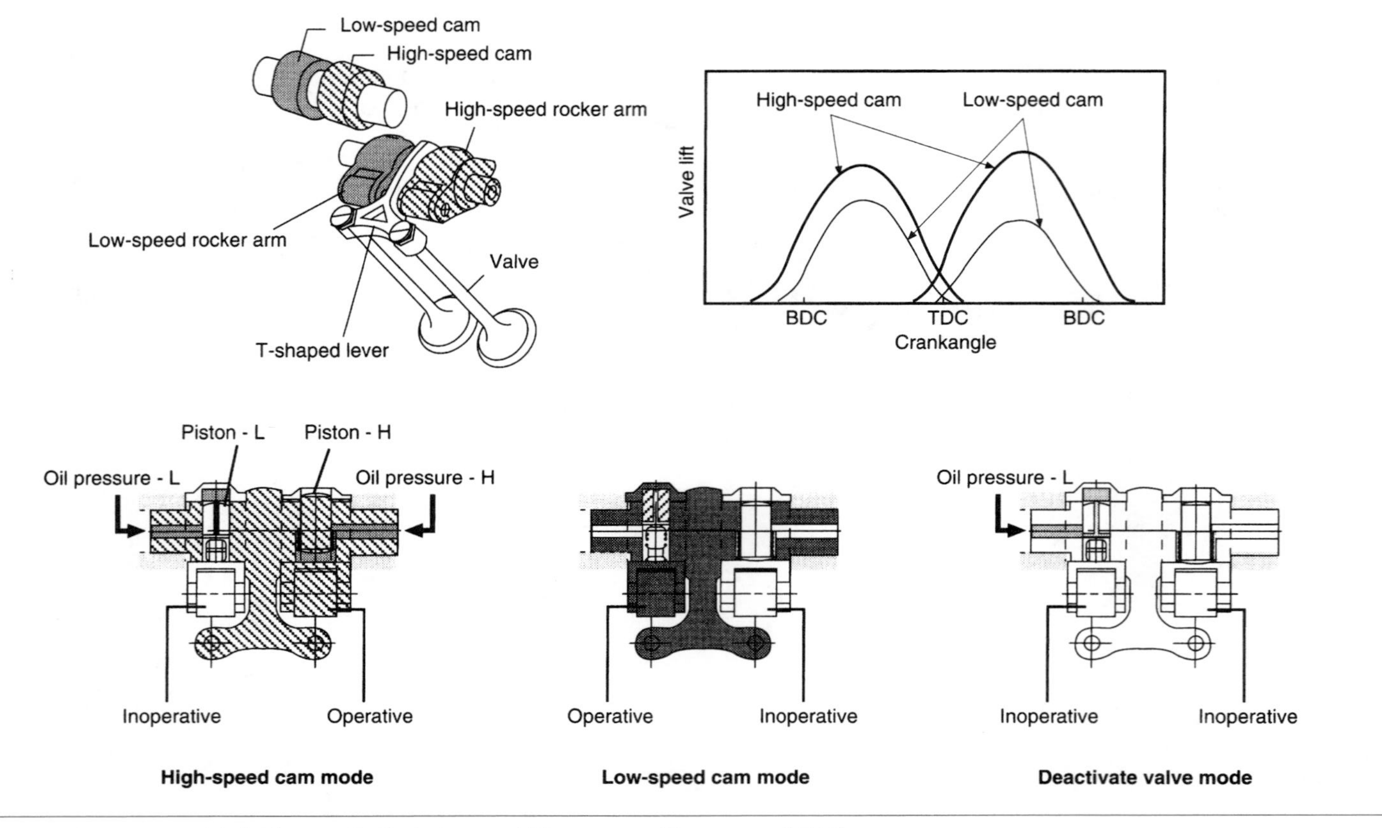


Figure 7-33 Variable camshaft timing and lift system. (Courtesy of SAE)

In the high-speed mode, the PCM operates the control solenoids to supply oil pressure to both the pistons in the rocker arms. When this condition occurs, the oil pressure on the low-speed piston forces this piston into the rocker arm shaft so it no longer engages the low-speed rocker arm to the T-shaped lever. Oil pressure is supplied to the high-speed piston in the rocker arms, and these pistons are forced out of the rocker arm shafts so they engage the high-speed rocker arms to the T-shaped lever. Under this condition, the high-speed camshaft lobes operate the valves in each cylinder through the high-speed rocker arms and T-shaped levers. The high-speed camshaft lobes provide longer valve opening and higher valve lift to provide improved engine performance (refer to Figure 7-33).

## A BIT OF HISTORY

Quotes from a 1954 Buick owner's manual.

1. Every 5,000 miles the distributor points should be cleaned and spaced. The proper gap setting is 0.0125 to 0.0175 inch.
2. The spark plugs should also be cleaned and gapped every 5,000 miles. Standard gap adjustment is 0.030 to 0.035 inch.
3. At 5,000 mile intervals, the timing should be checked and adjusted if necessary. This should be done with a timing light or syncroscope.
4. The voltage and current regulator should be adjusted properly in order to avoid burned ignition points or burned-out light units and radio tubes.

This recommended automotive service is much different from the service required on modern vehicles that do not have ignition points or timing adjustments; spark plug replacement intervals are often 100,000 miles, solid-state voltage regulators are not adjustable, and radios have transistors in place of tubes.

## Summary

- Rotation of the camshaft is controlled through a gear or sprocket combination. The camshaft rotates at half the crankshaft speed.
- The valve train components work together to open and close the valves. The common components include: camshaft, lifters or lash adjusters, pushrods, rocker arms or followers, valves, and valve timing components.
- Simply stated, the camshaft is used to control valve opening, the rate of opening and closing, and how long the valve will be open.
- The lobes of the camshaft push the valves open as the camshaft is rotated. The height of the lobe determines the amount of valve lift or opening. The design of the lobe will also determine the duration of valve opening.
- Lift can be expressed in two methods. The first is cam lift, which is the measured lift of the cam lobe. The second is valve lift. This is the cam lift multiplied by the rocker arm ratio to provide a specification for the amount of valve movement.
- The camshaft lobes generally have a very slight taper, which works with offsetting the centerlines of the lifter to provide for lifter rotation.
- Lifters are mechanical (solid) or hydraulic connections between the camshaft and the valves. Lifters follow the contour of the cam lobes to lift the valve off its seat.
- Hydraulic lifters use oil to automatically compensate for the effects of engine temperature and valve train wear.
- OHV engines with the camshaft located in the block use pushrods to transfer motion from the lifter to the rocker arms.
- The rocker arm is a pivoting lever used to transfer the motion of the pushrod to the valve stem. It changes the direction of the camshaft lift and spring closing forces, and it provides a leverage during valve opening.
- OHC or DOHC engines have the camshafts mounted on the cylinder heads. The camshaft lobes may contact the rocker arms directly, and a lash adjuster may be positioned under one end of the rocker arm.

### Terms to Know

Advanced camshaft
Asymmetrical
Cam lift
Camshaft
Clearance ramps
Duration
Free-wheeling engine
Interference engine
Lifter
Lobe lift
Lobes
Pedestal-mounted rocker arm
Pump-up
Pushrods
Retarded camshaft
Rocker arm
Rocker arm ratio
Shaft-mounted rocker arms
Single-pattern
Split overlap
Stud-mounted rocker arm
Valve float
Valve lift
Valve overlap
Valve train
Zero lash

## Review Questions

### Short Answer Essays

1. What is the function of the valve train?
2. Describe three timing belt and chain configurations on DOHC engines.
3. Explain the purpose and function of the camshaft.

4. Describe the relationship among the camshaft lobe design and lift, duration, and overlap.

5. Explain the purpose of the lifters.

6. Explain the purpose of pushrods.

7. List and explain the common methods of mounting the rocker arms.

8. Explain rocker arm ratio.

9. What is meant by the term *duration* as it relates to the camshaft?

10. If the rocker arm ratio is 1.5:1 and the cam lobe lift is 0.335 inch, what is the actual valve opening?

## Fill-in-the-Blanks

1. The camshaft rotates at ____________________ the speed of the crankshaft.

2. For proper valve timing and engine operation, the camshaft and the crankshaft must remain in the same ____________________ position to each other.

3. The camshaft is used to control valve ____________________, the ____________________ of opening and closing, and how ____________________ the valve will be open.

4. The ____________________ of the lobe determines the amount of valve lift or opening.

5. ____________________ is the length of time, expressed in degrees of crankshaft rotation, the valve is open.

6. The camshaft lobes generally have a very slight taper. This taper, along with the ____________________ of the lifter and the lobe, provides for lifter rotation.

7. Lifters are mechanical or hydraulic connections between the ____________________ and the ____________________.

8. OHV engines with the camshaft located in the block use ____________________ to transfer motion from the lifter to the rocker arms.

9. The design of the rocker arm is such that the side to the valve stem is usually ____________________ than the side to the pushrod.

10. When the valve is opened 50 percent of its travel, the rocker arm-to-stem contact should be ____________________ on the valve stem.

## Multiple Choice

1. The camshaft drive is being discussed.
*Technician A* says the crankshaft rotates at half the speed of the camshaft.
*Technician B* says the camshaft sprocket has twice as many teeth as the crankshaft sprocket.
Who is correct?
**A.** A only
**B.** B only
**C.** Both A and B
**D.** Neither A nor B

2. Camshafts are being discussed.
*Technician A* says camshafts control the rate of valve opening.
*Technician B* says clearance ramps are ground into the lobe to remove clearance in the valve train at a slower rate than the rate of lift.
Who is correct?
**A.** A only
**B.** B only
**C.** Both A and B
**D.** Neither A nor B

3. *Technician A* says duration is the amount the valve is moved off its seat.
   *Technician B* says lift is the length of time the valve is open.
   Who is correct?
   **A.** A only **C.** Both A and B
   **B.** B only **D.** Neither A nor B

4. *Technician A* says the longer the valve remains open after reaching the 0.300 in. (8 mm) lift, the better the airflow.
   *Technician B* says lift has more effect on engine performance and efficiency than any other valve timing specification.
   Who is correct?
   **A.** A only **C.** Both A and B
   **B.** B only **D.** Neither A nor B

5. *Technician A* says the camshaft lobes are generally ground flat across the surface.
   *Technician B* says offsetting the centerlines of the lifter and the lobe is done to rotate the lifter.
   Who is correct?
   **A.** A only **C.** Both A and B
   **B.** B only **D.** Neither A nor B

6. Lifters are being discussed.
   *Technician A* says solid lifters must provide room for expansion by a specified clearance between valve train components.
   *Technician B* says hydraulic lifters use oil to automatically compensate for the effects of engine temperature and valve train wear.
   Who is correct?
   **A.** A only **C.** Both A and B
   **B.** B only **D.** Neither A nor B

7. *Technician A* says the hydraulic lifter is filled with oil only when the valve is in the closed position.
   *Technician B* says the process of filling and leakdown automatically adjusts the hydraulic lifter.
   Who is correct?
   **A.** A only **C.** Both A and B
   **B.** B only **D.** Neither A nor B

8. *Technician A* says lifter pump-up can cause a valve to remain open.
   *Technician B* says lifter pump-up can only occur when the engine is operating at too low rpm for the load.
   Who is correct?
   **A.** A only **C.** Both A and B
   **B.** B only **D.** Neither A nor B

9. *Technician A* says rocker arms are designed so the longer side is the pushrod end.
   *Technician B* says rocker arm ratio is an expression of the leverage advantage of the rocker arm.
   Who is correct?
   **A.** A only **C.** Both A and B
   **B.** B only **D.** Neither A nor B

10. *Technician A* says rocker arm geometry can be affected by cylinder head and block deck resurfacing.
    *Technician B* says rocker arm geometry can be affected by changes in camshaft lift.
    Who is correct?
    **A.** A only **C.** Both A and B
    **B.** B only **D.** Neither A nor B

# The Cylinder Block Assembly

Upon completion and review of this chapter, you should be able to:

- ❑ Explain the purpose of the cylinder block.
- ❑ Describe the purpose of cylinder sleeves.
- ❑ Name and describe the components of the crankshaft.
- ❑ Explain the relationship of crankshaft throw to stroke and firing impulses.
- ❑ Describe the forces applied to the crankshaft resulting in wear or damage.
- ❑ Explain the purpose of the harmonic balancer.
- ❑ Explain the purpose of the flywheel.
- ❑ Describe the function of main and rod bearings.
- ❑ Explain the use of undersize and oversize bearings.
- ❑ Explain the function of the pistons.
- ❑ Describe the methods used in piston construction to control heat.
- ❑ List the advantages of cast, forged, and hypereutectic cast pistons.
- ❑ Explain the purposes of compression rings and oil rings.
- ❑ Describe the forces applied to the piston pin.
- ❑ Explain the function of the connecting rods.

## Introduction

The block assembly may be referred to as a **short block.** This assembly includes the cylinder block, crankshaft, connecting rods, pistons, and bearings (Figure 8-1). Each of these components has a specific purpose. A short block may also include the camshaft, timing gear, and balance shaft. A **long block** is basically a short block with cylinder heads.

The purpose and function of the flywheel and harmonic balancer are also discussed in this chapter, because these components are related to the block assembly. When there is a major engine failure, shops either rebuild or replace the engine. Most often the short block is repaired or replaced as an assembly.

A **short block** is a block assembly without the cylinder heads and some other components.

A **long block** is an engine block on which the cylinder heads and other components are mounted.

## Cylinder Blocks

The **cylinder block** makes up the lower section of the engine. The cylinder block supports many of the other engine components. It houses the cylinders, where combustion of the air/fuel mixture takes place. The upper section of the engine, known as the cylinder head, bolts to the top of the cylinder block. The head is also part of the combustion chamber and contains the valve train components.

The cylinder block (Figure 8-2) is normally a one-piece casting, machined so that all the parts contained in it fit properly. Blocks may be cast from several different materials: iron, aluminum, or possibly, in the future, plastic.

In recent years, some manufacturers have designed two-piece blocks utilizing a **bedplate** or a one-piece main bearing cap (Figure 8-3). The two-piece design provides a stronger lower end because it ties all of the main caps together to improve block stiffness and reduce engine vibration.

The word cast refers to how the block is made. To cast is to form molten metal into a particular shape by pouring or pressing it into a mold. This molded piece must then undergo a number of machining operations to make sure all the working surfaces are smooth and true. The top of the block must be perfectly smooth so that the cylinder head can seal it. The base or bottom of the block is also machined to allow for proper sealing of the oil pan. The cylinder bores must be smooth and have the correct diameter to accept the pistons.

The **cylinder block** is the main structure of the engine. Most of the other engine components are attached to the block.

The **bedplate** is the lower part of a two-piece engine block.

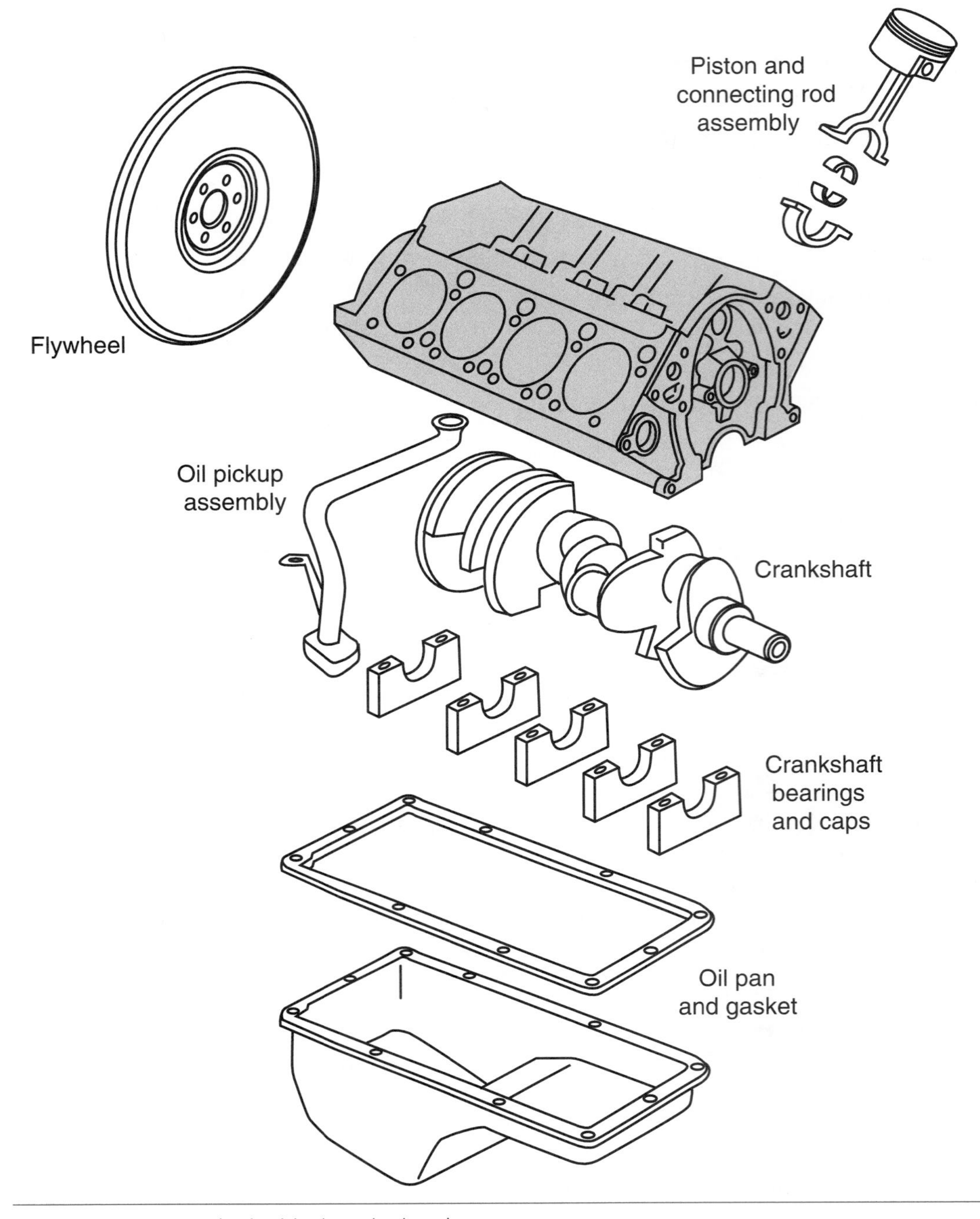


**Figure 8-1** Cylinder block and related components.

When main bearing bores are **align bored,** they are in perfect alignment with each other in a straight line.

The main bearing area of the block must be **align bored** to a diameter that will accept the crankshaft. When main bearing bores are align bored, they are in perfect alignment with each other in a straight line. If main bearings are not align bored, rapid and uneven main bearing wear will occur. Camshaft bearing surfaces must also be align bored. The word bore means to drill or machine a hole. Align boring cuts a series of holes in a straight line.

Cast-iron blocks offer great strength and controlled warpage. With the increased concern for improved gasoline mileage, however, car manufacturers are trying to make the vehicle lighter. One way to do this is to reduce the weight of the block. Aluminum is often used to reduce this weight. Certain materials are added to aluminum to make the aluminum stronger and less likely to warp from the heat of combustion. Aluminum blocks normally have a sleeve or steel liner

Figure 8-2 Assembled cylinder block. (Courtesy of Jasper Engine and Transmission Exchange, Inc.)

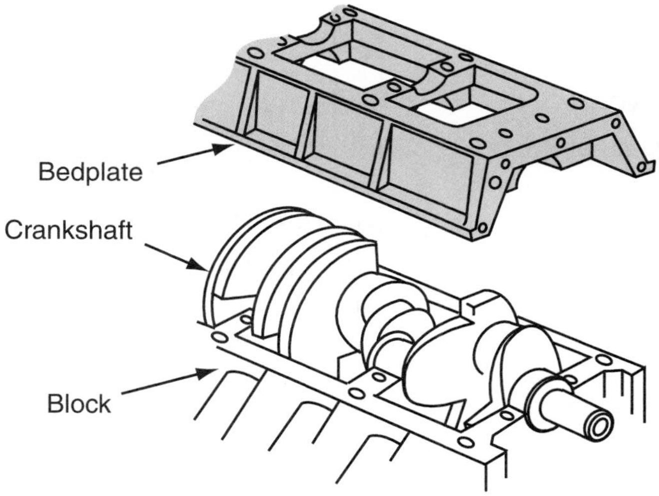

Bedplate
Crankshaft
Block

Figure 8-3 Cylinder block with bedplate.

placed in them to serve as cylinder walls. Steel liners are placed in the mold before the metal is poured. After the metal is poured, the steel liner cannot be removed.

There are several forces working against the lower engine block, including:

1. Vertical bending
2. Horizontal bending 90 degrees to the cylinder bore
3. Torsional bending along the crankshaft axis
4. Individual main bearing cap flutter coinciding with individual cylinder firing

## Lubrication and Cooling

A cylinder block contains a series of oil passages that allows engine oil to be pumped through the block and crankshaft and on to the cylinder head. The oil lubricates, cools, seals, and cleans engine components (Figure 8-4).

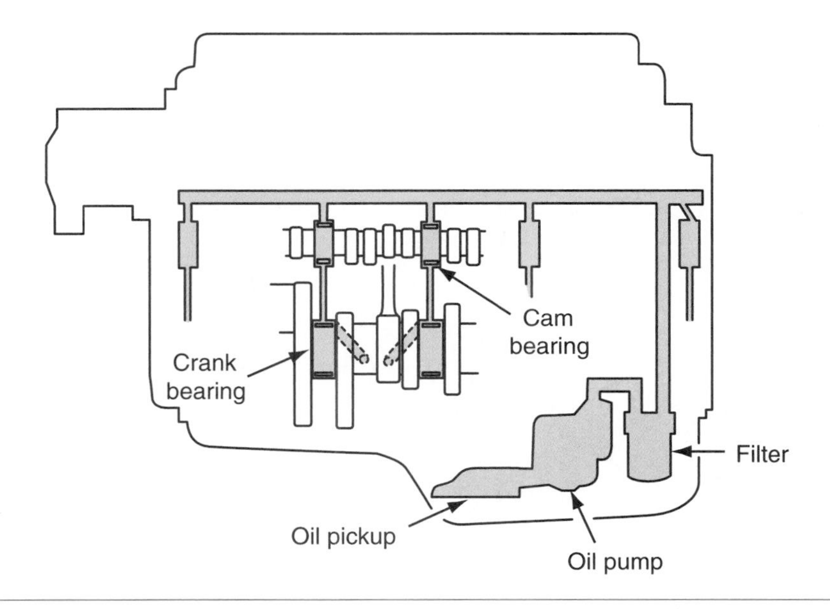


Figure 8-4 Cylinder block lubrication system.

Some of the heat generated by an engine is absorbed by the block and cylinder heads. This absorbed heat is wasted power and must be removed before it damages the engine. This is the job of the cooling system. Water jackets are also cast in the block around the cylinder bores. Coolant circulates through these jackets to transfer heat away from the block to the coolant.

**Shop Manual**
Chapter 8, page 355

## Core Plugs

**Core plugs** are also called expansion plugs.

All cast cylinder blocks use **core plugs.** These are also called expansion plugs. During the manufacturing process, sand cores are used. These cores are partly broken and dissolved when the hot metal is poured into the mold; however, holes must be placed in the block to get the sand out after the block is cast. These core holes are machined and core plugs are placed into them to seal them (Figure 8-5).

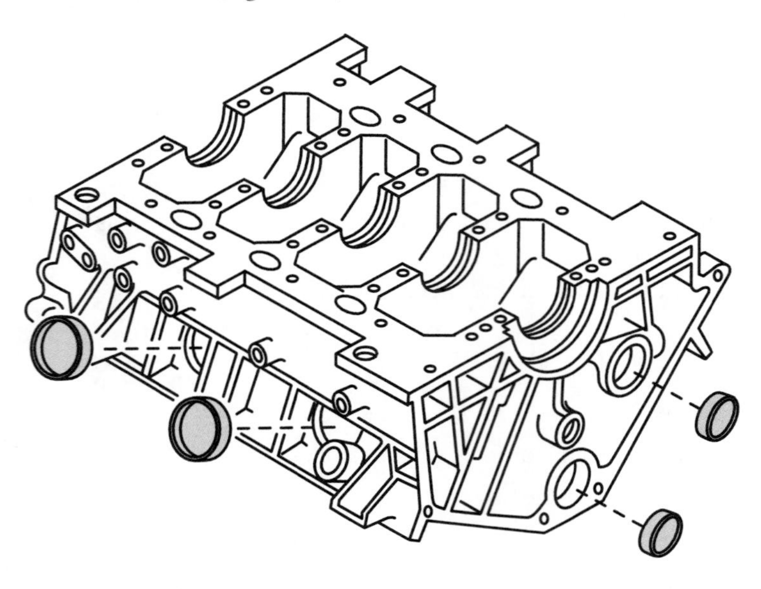

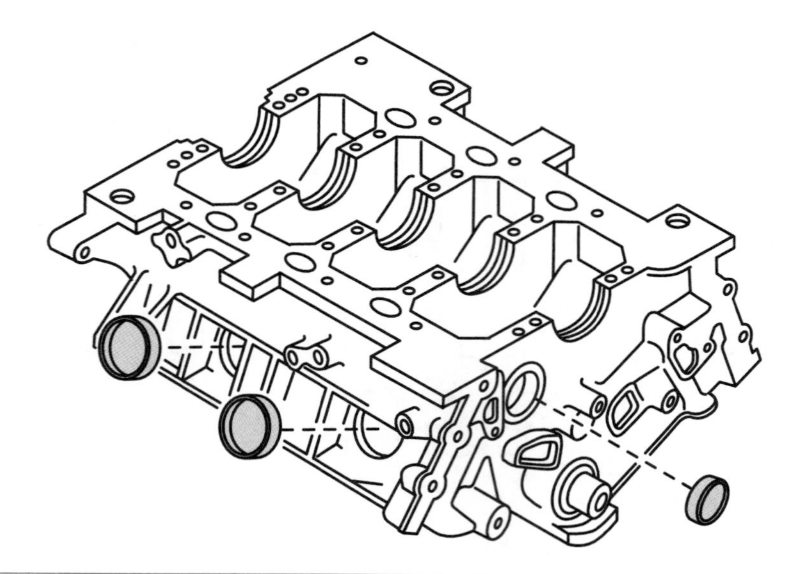

Figure 8-5 Cylinder block core and gallery plugs.

## Cylinder Sleeves

Some engines, such as those made with aluminum, have cylinder sleeves (Figure 8-6). Some engines have sleeves that can be replaced if the cylinder walls are damaged. Most blocks must be bored out and larger pistons or standard-sized sleeves must be installed. There are two types of sleeves: wet and dry. Both types are pressed into the block. The dry sleeve is supported from top to bottom by the block. The wet sleeve is supported only at the top and bottom. Coolant touches the center part of a wet sleeve, and this type of sleeve is sealed at the top and bottom.

A cylinder sleeve may be called a liner.

Wet sleeves are surrounded by engine coolant.

# Crankshafts

Crankshafts are generally constructed from cast iron, nodular iron, or forged steel (Figure 8-7). As the pistons are forced downward on the power stroke, pressure is applied to the crankshaft, causing it to rotate. The crankshaft transmits this torque to the drivetrain and ultimately to the drive wheels.

The saddle is the portion of the crankcase bore that holds the bearing half in place.

## Crankshaft Journals

**Main bearing journals** are machined along the centerline of the crankshaft and support the weight and forces applied to the crankshaft (Figure 8-8). Offset from the crankshaft centerline are the **connecting rod bearing journals.** Engine design determines the amount of offset (Figure 8-9). The offset places weight and pressure off the center of the crankshaft and provides leverage for the piston assembly to turn the crankshaft.

**Crankshaft throw** determines the length of the piston stroke. The crankshaft throw is the distance from the center of the main bearing journals to the center of the connecting rod journals. The piston stroke is double the crankshaft throw. The throws are set so pistons in opposite banks move in opposite directions. This design cancels some of the forces applied to the crankshaft. In addition, most crankshaft throws are arranged to provide equal intervals between firing impulses. Since a piston of a 4-stroke engine has only one firing impulse every 720 degrees of crankshaft revolution, a 4-cylinder engine would have the throws 180 degrees apart. A typical V8 engine has four crank throws since two connecting rods are attached to each journal. The design of the "V" engine is such that the angle between the cylinder banks is a multiple of the angle of the throw; for example, most V8 engines have a 90-degree angle between cylinder banks and the throws are set 90 degrees apart. Most V6 engines are designed with the cylinder banks set 60 degrees apart, with 120 degrees between throws. This is possible because 60 is a multiple of 120. An exception to this rule is the General Motors odd-fire engine. This V6 engine is designed with cylinder banks set at 90 degrees, with three crank throws set at 120 degrees. The engine is fired from every other throw, resulting in unequally spaced firing impulses. The two piston assemblies sharing common journals are fired in succession. When the first cylinder is fired, the crankshaft rotates 90 degrees before the second cylinder is fired. After the second cylinder is fired, the crankshaft has to rotate 150 degrees until the third cylinder is fired. This cycle is then repeated as the throws are alternated.

**Main bearing journals** are machined along the centerline of the crankshaft and support the weight and forces applied to the crankshaft.

**Connecting rod bearing journals** are offset from the crankshaft centerline.

**Crankshaft throw** is the measured distance between the centerline of the rod bearing journal and the centerline of the crankshaft.

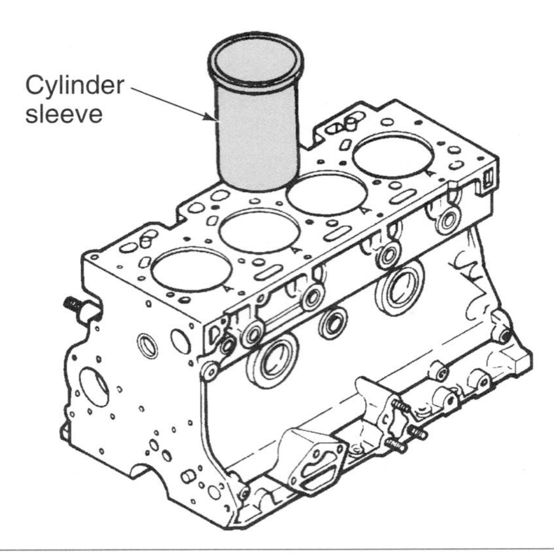


**Figure 8-6** Sleeves can be pressed into the cylinder block to provide a wear surface for the piston rings.

**Figure 8-7** Typical crankshaft construction. (Courtesy of JTG Associates)

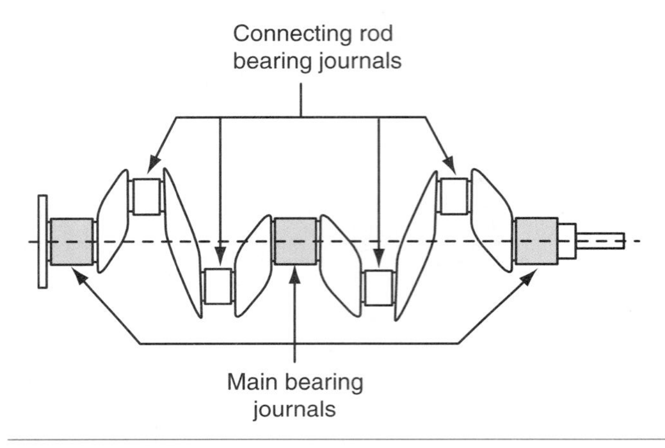


**Figure 8-8** Main bearing journals are the centerline of the crankshaft.

**Fillets** are small rounded corners machined on the edge of the journals to increase strength.

V6 engines with 90-degree banks can be even firing by redesigning the crankshaft to have one connecting rod per journal and two journals per throw (Figure 8-10). This splayed crankshaft design offsets the journals of the same throw 30 degrees.

The **fillet** on the journals is used to increase the strength of the crankshaft (Figure 8-11). This joint area is subject to high stress and is a common location for cracks. Since sharp corners are weak, manufacturers call for a curved fillet of a specific radius. The size of the fillet is increased for applications requiring severe duty. Most fillets are formed by grinding the joint. Some high-performance or turbocharged engines use a process of rolling the fillet. This is done by rolling hardened steel balls under pressure on the joint. This causes the metal at the joint to be compacted and tightens the grain structure, thus strengthening the crankshaft by as much as 30 to 40 percent. The fillet is not usually a wear area unless the bearing is worn. If the journal requires machining, the fillet should be reground to the proper radius.

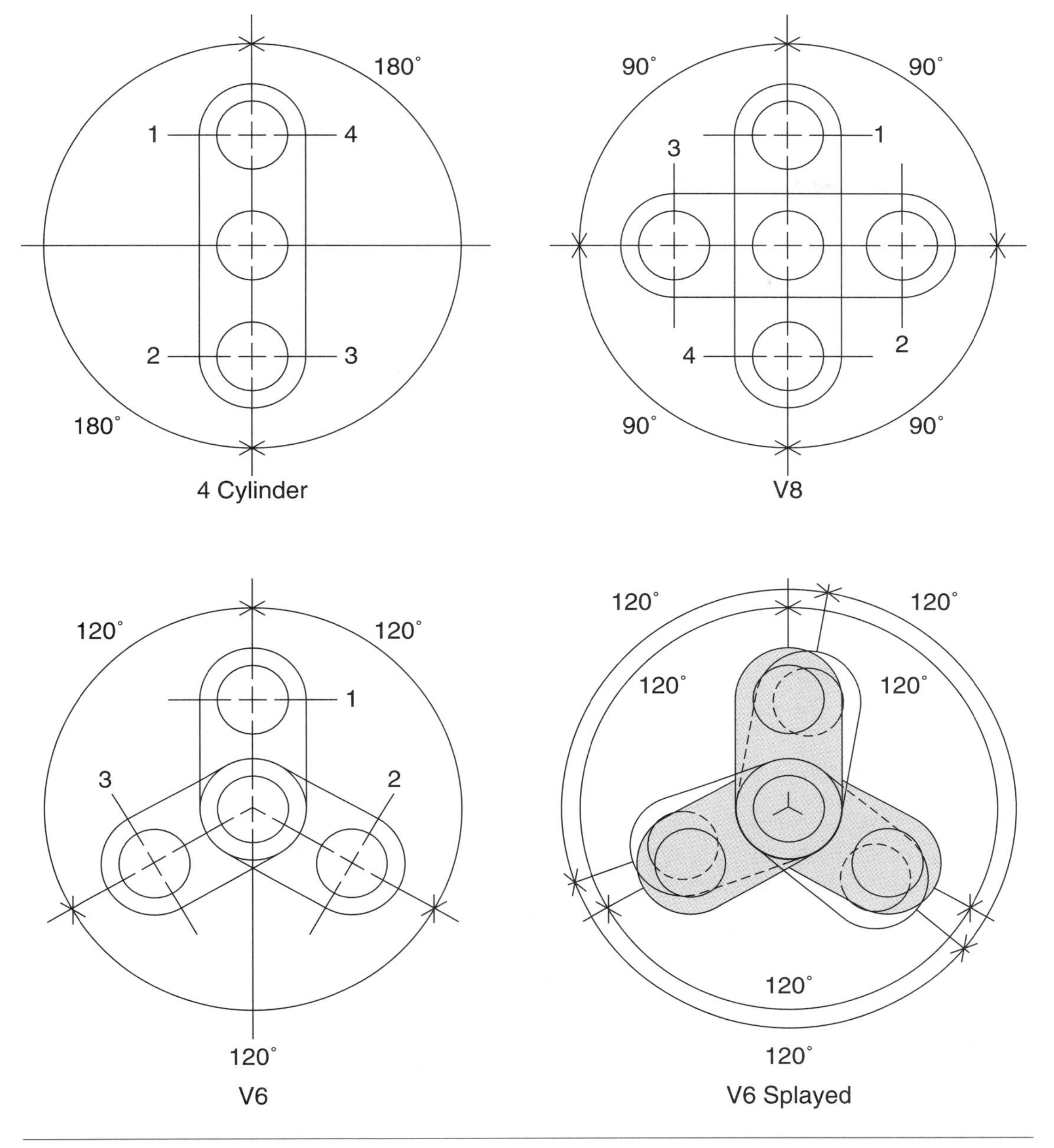


**Figure 8-9** Typical connecting rod throw offsets.

The crankshaft is also subject to lateral thrust forces. These come from clutch engagement and automatic transmission loads. The lateral movement of the crankshaft is controlled by a thrust bearing and a thrust surface on the crankshaft. The thrust bearing is usually attached vertically to the sides of one main bearing insert.

For smooth engine operation, the crankshaft is balanced by **counterweights** positioned opposite the connecting rod journals. The counterweight overcomes the total weight of the rod bearing journal, bearings, connecting rod, piston pin, and piston. There is normally no wear on the counterweights; however, if a connecting rod breaks, it can strike the counterweight, resulting in the removal of material. This may require balancing of the crankshaft before it can be reused.

**Counterweights** are positioned opposite the connecting rod journals and balance the crankshaft.

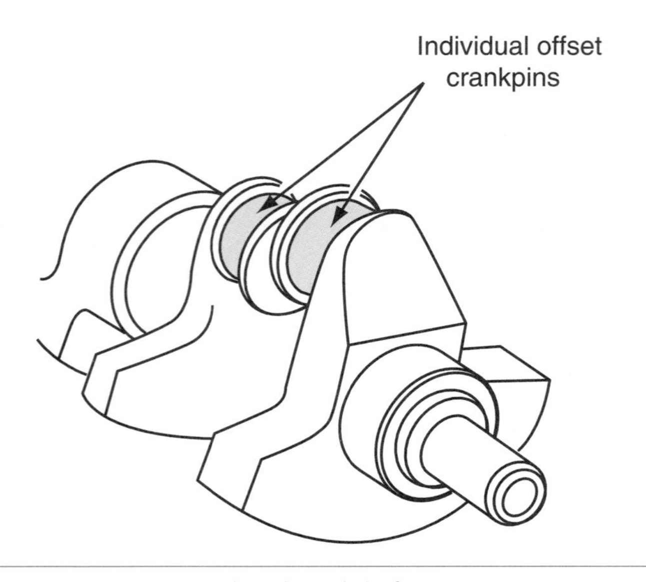

Figure 8-10 Splayed crankshaft.

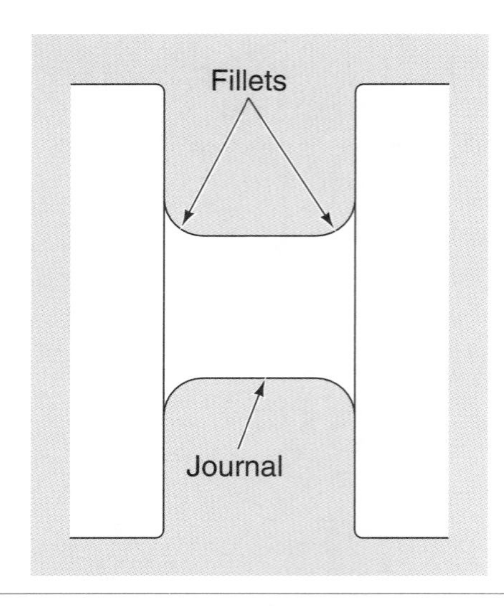


Figure 8-11 The fillet strengthens the crankshaft.

## Crankshaft Construction

Many manufacturers construct their crankshafts from gray cast iron. This construction is strong enough for most applications. Some engine applications require additional strength above what gray cast iron can provide; in these instances, nodular iron is often used. A cast-iron crankshaft is identifiable by its straight cast mold parting line (Figure 8-12).

In other applications, such as some turbocharged engines, a forged medium-carbon steel crankshaft (carbon content between 0.30 and 0.60 percent) is used. High-performance engines may use a crankshaft using chromium and molybdenum (sometimes called chrome-moly) alloys to increase its strength. The forging process condenses the grain of the steel and increases its strength. Because of the increased density of the steel, these crankshafts weigh more than a cast-iron crankshaft. To lighten the weight, some manufacturers hollow the rod bearing journals.

Forged crankshafts designed for use in V8 and in-line 6-cylinder engines must be twisted to achieve the desired throw offset. This process is done immediately after the crankshaft is forged, while the metal is still hot. A forged crankshaft can be identified by the staggered die parting lines because of this twisting (Figure 8-13).

Case hardening crankshafts protects the journals from wear and fractures. Many manufacturers use a case-hardening process called ion nitriding. The component is placed into a pressur-

Figure 8-12 Casting lines on a cast-iron crankshaft.

Figure 8-13 Casting lines on a forged crankshaft.

ized chamber filled with hydrogen and nitrogen gases. An electrical current is then applied through the component. This changes the molecular structure of the metal and allows the induction of the gases into the surface area of the journal.

## The Harmonic Balancer

During normal engine operation, the crankshaft is subjected to twisting and vibrations. When a piston is approaching top dead center (TDC) on the compression stroke, the portion of the crankshaft connected to the piston attempts to slow down. At the same time, other pistons are forcing the crankshaft to increase in speed. This results in torsional vibrations. **Harmonic balancers** (also referred to as vibration dampers) are one method used to control and compensate for **torsional vibrations.** The vibration damper is constructed of an inertia ring and a rubber ring. The two are bonded together and attached to the front of the crankshaft (Figure 8-14).

**Shop Manual**
Chapter 8, page 356

**Harmonic balancers** are used to control and compensate for torsional vibrations.

**Torsional vibration** is the result of the crankshaft twisting and snapping back during each revolution.

As the crankshaft twists, the inertia ring has a slow-down effect. The inertia and the rubber rings absorb the torsional vibrations. Many of today's engines use dual-mode harmonic balancers (Figure 8-15). The first mode eliminates torsional vibration, and the other mode eliminates vertical vibration.

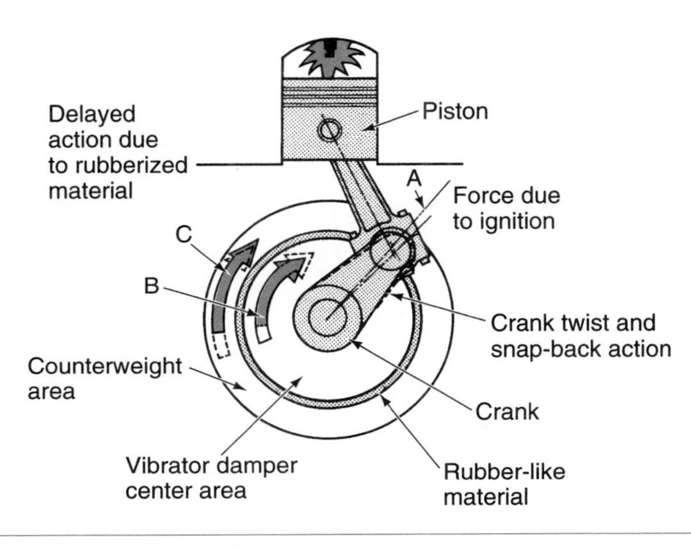


Figure 8-14 Harmonic balancer action.

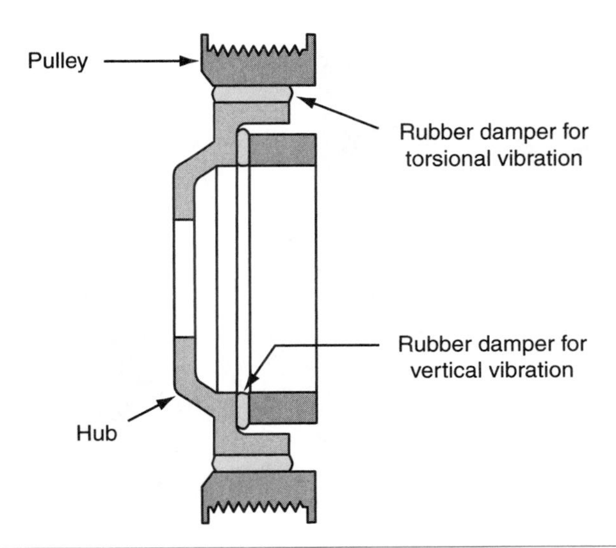


Figure 8-15 Dual-mode harmonic balancer.

Many harmonic balancers are pressed onto the end of the crankshaft, and a square key fits in matching keyways in the crankshaft and balancer to prevent balancer rotation. A bolt and washer usually retain the harmonic balancer on the end of the crankshaft.

## The Flywheel

The **flywheel** is located at the end of the crankshaft and assists the crankshaft in rotating once power has been applied to the piston.

The **flywheel** is located at the end of the crankshaft (Figure 8-16). It is used to assist the crankshaft in rotating once power has been applied to the piston. It acts as an inertia ring to keep the crankshaft rotating through nonproductive strokes. It is also used to absorb vibrations and power impulses. The most common materials used to construct flywheels are cast iron, nodular iron, steel, and aluminum with steel plasma.

The term flywheel is usually used if the engine is mated to a manual transmission. The machined surface of the flywheel is where the clutch assembly is attached. An engine that is mated to an automatic transmission generally has a flex plate with a torque converter attached to it. The torque converter is filled with transmission fluid and is as heavy as the flywheel and clutch assembly used with a manual transmission. The torque converter is capable of performing the same basic functions as the flywheel.

## Bearings

Bearings are used to carry the loads created by rotational forces. The bearings used to support the crankshaft in the engine block are called main bearings. Bearings are also used around the connecting rod where it is attached to the crankshaft. In addition, many engines use bearings to support the camshaft.

There are many different materials used in the construction of bearings. The main concern of the manufacturers is to achieve a bearing construction that will meet four primary requirements:

1. *Surface action*—The ability of the bearing to withstand metal-to-metal contact without being damaged.
2. *Embedability*—The ability of the bearing material to tolerate the presence of foreign material.
3. *Fatigue resistance*—The ability of the bearing to withstand the stress placed upon it.
4. *Corrosion resistance*—The ability of the bearing to withstand the corrosive acids produced within the engine.

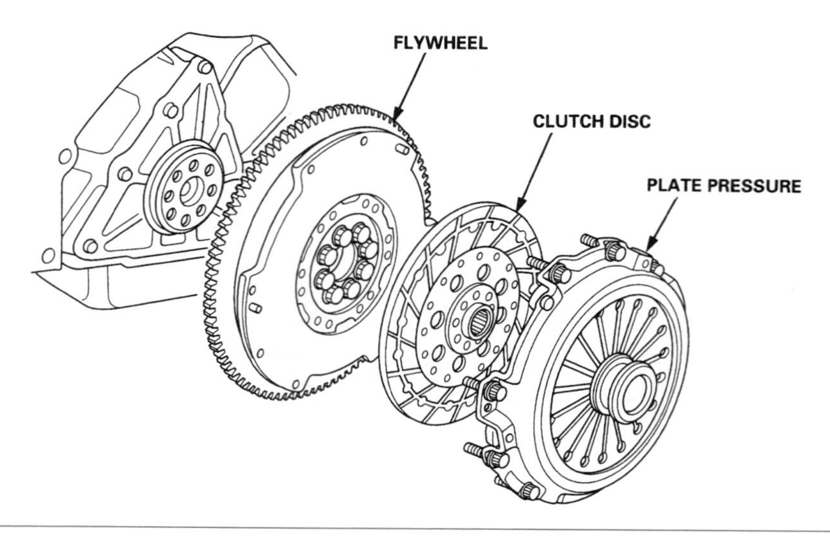


Figure 8-16 The flywheel is attached to the rear of the crankshaft. (Courtesy of American Honda Motor Co., Inc.)

Since no single material is capable of meeting all of these requirements, most bearings are constructed using multilayers of different materials. The backing is usually made of steel, then additional lining material is added. The most common materials used include:

1. Babbitt
2. Sintered copper-lead
3. Cast copper-lead
4. Aluminum

To assist the technician in the selection of bearings, most manufacturers use a part number suffix to identify the bearing lining material; for example, a Federal Mogul bearing with the part number 8-3400CP indicates the bearing is steel-backed copper-lead alloy with overplate. The following chart provides the codes used:

| Part Number Suffix | Lining Material and Application |
|---|---|
| AF | Steel-backed aluminum alloy thrust washer |
| AP | Steel-backed aluminum alloy with overplate connecting rod or main bearings |
| AT | Solid aluminum alloy connecting rod or main bearings |
| B | Bronze-backed babbitt connecting rod or main bearings |
| BF | Steel-backed bronze or babbitt thrust washers |
| CA | Steel-backed copper alloy connecting rod or main bearings |
| CH | Steel-backed copper-lead alloy with overplate connecting rod and main bearings for performance applications |
| CP | Steel-backed copper-alloy with overplate connecting rod and main bearings |
| DR | Steel-backed babbitt, steel-backed copper-lead alloy, or solid aluminum camshaft bushing |
| F | Bronze-backed babbitt, bronze or aluminum thrust washers |
| RA | Steel-backed aluminum alloy connecting rod and main bearings |
| SA, SB, SBI, SH, SI, SO | Steel-backed babbitt connecting rod and main bearings |
| TM | Steel-backed copper alloy with thin overplate connecting rod and main bearings |
| W | Camshaft thrust washer |

Rod, main, and thrust bearings are usually two-piece insert design. Figure 8-17 provides common nomenclature used with insert-type bearings. Most main and connecting rod bearings are designed to have a certain amount of spread. Bearing spread is provided by manufacturing the bearing inserts so they have a slightly larger curvature compared to the curvature of the rod bearing or cap. This design assures positive positioning of the bearing against the total bore area. When the bearing is installed into the bore, a light pressure is required to set the bearing (Figure 8-18).

In addition to **bearing spread,** most main and connecting rod bearings are designed to provide **bearing crush** (Figure 8-19). When the bearing half is installed in the connecting rod or cap, the bearing edges protrude slightly from the rod or cap surfaces. When the two halves are assembled, the crush exerts radial pressure on the bearing halves and forces them into the bore. Proper crush is important for heat transfer. If the bearing is not tight against the bore, it will act as a heat dam and not transfer the heat to the engine block.

To assure proper installation of the bearing in the bore, most manufacturers use a **locating tab** (Figure 8-20). A protruding tab at the parting of the bearing halves is set into a slot in the housing bore. When properly installed, the tab will locate the bearing, preventing it from shifting sideways in the bore.

**Bearing spread** means the distance across the outside parting edges is larger than the diameter of the bore.

**Bearing crush** refers to the extension of the bearing half beyond the seat.

The **locating tab** is a tab sticking out of bearing inserts that fits in a groove in the block, main bearing cap, connecting rod, or connecting rod cap to prevent insert rotation.

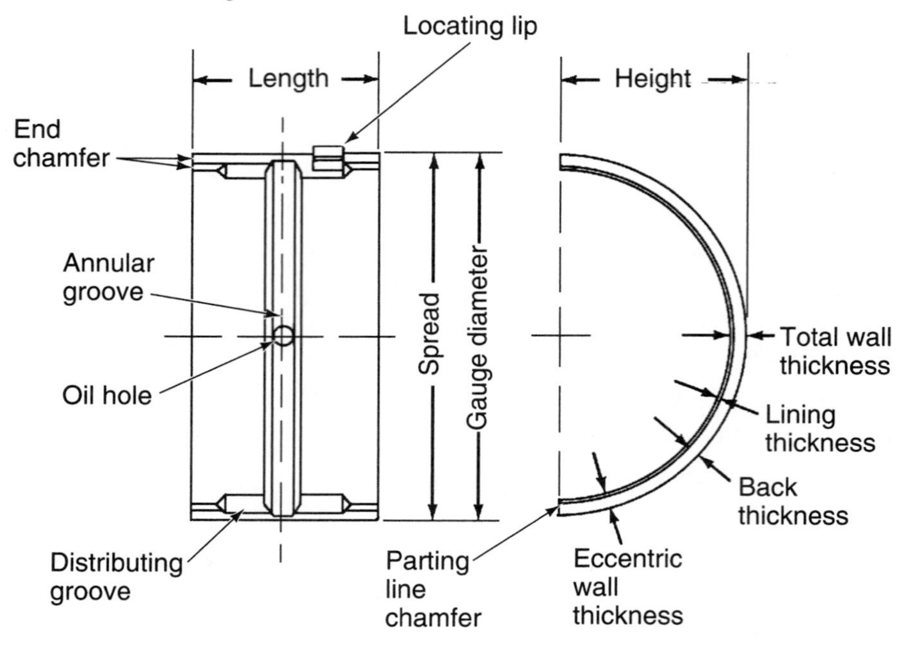


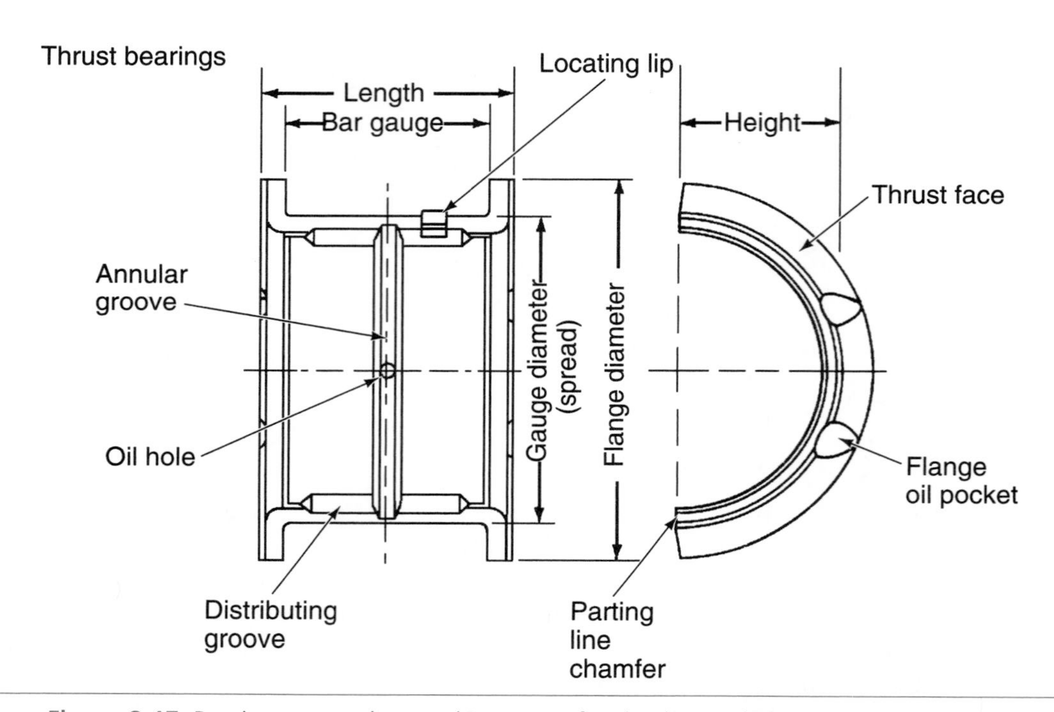


Figure 8-17 Bearing nomenclature. (Courtesy of Federal-Mogul Corporation)

## Main Bearings

Main bearings carry the load created by the movement of the crankshaft. Most of these bearings are insert design. The advantages of this design include ease of service, variety of materials, and controlled thickness.

Split bearings surround each main bearing journal. The bearings are supplied oil under pressure from the oil pump through oil passages drilled in the engine block and crankshaft (Figure 8-21). The crankshaft does not rotate directly on the main bearings; instead, it rides on a thin film of oil

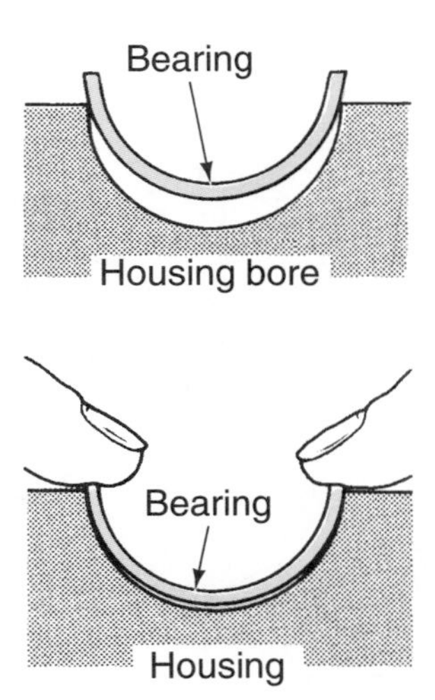


Figure 8-18 Designing the bearing with a certain amount of spread assures positive positioning of the bearing against the total bore area.

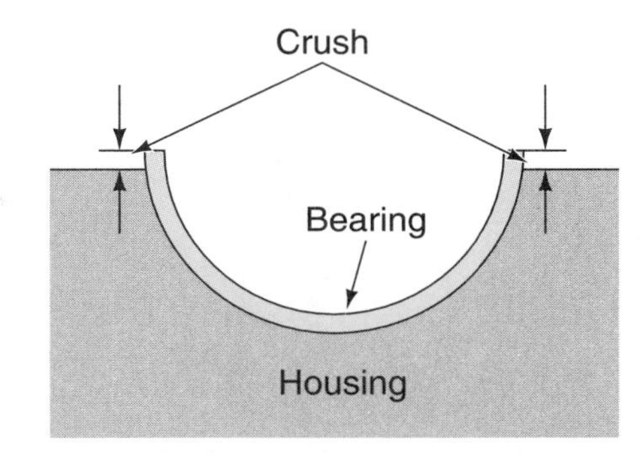


Figure 8-19 Bearing crush exerts radial pressure on the bearing halves and forces them into the bore.

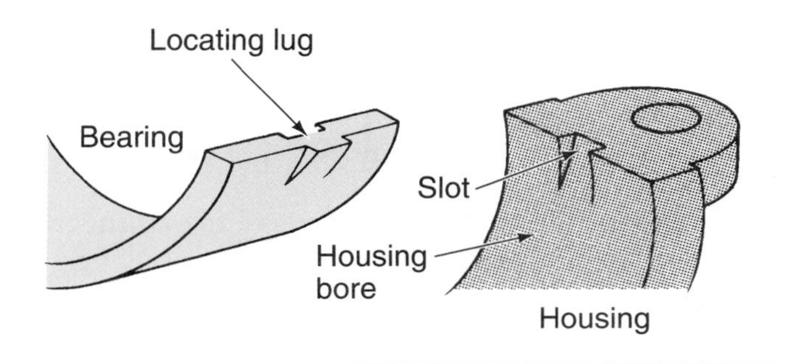


Figure 8-20 A protruding tab at the parting of the bearing halves is mated to a slot in the housing bore. The tab locks the bearing to prevent it from spinning or shifting sideways.

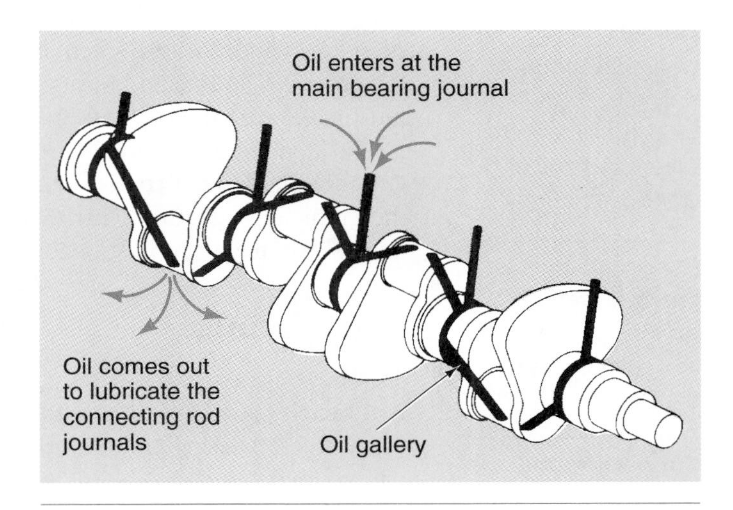


Figure 8-21 Internal passages in the crankshaft send lubrication to the connecting rod bearings.

trapped between the bearing and the crankshaft (Figure 8-22). If the journals are worn and become out-of-round, tapered, or scored, the oil film is not formed properly. This will result in direct contact between the crankshaft and bearing and will eventually damage the bearing and/or crankshaft. Soft materials are used to construct the bearings in an attempt to limit wear of the crankshaft.

**Oil Grooves.** Providing an adequate oil supply to all parts of the bearing surface, particularly in the load area, is an absolute necessity. In many cases, this is accomplished by the oil flow through the bearing oil clearance. In other cases, however, engine operating conditions are such that this oil distribution method is inadequate. When this occurs, some type of oil groove must be added to the bearing. Some oil grooves are used to assure an adequate supply of oil to adjacent engine parts by means of oil throw-off.

**Oil Holes.** Oil holes allow for oil flow through the engine block galleries and into the bearing oil clearance space. Connecting rod bearings receive oil from the main bearings by means of oil

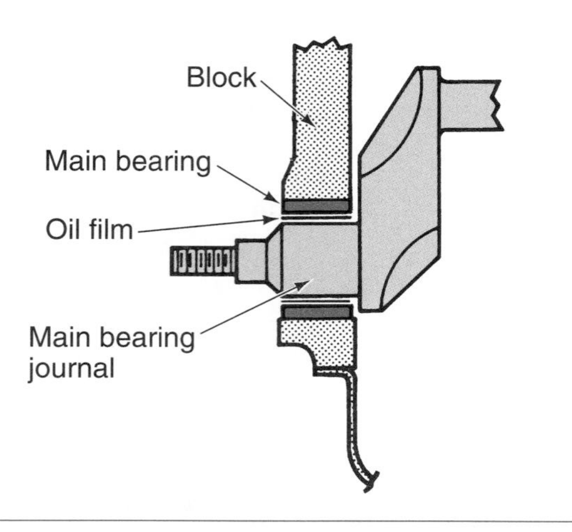


Figure 8-22 The crankshaft rotates on a thin film of oil.

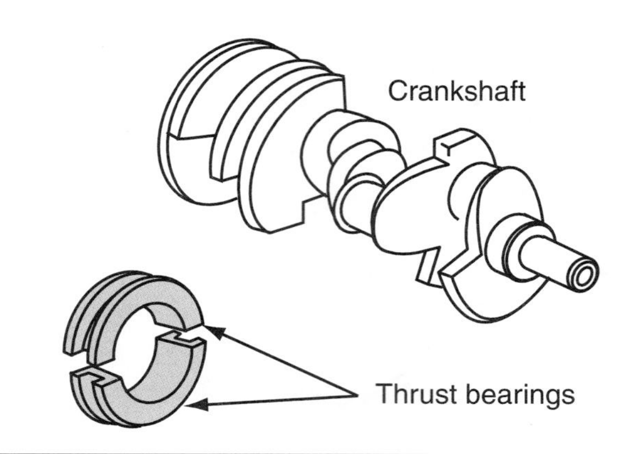


Figure 8-23 Flanged thrust bearing.

The **thrust bearing** prevents the crankshaft from sliding back and forth by using flanges that rub on the side of the crankshaft journal.

Crankshaft end play is the measure of how far the crankshaft can move lengthwise in the block.

**Connecting rod bearings** are insert-type bearings retained in the connecting rod and mounted between the large connecting rod bore and the crankshaft journal.

passages in the crankshaft. Oil holes are also used to meter the amount of oil supplied to other parts of the engine; for example, oil squirt holes in connecting rods are often used to spray oil onto the cylinder walls. When the bearing has an oil groove, the oil hole normally is in line with the groove. The size and location of oil holes is critical. Therefore, when installing bearings, you must make sure the oil holes in the block line up with holes in the bearings.

The engine will usually use a straight shell bearing design on all but one crankshaft journal. One double-flanged **thrust bearing** is used to control any horizontal movement (end play) of the crankshaft (Figure 8-23). The thrust bearing may have the flange formed as a part of the insert, or it may use thrust bearing inserts (Figure 8-24).

## Rod Bearings

Split bearings surround each connecting rod journal. Like the main bearings, the **connecting rod bearings** do not rotate directly on the crankshaft. They rotate on a thin film of oil trapped between the bearing and the crankshaft. Journals that are worn, out-of-round, tapered, or scored will not maintain this oil film, resulting in damage to the bearing and/or crankshaft.

Most main and rod bearings are designed with a bearing crown to maintain close clearances at the top and bottom of the bearing. This is the area where most of the loads are applied. The crown also allows increased oil flow at the sides of the bearing (Figure 8-25). In addition, most bearings have a beveled edge to provide room for the journal fillet.

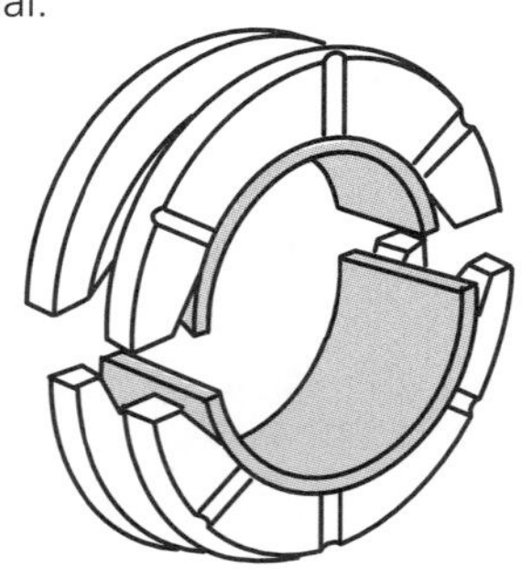

Figure 8-24 Multipiece thrust bearing.

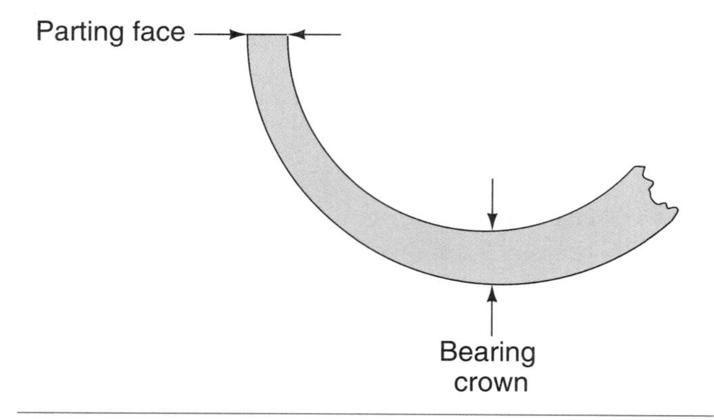


Figure 8-25 Bearing crown maintains close clearances at the top and bottom of the bearing where most of the loads are applied.

**Oil Clearance.** There must be a gap or clearance between the outside diameter of the crankshaft journals and the inside diameter of its bearings. This clearance allows for the building and maintenance of the oil film. During an engine rebuild, if there is little or no wear on the journals, the proper oil clearance can be restored with the installation of standard-size replacement bearings; however, if the crankshaft is worn to the point where the installation of standard-size bearings will result in excessive oil clearance space, a bearing with a thicker wall must be used.

**Undersize bearings** have the same outside diameter as standard bearings, but the bearing material is thicker to fit an undersize crankshaft journal.

**Bearing Sizing.** Undersize and **oversize bearings** are available to correct oil clearances if grinding of the journal or bore is required. **Undersize bearings** are used to maintain correct oil clearance when the crankshaft journal has been ground to provide a new journal surface. Undersize bearings are available in 0.001, 0.002, 0.009, 0.010, 0.020, and 0.030 inch sizes. Metric sizes are available in 0.050, 0.250, 0.500, and 0.750 mm. Oversize bearings are used if the bearing bore diameter has been increased. Available oversize bearings include 0.010, 0.020, 0.030, and 0.040 in. (0.250, 0.500, 0.750, and 1.000 mm).

**Oversize bearings** are thicker than standard to increase the outside diameter of the bearing to fit an oversize bearing bore. The inside diameter is the same as standard bearings.

**AUTHOR'S NOTE:** It has been my experience that many connecting rod and main bearing failures are caused by contaminated engine oil or lack of lubrication. When overhauling an engine with failed connecting rod and/or main bearings, always be sure the oil is not contaminated with antifreeze. This type of contamination causes bearing corrosion and erosion, and the engine oil can be analyzed for contaminants by a laboratory. Always be sure the engine lubrication system including the oil pump is in satisfactory condition, and inform the customer about the importance of oil and filter changes at the manufacturer's specified intervals.

## Camshaft and Balance Shaft Bearings

In an overhead valve (OHV) engine, the circular bearings are pressed into openings in the block (Figure 8-26). The camshaft journals are positioned in these bearings. The specified bearing clearance must be present between the camshaft journals and the bearings. Oil from the lubrication system is supplied through passages in the block to an opening in each camshaft bearing. The opening in the camshaft bearing must be properly aligned with the oil passage in the block. Excessive camshaft bearing clearance causes low oil pressure. During an engine rebuild, the camshaft bearings should be replaced. To make camshaft removal and replacement easier, the camshaft bearings on many engines are progressively larger from the back to the front of the block. Special tools are available to remove and replace the camshaft bearings. Balance shaft bearings are similar to camshaft bearings.

**Shop Manual**
Chapter 8, page 387

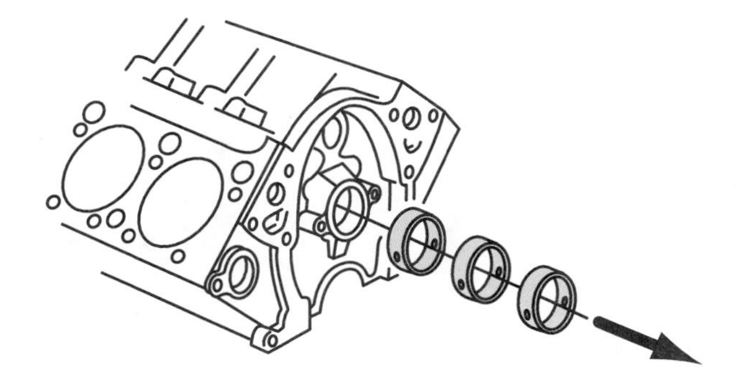

Figure 8-26 Camshaft bearings in the cylinder block.

## A BIT OF HISTORY

Isaac Babbitt invented his bearing material in 1839 using a mixture of tin, lead, and antimony. This tin-based babbitt was later replaced with a lead-based material. When lead-based babbitt was used in the first automotive engines, it was melted and then poured into the block and connecting rods. After the babbitt cooled, the correct oil clearance was achieved by hand scraping the excess material. Final adjustment of the bearings was accomplished by using shims between the bearing caps. Due to the nature of these early bearings, they often smeared under heavy loads. To decrease the oil clearance as the bearing wore, the caps were refilled. This operation was considered owner maintenance, and babbitt material was carried in the vehicle to make the necessary roadside repairs.

# Balance Shafts

The **balance shaft** is a shaft driven by the crankshaft that is designed to reduce engine vibrations.

Many engine designs tend to have inherent vibrations. The in-line 4-cylinder engine is an example of this. Some engine manufacturers use a balance shaft to counteract this tendency (Figure 8-27). Four-cylinder engine vibration occurs when the piston is moving down during its power stroke. When the engine completes one revolution, two vibrations occur. The **balance shaft** rotates at twice the speed of the crankshaft. This creates a force that counteracts crankshaft vibrations.

**Shop Manual**
Chapter 8, page 390

# Pistons

The **piston,** when assembled to the connecting rod, is designed to transmit the power produced in the combustion chamber to the crankshaft.

The **piston,** when assembled to the connecting rod, is designed to transmit the power produced in the combustion chamber to the crankshaft (Figure 8-28). The piston must be able to withstand severe operating conditions. Stress and expansion problems are compounded by the extreme

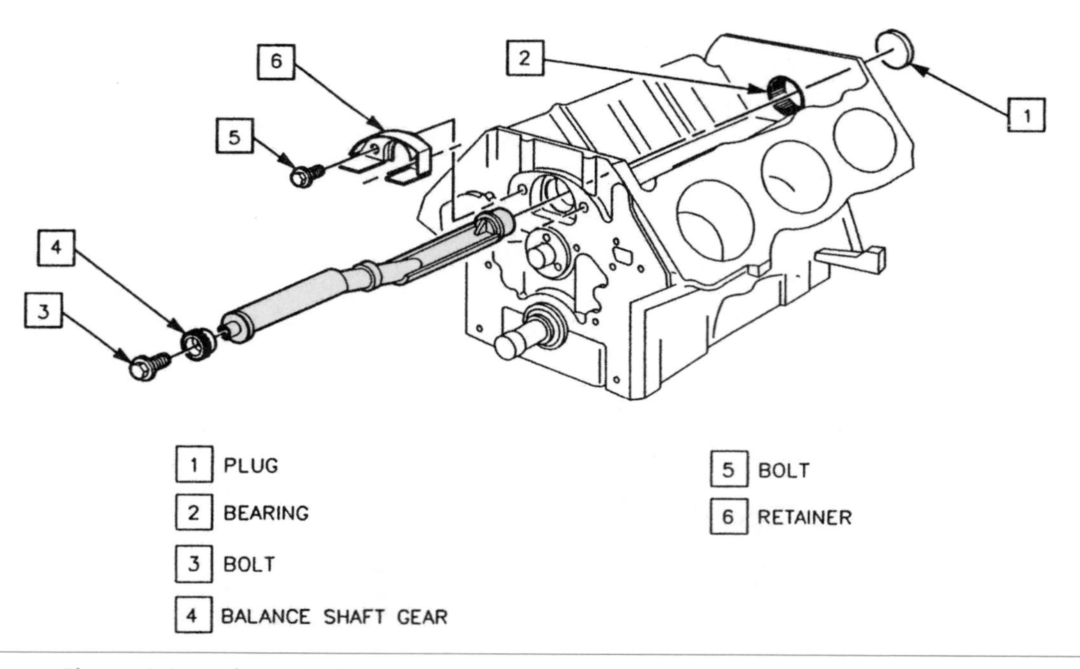


**Figure 8-27** Balance shaft.

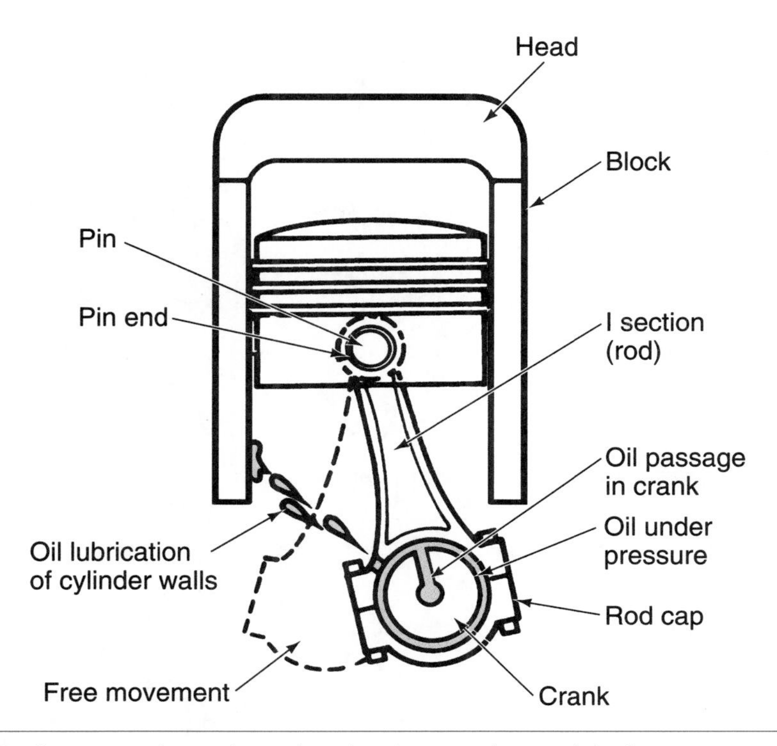


Figure 8-28 The connecting rod attaches the piston to the crankshaft.

temperatures to which the top of the piston is exposed. In addition, the rapid movement of the piston creates stress and high pressures. To control these stress conditions, most pistons are constructed from aluminum. The lightweight aluminum operates very efficiently in high rpm engines.

The piston has several parts, including:

1. *Land*—Used to confine and support the piston rings in their grooves.
2. *Heat dam*—Used on some pistons to reduce the amount of heat flow to the top ring groove. A narrow groove is cut between the top land and the top of the piston. As the engine is run, carbon fills the groove to dam the transfer of heat.
3. *Piston head or crown*—The top of the piston against which the combustion gases push. Different head shapes are used to achieve the manufacturers' desired results (Figure 8-29). Most engines use flat-top piston heads. If the piston comes close to the valves, the recessed piston head is used to provide additional clearance. Different types of domed and wedged piston heads are used to increase compression ratios.
4. *Piston pins*—Connect the piston to the connecting rod. Three basic designs are used: piston pin anchored to the piston and floating in the connecting rod; piston pin anchored to the connecting rod and floating in the piston; and piston pin full floating in the piston and connecting rod.
5. *Skirt*—The area between the bottom of the piston and the lower ring groove and at a 90-degree angle to the piston pin. The skirt forms a bearing area in contact with the cylinder wall. To reduce the weight of the piston and connecting rod assembly, many manufacturers use a slipper skirt (Figure 8-30). The piston skirt surface is etched to allow it to trap oil (Figure 8-31). This helps lubricate the piston skirt as it moves within the cylinder and prevents scuffing of the skirt.
6. *Thrust face*—The portion of the piston skirt that carries the thrust load of the piston against the cylinder wall.
7. *Compression ring grooves*—The upper ring grooves used to hold the compression rings.
8. *Oil ring groove*—The bottom ring groove used to hold the oil ring.

Slipper skirts allow shorter connecting rods to be used. Part of the skirt is removed to provide additional clearance between the piston and the counterweights of the crankshaft. Without this recessed area, the piston would contact the crankshaft when the shorter connecting rods are used.

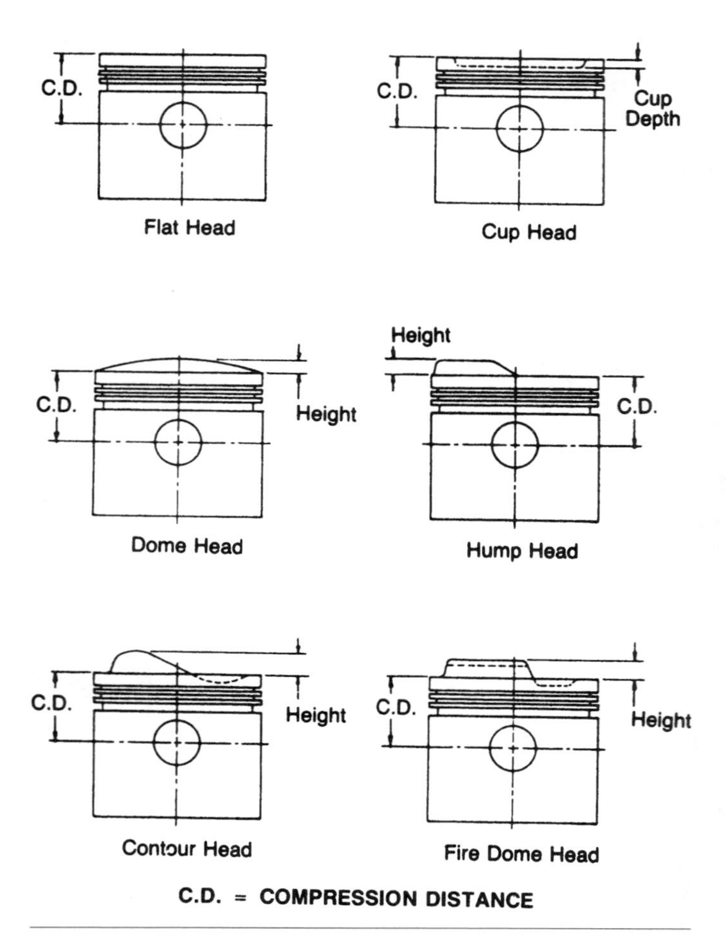


Figure 8-29 Some of the common piston head designs used in automotive engines. (Courtesy of TRW, Incorporated)

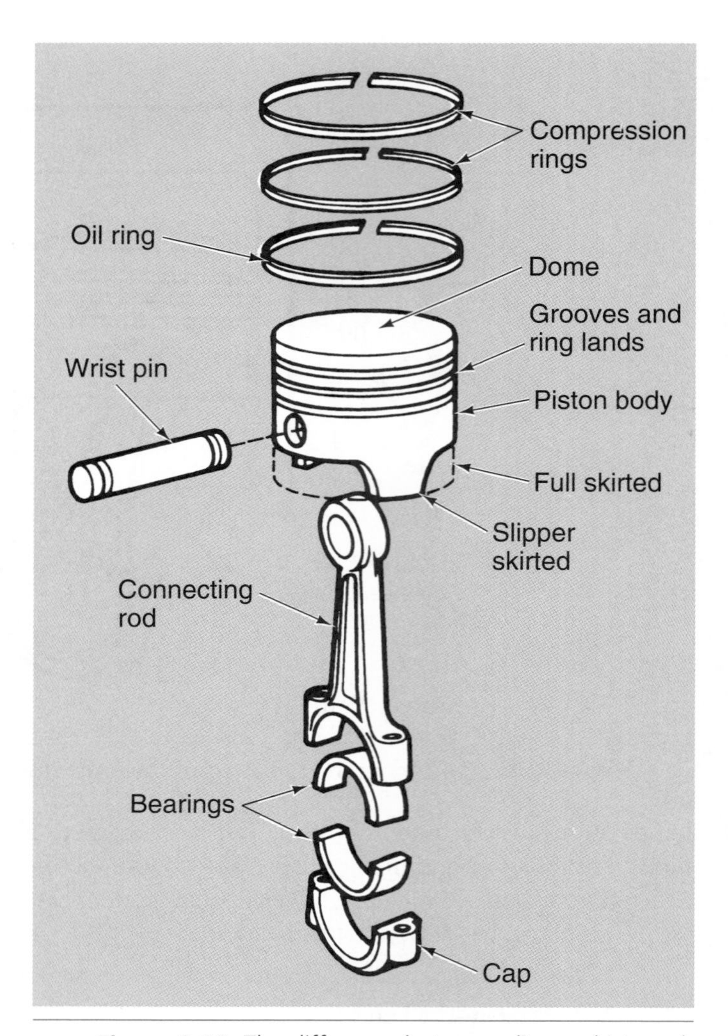


Figure 8-30 The difference between slipper skirts and full skirts. The dotted line indicates where the full skirt would be.

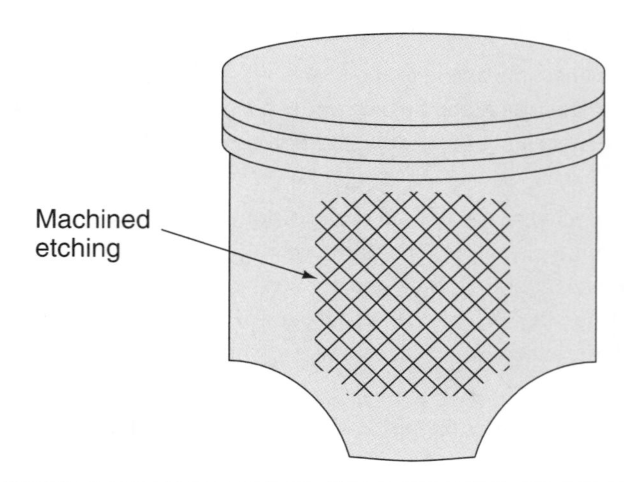


Figure 8-31 Most piston skirts have an etching machined into them to help trap oil to lubricate the skirt. Scuffing prevents oil from being trapped, resulting in increased temperatures and wear.

## Heat Control

When combustion occurs, high temperatures and pressures are applied to the top of the piston. Some of this heat is transmitted to the piston body, causing the piston to expand. To control piston expansion, most pistons are cam ground. **Cam ground pistons** are made to be an oval shape when the piston is cold. The larger diameter of the piston is across the thrust surfaces. The thrust surfaces are perpendicular to the piston pin (Figure 8-32). On a cold piston, the thrust surfaces have a close fit to the cylinder wall while the surface around the piston pin has greater clearance. As the piston warms, it will expand along the piston pin. As it expands, the shape of the piston becomes more round. Since the lower portion of the piston skirt is not subjected to as much heat as the top, the piston skirt is tapered. The diameter at the bottom is usually larger than the top of the piston. As the piston warms, the upper part will expand and the skirt will become straight.

Thermal expansion of aluminum is about twice as much as iron.

**Cam ground pistons** are oval or cam shaped to allow for expansion. As the piston warms, it will become round.

Many manufacturers design their cast-aluminum pistons with steel struts located in the skirt area (Figure 8-33). The strut works to control the amount and rate of expansion. The struts are cast into the piston during manufacturing. Since steel and aluminum expand at different rates for

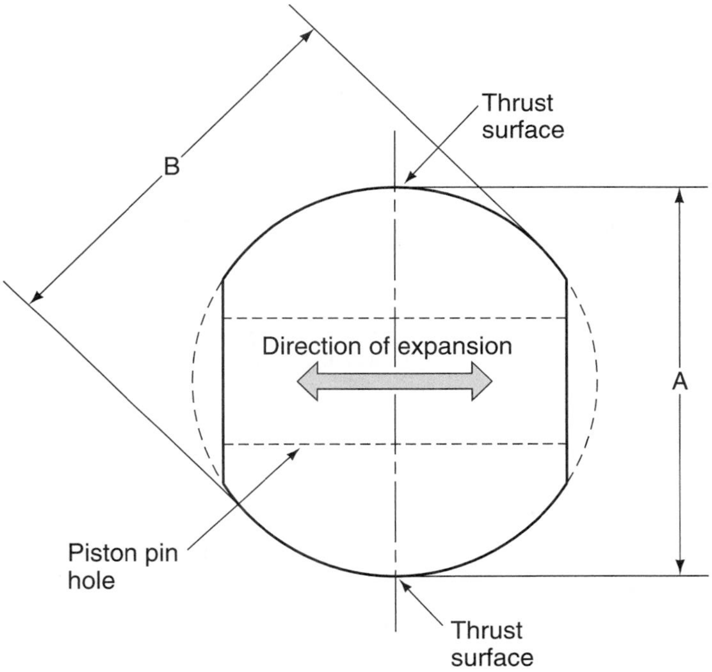

Figure 8-32 Cam ground pistons have the largest diameter across the thrust surfaces (A).

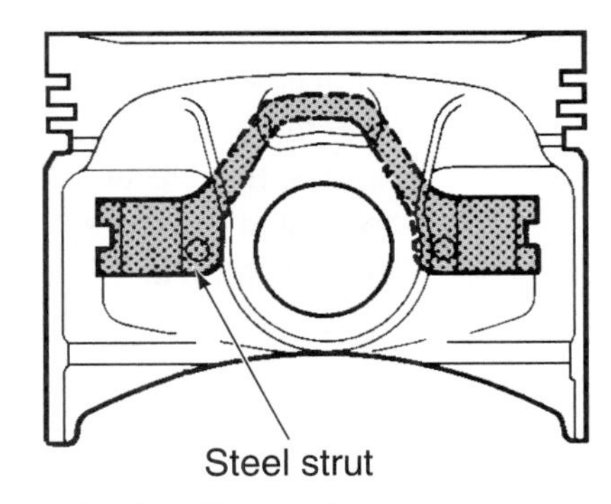


Figure 8-33 Some manufacturers use a steel strut to help control expansion and add strength to the piston. (Courtesy of Hyundai Motor America)

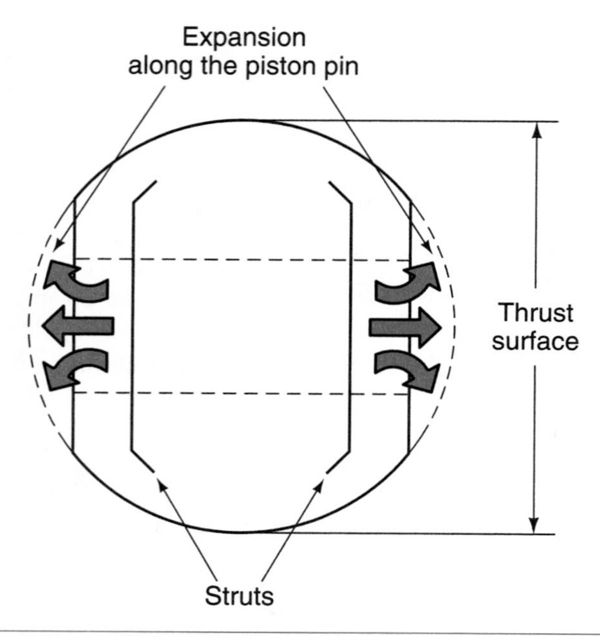


Figure 8-34 Steel struts in the piston keep the piston from expanding across the thrust surfaces.

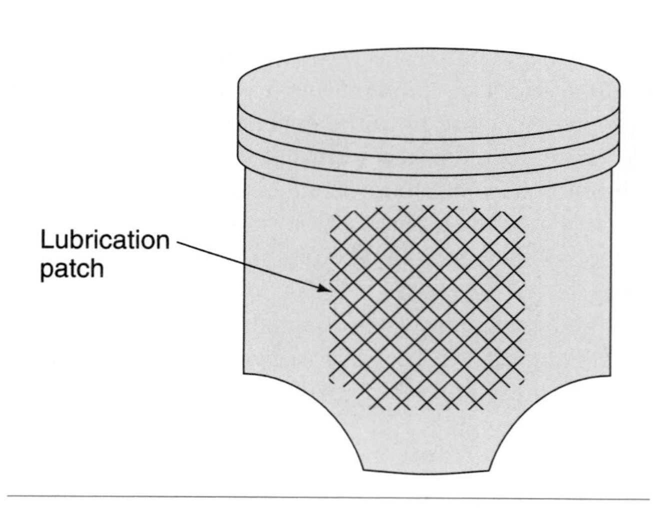


Figure 8-35 The piston may have a lubrication patch to prevent scuffing and piston slap.

Piston offset is used to provide more effective downward force onto the crankshaft by increasing the leverage applied to the crankshaft.

The skirt that pushes against the cylinder wall during the power stroke is the **major thrust surface.** The skirt that pushes against the cylinder wall during the compression stroke is the **minor thrust surface.**

An **offset pin** is a piston pin that is not mounted on the vertical center of the piston.

The pin boss accepts the piston pin to attach the piston to the connecting rod.

the same temperature, the struts keep the skirt from expanding as much (Figure 8-34). This allows the manufacturer to use tighter clearances between the piston and cylinder to prevent the piston from rocking in the cylinder and generating noise.

In addition, manufacturers may use a patch material on the outside of the skirt to improve lubrication so tighter clearances between the piston and cylinder are possible (Figure 8-35).

## Piston Pin Offset

Most manufacturers offset the piston pin to reduce piston slap (Figure 8-36). The connecting rod is angled to different sides on the compression and power strokes, causing the piston to rock from one skirt to the other at TDC. During the compression stroke, not as much force is exerted on the piston skirt as during the power stroke. Offsetting the piston pin hole toward the major thrust surface about 0.062 in. (1.57 mm) reduces the tendency of the piston to slap the cylinder wall as it rocks from the minor to the **major thrust surface.** During the compression stroke, the connecting rod pushes the minor thrust surface against the cylinder wall (Figure 8-37). The **offset pin** causes more of the combustion pressure to be exerted on the larger half of the piston head (Figure 8-38). This uneven pressure application causes the piston to tilt so the top of the piston contacts the cylinder wall on the **minor thrust surface** and the bottom of the piston contacts the cylinder wall on the major thrust surface. When the piston begins its downward movement, its upper half slides into contact with the major thrust surface (Figure 8-39).

To assist in the installation of the piston, most manufacturers have a groove or other marking to indicate the side of the piston that faces the front of the block. On V-type engines, the pin offset for each bank is on opposite sides. This is because crankshaft rotation determines the major thrust surface side. If the pistons are installed in the wrong direction, the engine will not rotate.

## Piston Speed

As the piston moves up and down in the cylinder, it is constantly changing speeds. At top and bottom dead center, the speed of the piston is zero. It then immediately accelerates to maximum speed. This action places heavy loads on the piston pin and pin boss.

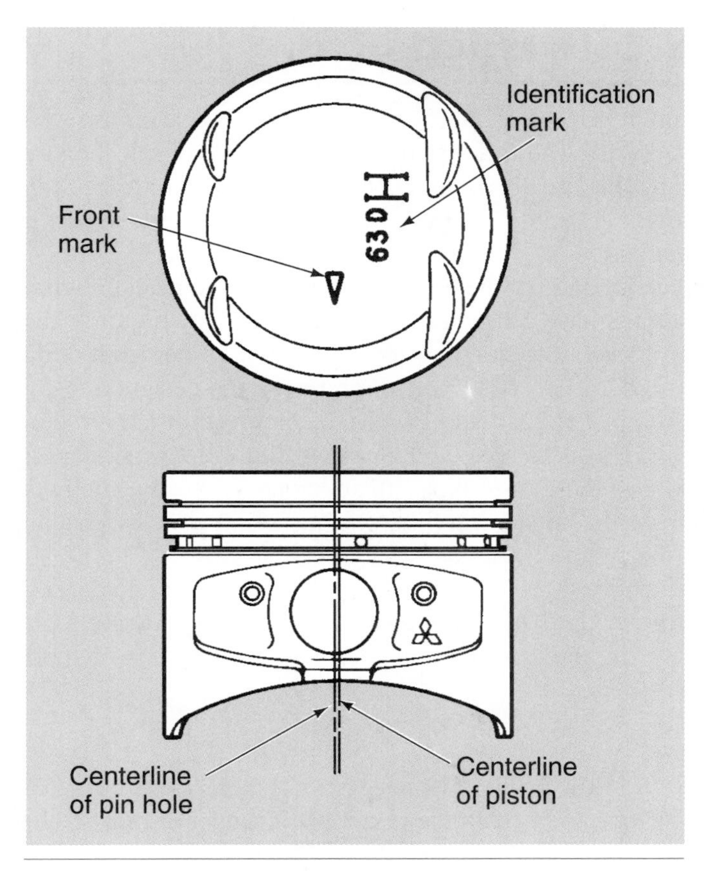


Figure 8-36 Some pistons are slightly offset on the piston pin. (Courtesy of Hyundai Motor America)

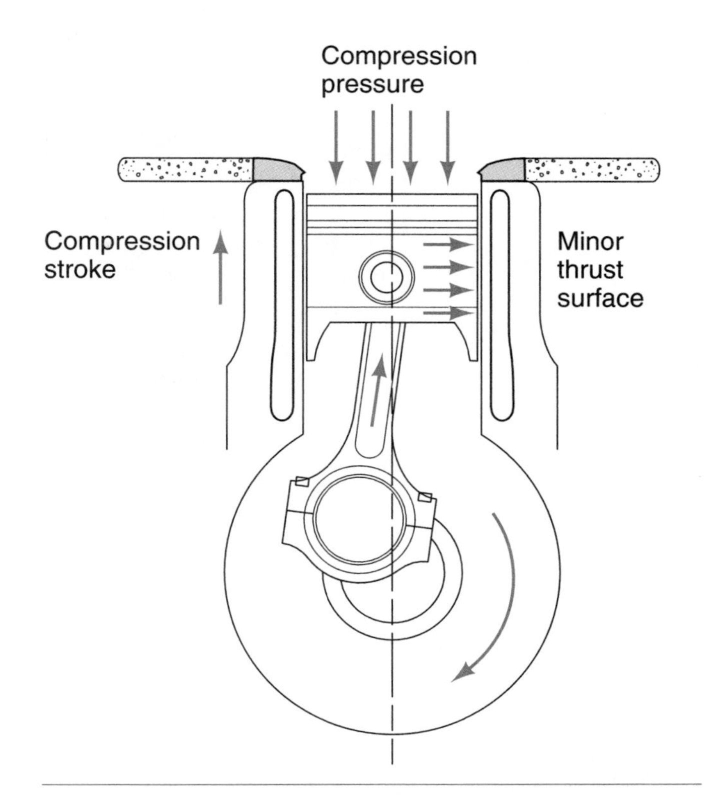


Figure 8-37 During the compression stroke, the connecting rod forces the piston against the cylinder wall.

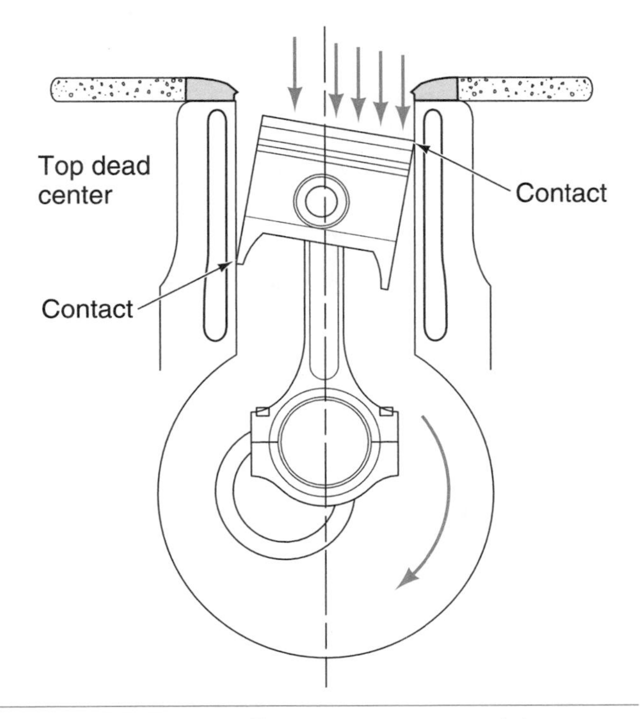


Figure 8-38 The offset pin causes more of the combustion pressure to be exerted on the larger half of the piston head, causing the piston to tilt. Illustration is exaggerated for clarity.

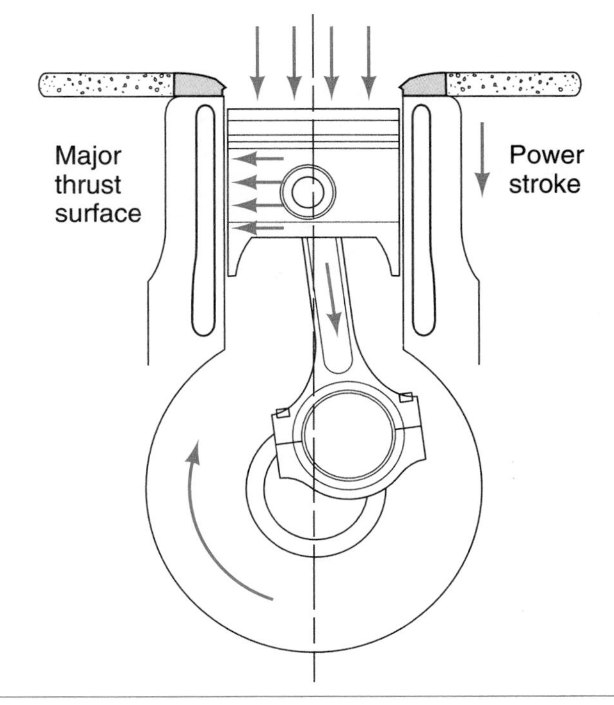


Figure 8-39 When the piston begins its downward movement, its upper half slides into contact with the major thrust surface.

# Piston Designs and Construction

If it is determined that the pistons require replacement, there are several piston designs available from manufacturers and aftermarket suppliers. The technician must be capable of selecting the correct design for the engine application. This usually means replacing the piston with the same design as removed; however, if the engine is to be modified, piston selection is one aspect in increasing engine performance and durability.

It is possible to change the compression ratio of the engine by changing the head design of the piston. Most original equipment pistons have a flat head. These pistons may require valve reliefs machined into the head to prevent valve-to-piston contact (Figure 8-40). The depth of the relief is determined by camshaft timing, duration, and lift. A dished piston is designed to put most of the combustion chamber into the piston head and lower the compression ratio (Figure 8-41). A domed piston will fill the combustion area and increase the compression ratio (Figure 8-42). Other designs are variations of these three types. In addition to changing compression ratio, piston head design can increase the efficiency of the combustion process by creating a turbulence to improve the mixing of the air and fuel.

Another consideration when replacing pistons is construction type. Weight and expansion rates of the piston are important aspects in today's high rpm engines. The use of aluminum pistons reduces the weight of the assembly and the load on the crankshaft. Construction methods

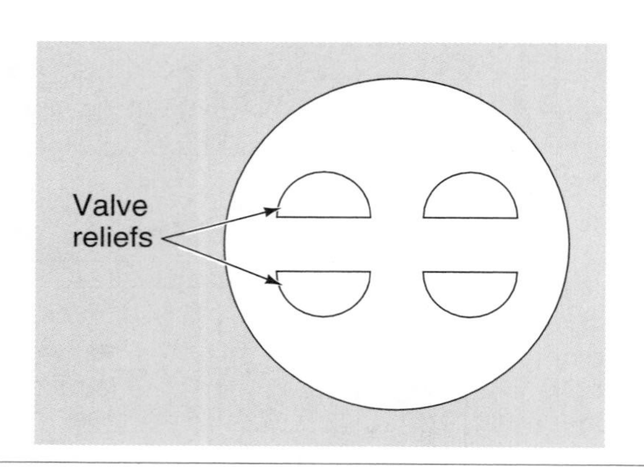


**Figure 8-40** Flat piston head with recessed valve reliefs.

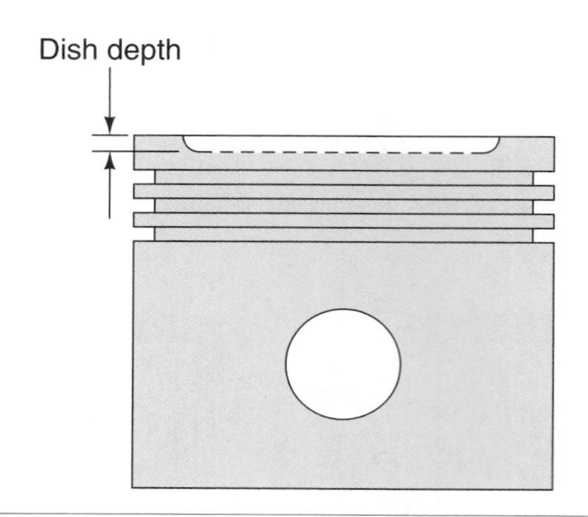


**Figure 8-41** Dished piston design.

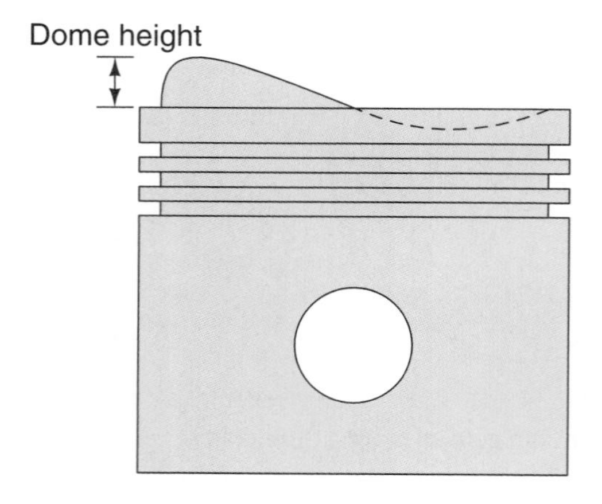


**Figure 8-42** Domed piston design.

also affect the piston's expansion rate and amount. Most original equipment pistons are made of cast aluminum. These are very capable of handing the requirements of most original equipment engines. The advantage of cast pistons is their low expansion rate. The expansion rate is low because the grain structure of cast is not as dense as other manufacturing processes. This low expansion rate allows the engine manufacturer to tighten the piston clearance tolerances, helping to prevent piston slap and rattle when the engine is cold.

Forged pistons are stronger due to their dense grain structure (Figure 8-43). The tighter grain structure allows the piston to run about 20 percent cooler than a cast piston (Figure 8-44). Depending on the alloy used in the aluminum, the expansion rate may be greater than cast. This requires additional piston clearance and may result in cold engine piston rattle. Alloys have been developed in recent years that reduce the expansion rate of forged pistons. Consequently, it is important to follow the piston manufacturer's specifications for piston clearance.

A third alternative is hypereutectic cast pistons. Most cast pistons are constructed with about 9.5 percent silicon content, while forged pistons have between 0.10 and 10 percent, depending on the aluminum used. Silicon is added to help provide lubrication and increase strength.

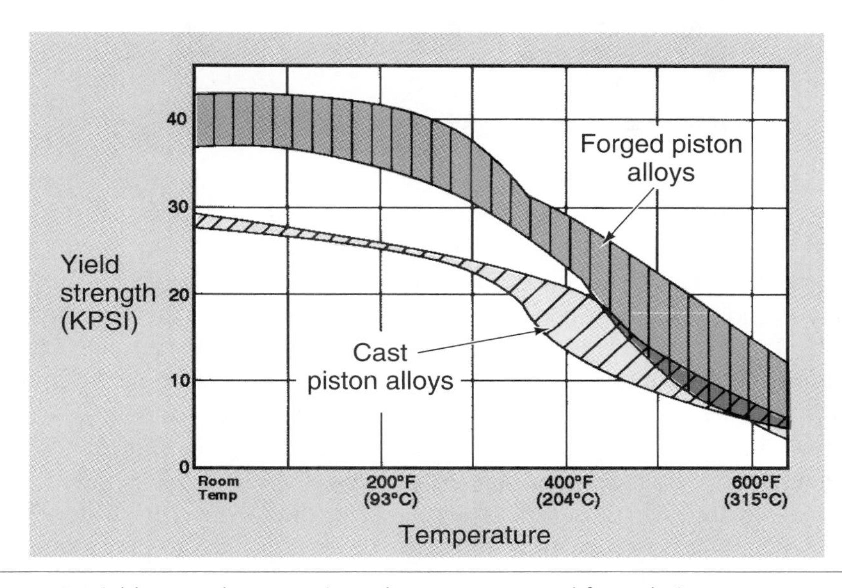


Figure 8-43 Yield strength comparisons between cast and forged pistons. (Courtesy of TRW, Incorporated)

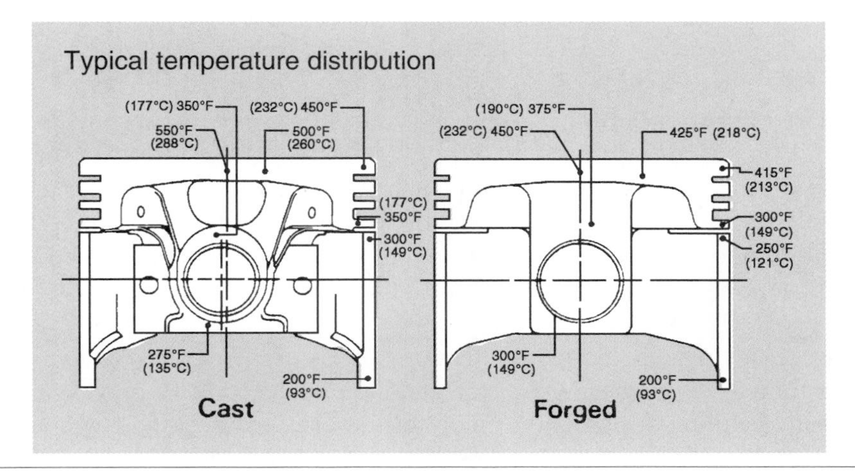


Figure 8-44 Temperature comparisons between cast and forged pistons. (Courtesy of TRW, Incorporated)

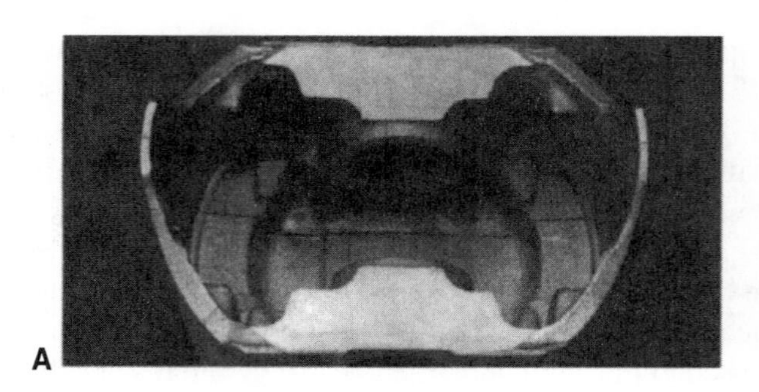


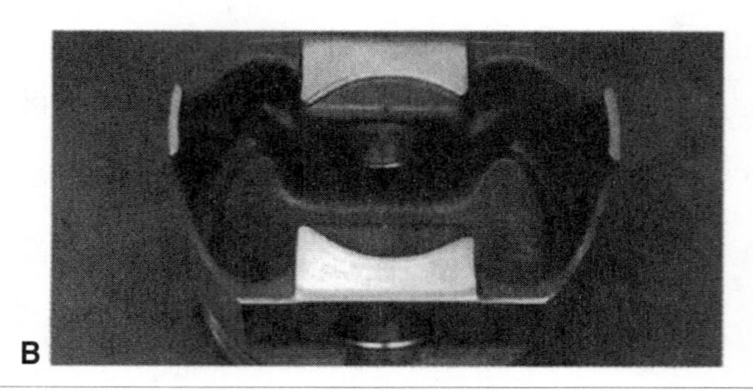


Figure 8-45 (A) Cast piston will have mold dividing lines, while the forged piston (B) will have the appearance of flow lines. (Courtesy of TRW, Incorporated)

Hypereutectic pistons contain between 16 and 22 percent silicon. Aluminum can only dissolve about 12 percent silicon and hold it in suspension. The silicon is added when the aluminum is molten and dissolves. At the 12 percent level, the silicon remains dissolved when the piston cools. Normally, silicon added above this saturation point will not dissolve, and settles at the bottom of the mold. Hypereutectic casting keeps the undissolved silicon dispersed throughout the piston by closely monitoring and controlling the heating and cooling rates during the manufacturing process.

Hypereutectic casting increases the temperature fatigue resistance of the piston. This allows the piston to be made thinner than most cast pistons, reducing the weight of the piston. The strength of hypereutectic pistons falls between cast and forged. The drawback to this piston construction is that silicon can reduce the piston's ability to conduct heat away from the combustion chamber. Although its temperature fatigue is increased, the piston runs hotter than cast and may cause detonation.

It is possible to tell if the piston is forged or cast by looking at the underside of the piston head. Cast pistons have mold dividing lines, while forged pistons do not have these parting lines (Figure 8-45).

**Shop Manual**
Chapter 8, page 401

Blowby refers to compression pressures that escape past the piston.

Compression rings form a seal between the piston and the cylinder wall.

## Piston Rings

For the engine to produce maximum power, compression and expansion gases must be sealed in the combustion chamber. To control blowby, the piston is fitted with compression rings. The rings seal the piston to the cylinder wall, sealing in combustion pressures. In addition, oil rings are used to prevent oil from reaching the top of the piston. Oil in this area will result in blue smoke being expelled from the exhaust system.

**Compression rings** are located on the upper lands of the piston (Figure 8-46). The compression rings form a seal between the piston and the cylinder wall. To provide a more positive seal, many manufacturers design one or more of the compression rings with a taper (Figure 8-47). The taper provides a sharp, positive seal to the cylinder wall. This style of ring must be installed in the correct position. A dot or other marking is usually provided to indicate the top of the ring.

In addition to tapering or chamfering the ring, the ring is designed to use combustion pressures to force the ring tighter against the cylinder wall and the bottom of the ring groove (Figure 8-48). The top ring provides primary sealing of the combustion pressures. The second ring is used to control the small amount of pressures that may escape past the top ring.

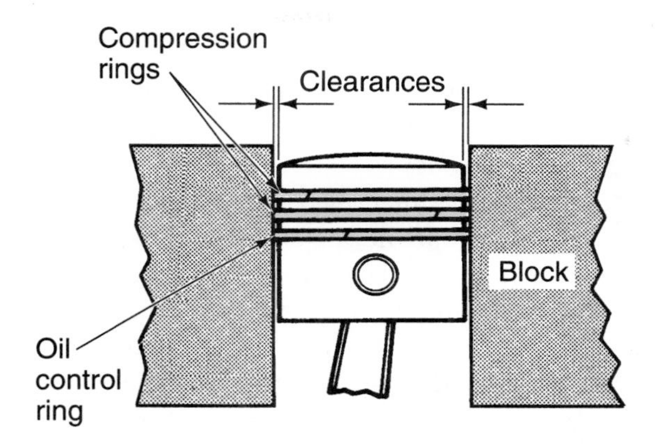


Figure 8-46 The top piston grooves are used to hold the compression rings. The rings fit against the cylinder wall. The piston should not touch the wall.

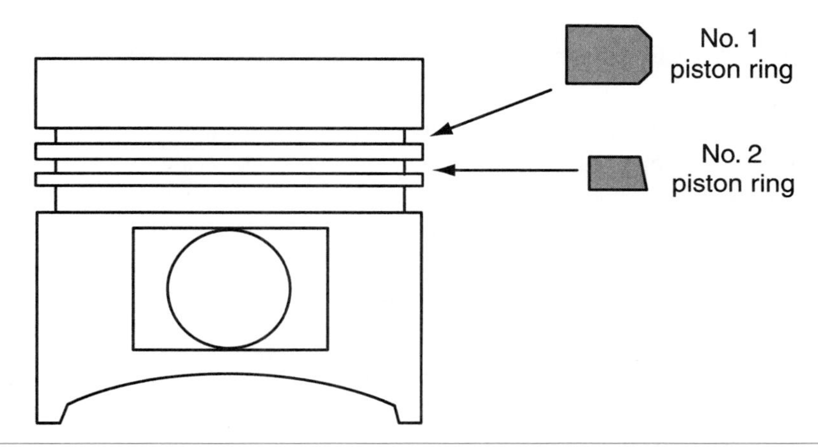


Figure 8-47 Some compression rings are designed with a chamfer or taper. These types of rings may be directional.

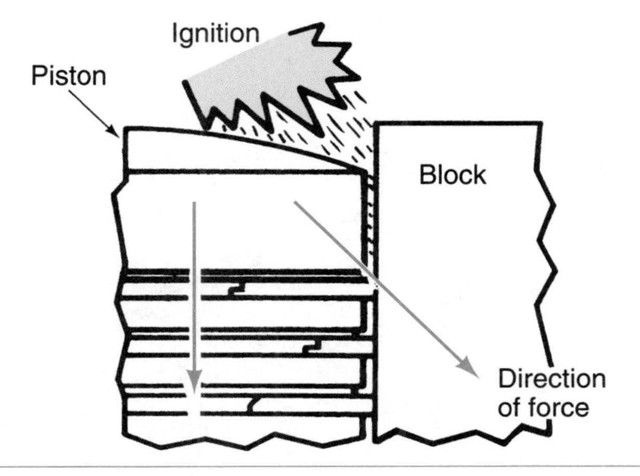


Figure 8-48 The rings are designed to use combustion pressures to exert additional outward forces of the rings against the cylinder wall.

Ring seating is accomplished by the movement of the piston in the cylinder. If the cylinder is properly conditioned, this movement laps the rings against the wall.

Wear-in is the required time needed for the ring to conform to the shape of the cylinder bore.

**Oil control rings** are piston rings that control the amount of oil on the cylinder walls.

The most common materials for constructing compression rings are steel and cast iron. The face of the ring is usually coated to aid in the wear-in process. This coating can be graphite, phosphate, iron oxide, molybdenum, or chromium. The bottom piston ring groove houses the oil control ring. There are two common styles of oil control rings: the cast-iron oil ring and the segmented oil ring. Both styles of oil rings have slots to allow the oil from the cylinder walls to pass through. The ring groove also has slots to provide a passage back to the oil sump.

Many engine manufacturers now install low-tension piston rings. These rings provide adequate combustion chamber sealing with reduced friction on the cylinder walls as the piston moves up and down. Since low-tension rings provide a reduced internal engine friction, they also provide a slight improvement in fuel economy.

Segmented **oil control rings** have three pieces (Figure 8-49). The upper and lower portions are thin metal rails that scrape the oil. The center section of the ring is constructed of spring steel, and is called the expander or spacer. The movement of the piston causes the oil ring to scrape oil from the cylinder wall and return it to the sump. The rings are usually replaced whenever an engine is rebuilt. Once again, the original ring design and construction should be selected if the engine is going to be used under normal conditions. There is a wide selection of aftermarket rings available for special-purpose engines. Some concerns to be aware of when selecting replacement rings include:

1. Molybdenum-coated cast rings are popular original equipment rings because of their resistance to scuffing. They seal quickly, provided the cylinder walls are properly finished.
2. Phosphate, iron oxide, and graphite are commonly used coatings on gray cast-iron rings. These rings help reduce scuffing during the break-in period and seat quickly. The drawback to this type of ring is its short service life.
3. Chromium-coated cast rings have a long service life, but they require a proper cylinder wall finish to seat the rings. In addition, chromium-coated rings are hard and may cause wear of the cylinder walls unless the block is cast with additional amounts of nickel to increase its strength and resistance to abrasives.
4. Ceramic-coated cast rings offer high resistance to scuffing and damage due to detonation. The disadvantage to these rings is that they create a heat dam and do not conduct heat away from the piston as well as other coatings.
5. Most steel rings have chromium or titanium nitride coatings. These offer long service life but require careful cylinder wall finishing to seat the rings. They may cause wear of cylinder walls on blocks made from softer alloys.

## Oversize Rings

If the cylinder bore diameter is increase to correct for wear, the piston and rings must be replaced. The piston and rings are oversize to fit the new bore diameter. Piston rings are avail-

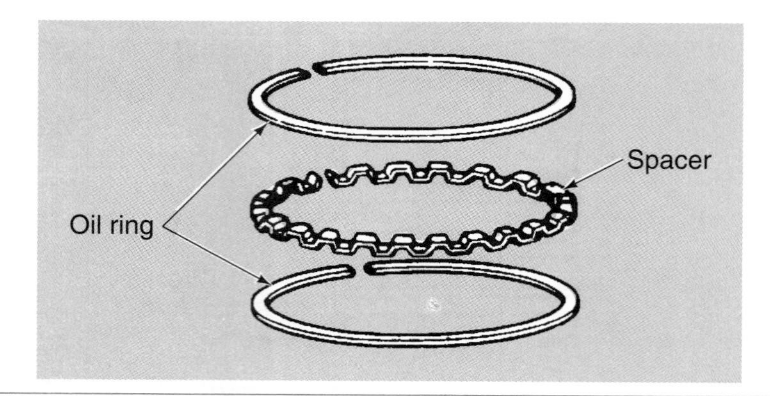


**Figure 8-49** Typical three-piece segmented oil control ring. (Courtesy of American Honda Motor Co., Inc.)

able in standard, 0.020, 0.030, 0.040, and 0.060 inch oversizes. If an odd size is needed, such as 0.010 or 0.050 inch oversize, use the ring one size smaller; for example, for 0.010 inch oversize, use a standard ring; for 0.050 inch oversize, use a 0.040 inch oversize. Metric oversizes are available in 0.50 mm, 0.75 mm, 1.0 mm, and 1.5 mm.

## Connecting Rods

**Shop Manual**
Chapter 8, page 395

**Connecting rods** are the link between the piston and the crankshaft.

Big end refers to the end of the connecting rod that attaches to the crankshaft. Small end refers to the end of the connecting rod that accepts the piston pin.

The **connecting rod** transmits the force of the combustion pressures applied to the top of the piston to the crankshaft (Figure 8-50). To provide the strength required to withstand these forces and still be light enough to prevent rough running conditions, the connecting rod is constructed from one of the following methods:

1. Forged from high-strength steel
2. Made of nodular steel
3. Made from cast iron
4. Made of sintered powdered metal

To increase its strength, the connecting rod is constructed in the form of an I (Figure 8-51). The small bore at the upper end of the rod is fitted to the piston pin. The piston pin can be pressed into the piston and a free fit in the connecting rod. In this type of mounting, a bushing is installed in the small rod bore. The piston pin can also be pressed into the small rod bore and a free fit in the piston. The piston moves on the pin surface without the use of bushings.

The larger bore connects the rod to the crankshaft journal. The large bore is a two-piece assembly. The lower half of this assembly is called the rod cap. The rod cap and connecting rod are constructed as a unit and must remain as a matched set. During cylinder block disassembly, the connecting rods and caps should be marked to assure proper match when installed. Most rod caps are machined from the connecting rods during the manufacturing process. Some manufacturers now use a new process of "breaking" the cap off of the connecting rod. This process uses

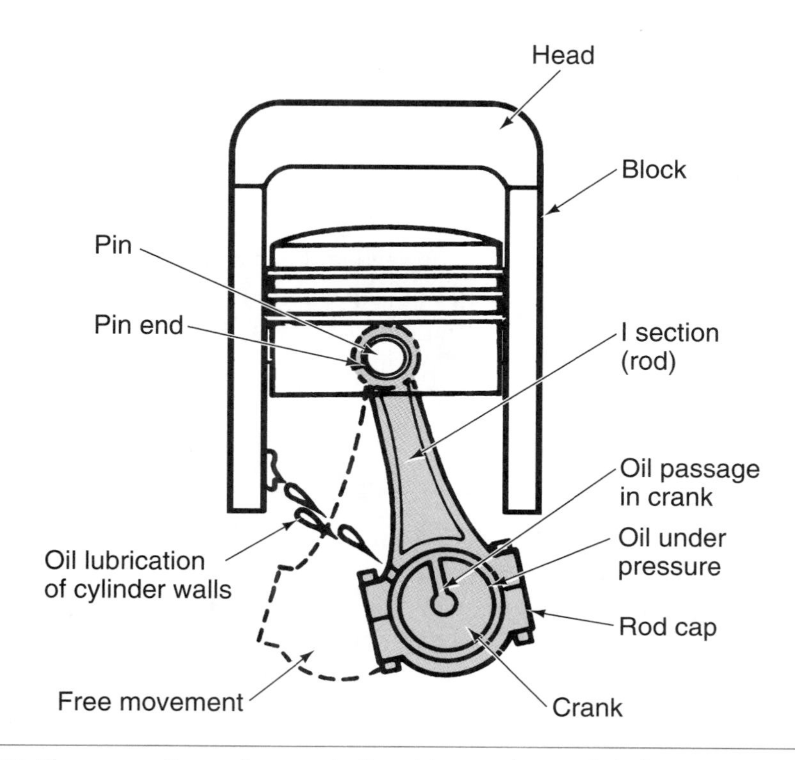


Figure 8-50 The connecting rod connects the piston to the crankshaft.

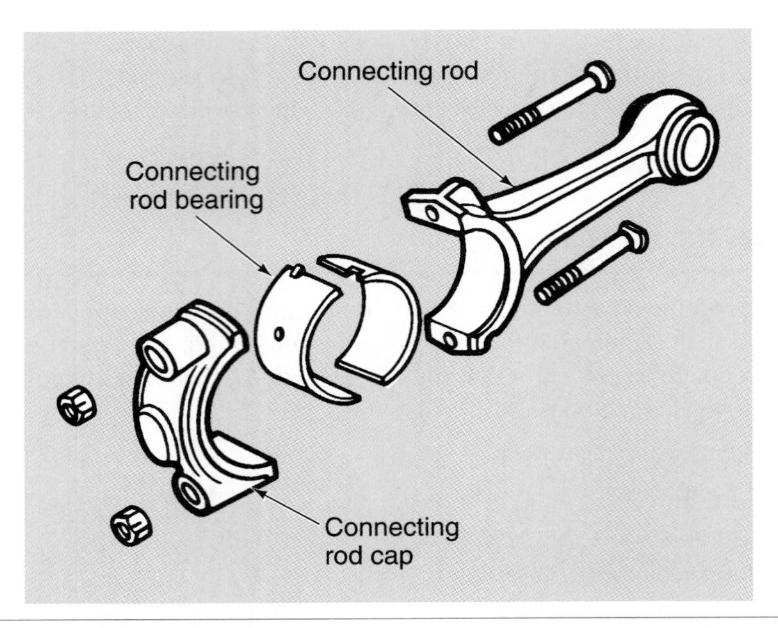


Figure 8-51 The I-beam construction adds strength to the connecting rod.

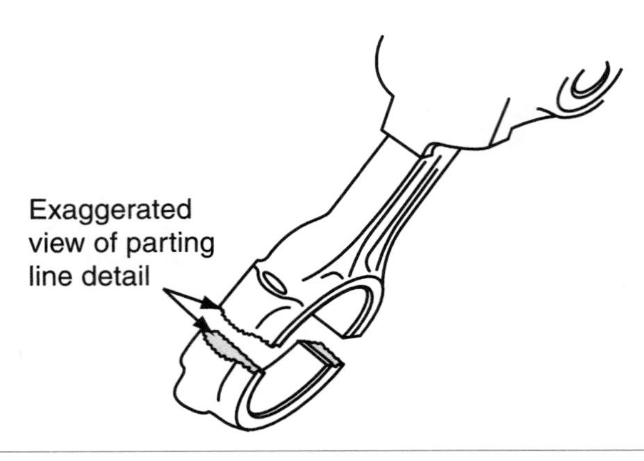


Figure 8-52 Fracturing, instead of machining, the connecting rod caps provides a perfect fit between cap and rod.

**Shop Manual**
Chapter 8, page 397

connecting rods constructed from sintered powdered metal. The connecting rod is then shot-peened to remove any flash and to increase surface hardness. After the rod bolt holes are drilled and tapped, the bolts are installed loosely. Next, the break area is laser scribed and the cap is fractured in a special fixture. This creates a rod and cap parting surface that is a perfect fit (Figure 8-52).

## Summary

Terms to Know

**Align bored**
**Balance shaft**
**Bearing crush**
**Bearing spread**
**Bedplate**
**Cam ground pistons**

- The cylinder block is the main structure of the engine. Most of the other engine components are attached to the block.
- Many aluminum and some cast-iron blocks use liners in their cylinders. Dry sleeves are thin-walled liners that are pressed into the block, while wet sleeves are thicker and are held in place by supporting flanges in the block. Wet sleeves are surrounded by coolant that is kept out of the crankcase by O-rings.

- ❑ The crankshaft converts the reciprocating movement of the pistons into rotary motion.
- ❑ Main bearing journals are machined along the centerline of the crankshaft and support the weight and forces applied to the crankshaft.
- ❑ Offset from the crankshaft centerline are the connecting rod bearing journals. The offset places weight and pressure off the center of the crankshaft and provides leverage for the piston assembly to turn the crankshaft.
- ❑ The fillet on the journals is used to increase the strength of the crankshaft.
- ❑ Bearings are used to carry the loads created by rotational forces. Main bearings carry the load created by the movement of the crankshaft. Connecting rod bearings carry the load of the power transfer from the connecting rod to the crankshaft.
- ❑ Oil is supplied to the rod bearing journals through holes in the crankshaft.
- ❑ Undersize bearings are used to maintain correct oil clearance when the crankshaft journal has been ground to provide a new journal surface.
- ❑ Oversize bearings are used if the bearing bore diameter has been increased.
- ❑ The piston, when assembled to the connecting rod, is designed to transmit the power produced in the combustion chamber to the crankshaft.
- ❑ To seal combustion pressures, the piston is fitted with compression rings. The rings seal the piston to the cylinder wall.

**Terms to Know**

Compression rings
Connecting rod
Connecting rod bearing journals
Connecting rod bearings
Core plugs
Counterweights
Crankshaft throw
Cylinder block
Fillet
Flywheel
Harmonic balancers
Locating tab
Long block
Main bearing journals
Major thrust surface
Minor thrust surface
Offset pin
Oil control rings
Oversize bearings
Piston
Short block
Thrust bearing
Torsional vibration
Undersize bearings

# Review Questions

## Short Answer Essays

1. List the purpose of the following major engine components:
   - **A.** Crankshaft
   - **B.** Engine block
   - **C.** Piston
   - **D.** Connecting rods
2. Explain the types of forces applied to the crankshaft.
3. List the parts of the piston and their purpose.
4. Explain the term align boring as it relates to the main bearing bores.
5. Explain the purpose of core plugs during the block manufacturing process.
6. Describe the purpose of the fillet on crankshaft journals.
7. Explain bearing crush and bearing spread, including the importance of these features.
8. Explain why the piston pins are offset in the pistons.
9. Describe cam-ground piston operation during engine warmup.
10. Describe the design and purpose of compression rings.

## Fill-in-the-Blanks

1. The __________ __________ is the main structure of the engine. Most of the other engine components are attached to it.

2. The __________ converts the reciprocating movement of the pistons into rotary motion.

3. The __________ __________ transmits the force of the combustion pressures applied to the top of the piston to the crankshaft.

4. __________ are used to carry the loads created by rotational forces. __________ __________ carry the load created by the horizontal movement of the crankshaft.

5. __________ bearings are used to maintain correct oil clearance when the crankshaft journal has been ground to provide a new journal surface. __________ bearings are used if the bearing bore diameter has been increased.

6. A __________ __________ assures proper installation of the bearing insert in the bore.

7. A harmonic balancer is installed on the front of the crankshaft to reduce __________ __________ .

8. The flywheel acts as an inertia ring to keep the crankshaft turning through __________ __________ .

9. Steel struts help to control piston __________ .

10. Some connecting rod caps are __________ off the connecting rod during the manufacturing process to provide an improved cap to rod fit.

## Multiple Choice

1. Crankshaft bearings are being discussed.
*Technician A* says undersize bearings are used if the journal diameter is increased.
*Technician B* says oversize bearings are used when the journals have been machined to a smaller diameter.
Who is correct?
**A.** A only
**B.** B only
**C.** Both A and B
**D.** Neither A nor B

2. *Technician A* says an engine bedplate contains the main bearing caps.
*Technician B* says an engine bedplate reduces engine vibrations.
Who is correct?
**A.** A only
**B.** B only
**C.** Both A and B
**D.** Neither A nor B

3. All of these statements about aluminum engine blocks are true EXCEPT:
**A.** Aluminum blocks are lighter than cast-iron blocks.
**B.** Aluminum blocks have greater strength than cast-iron blocks.
**C.** Aluminum blocks usually have steel cylinder liners.
**D.** Certain materials are usually added to the aluminum to provide increased block strength.

4. *Technician A* says crankshaft throw is the distance from the center of the main bearing journals to the center of the connecting rod journals.
*Technician B* says if the crankshaft throw is increased, it does not change the piston stroke.
Who is correct?
**A.** A only
**B.** B only
**C.** Both A and B
**D.** Neither A nor B

5. To make a 90-degree V6 engine even firing, the connecting rods on each crankshaft throw must be offset:
   **A.** 10 degrees.
   **B.** 30 degrees.
   **C.** 35 degrees.
   **D.** 45 degrees.

6. A harmonic balancer mounted on front of the crankshaft:
   **A.** Contains an inertia ring and a rubber ring.
   **B.** Controls crankshaft end play.
   **C.** Helps to keep the crankshaft turning during the nonproductive strokes.
   **D.** Is threaded onto the crankshaft.

7. *Technician A* says connecting rod and main bearing inserts may be made from aluminum.
   *Technician B* says connecting rod and main bearing inserts may be made from sintered copper-lead.
   Who is correct?
   **A.** A only
   **B.** B only
   **C.** Both A and B
   **D.** Neither A nor B

8. While discussing connecting rod bearings:
   *Technician A* says the bearing curvature is slightly greater than the curvature of the connecting rod bore.
   *Technician B* says when half of the bearing insert is installed in the rod cap, the bearing should protrude above the cap surfaces.
   Who is correct?
   **A.** A only
   **B.** B only
   **C.** Both A and B
   **D.** Neither A nor B

9. While discussing pistons:
   *Technician A* says the larger diameter of a cam ground piston is across the piston pin bores.
   *Technician B* says pistons are manufactured with a perfectly vertical skirt.
   Who is correct?
   **A.** A only
   **B.** B only
   **C.** Both A and B
   **D.** Neither A nor B

10. Common piston ring designs include:
   **A.** A three-piece oil ring.
   **B.** A top compression ring with no chamfer or taper.
   **C.** A top compression ring with an expander behind the ring.
   **D.** An oil ring below the piston pin.

CHAPTER 9

# Engine Seals, Sealants, and Gaskets

Upon completion and review of this chapter, you should be able to:

- ❑ Explain the purposes of engine gaskets.
- ❑ Explain the purpose of engine seals.
- ❑ Describe the construction of a conventional head gasket.
- ❑ Describe the construction and advantages of a rubber-coated embossed steel shim head gasket.
- ❑ Explain the purpose of a fire ring in a head gasket.
- ❑ Describe the forces applied to the head gasket in an engine with a cast-iron block and an aluminum cylinder head.
- ❑ Describe three types of intake manifold gaskets.
- ❑ List three different materials used in valve cover gaskets.
- ❑ Describe the construction of a typical lip seal.
- ❑ List three different types of rear main bearing seals.
- ❑ Explain the proper application of room-temperature vulcanizing (RTV) sealant.
- ❑ Explain the proper application of anaerobic sealant.

## Introduction

The mating surfaces of the engine require proper sealing to prevent leakage and for proper engine performance. **Gaskets** are used to prevent gas, coolant, oil, or pressures from escaping between two stationary parts (Figure 9-1). In addition, gaskets are used as spacers, shims, wear indicators, and vibration dampers. In recent years, many gaskets have been replaced with the use of sealants. The sealant is less expensive and easy to apply. **Seals** are used to prevent leakage of fluids around a rotating part.

**Gaskets** are used to prevent gas, coolant, oil, or pressure from escaping between two stationary parts.

**Seals** are used to prevent leakage of fluids around a rotating part.

## Gaskets, Seals, and Sealants

### Gaskets

There are three basic engine gasket classifications:

- Hard gaskets. These include gaskets made of metals or metals covering a layer of clay/fiber compound. Head gaskets, exhaust manifold, intake manifold, and exhaust gas recirculation (EGR) valve gaskets are examples of this type of gasket.
- Soft gaskets. These gaskets are made of rubber, cork, paper, and rubber-covered metal. Common usages include valve covers, oil pans, thermostat housings, water pumps, timing covers, and some intake manifolds.
- Sealants. These are usually a type of liquid material used to form gaskets. Silicones and anaerobics are examples of sealants.

**Shop Manual**
Chapter 9, page 421

The **cylinder head gasket** prevents compression pressures, gases, and fluids from leaking. It is located on the connection between the cylinder head and engine block.

**Cylinder Head Gaskets.** Perhaps the **cylinder head gasket** is subjected to the greatest demands. It must be capable of sealing combustion pressures up to 2,700 psi (18,616 kPa) and withstanding temperatures over 2,000°F (1,100°C). In addition, it must be able to seal coolant and oil under pressure. To complicate the matter further, the head gasket must be able to perform these functions while accommodating a shearing action resulting from expansion rates of the metals. Shearing results from the difference in thermal expansion between the cylinder head

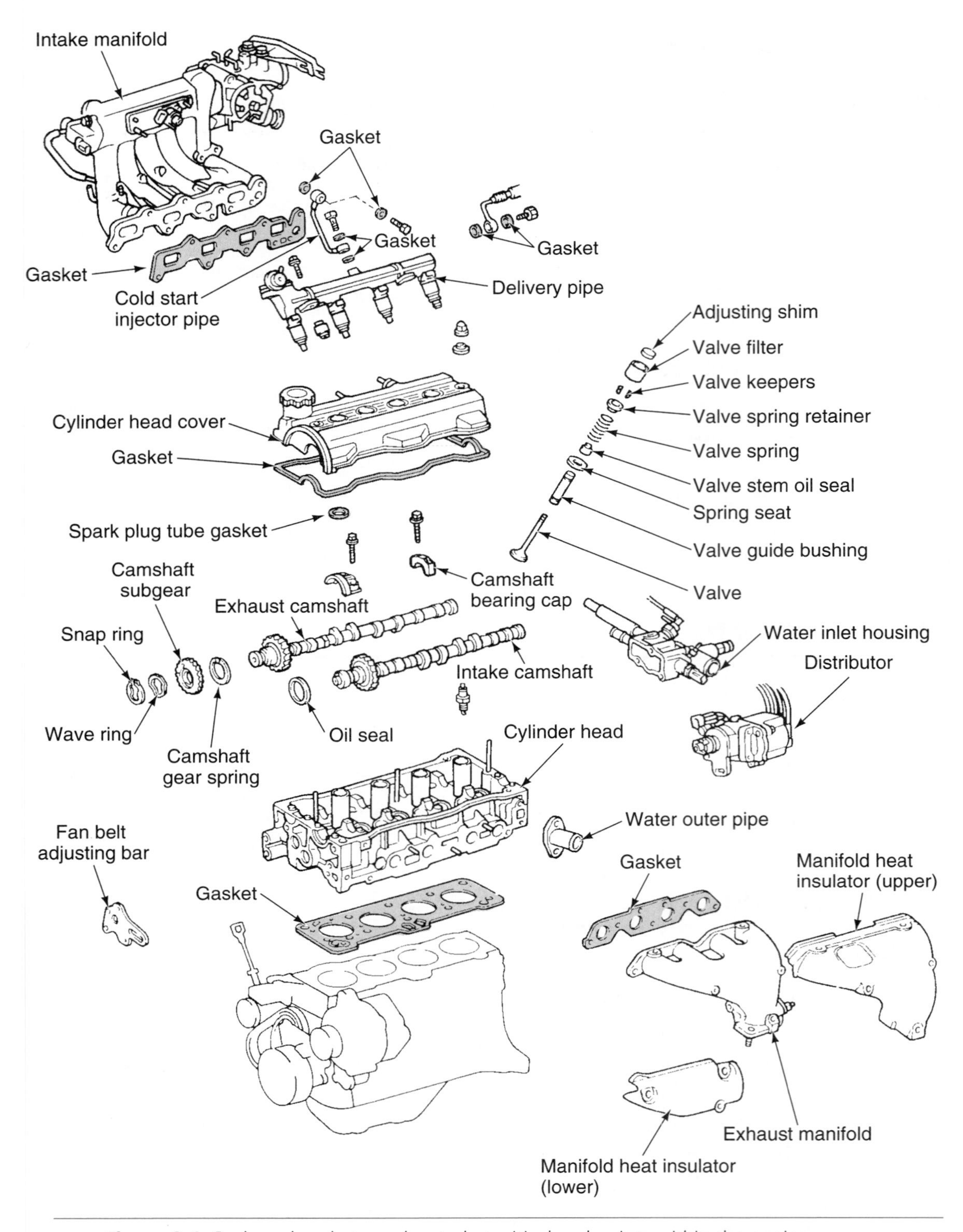


**Figure 9-1** Seals and gaskets are located at critical seal points within the engine.

and block. As the two metals expand and contract at different levels, a scrubbing stress is created (Figure 9-2). The head gasket cannot move during these times.

The head gasket consists of a core, facing, and coating. The core is usually made from solid or clinched steel. The facing is usually constructed of graphite and rubber fiber. These materials allow sufficient compression to conform to minor surface irregularities. The coating may be Teflon or silicone based and work with the facing to seal minor surface irregularities and resist

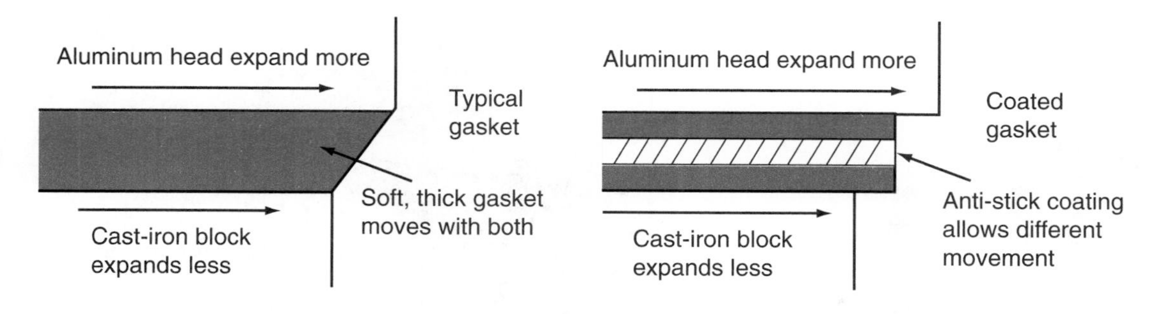


**Figure 9-2** The coating on the gasket material allows for different expansion rates between metals and still provides a good seal without distorting the gasket.

shearing (Figure 9-3). Some head gaskets use an elastomeric sealing bead to increase clamping forces around fluid passages (Figure 9-4).

One of the most recent developments in head gaskets is rubber-coated embossed (RCE) steel shim gaskets. A rubber coating is bonded to a steel shim gasket. The coating protects the shim from corrosive elements in the cooling and lubrication systems and provides good friction reduction. Because of their construction (Figure 9-5), RCE gaskets are also referred to as multilayer steel (MLS) gaskets. There are many advantages to the multilayer design. First, it has a uniform thickness that prevents bore distortion during cylinder head installation. Also, the MLS gasket is very resilient in that, once compressed, it "bounces" back for a good seal.

Most cylinder head gaskets have a metal **fire ring** around the combustion chamber opening (refer to Figure 9-3). The fire ring protects the gasket material from the high temperatures to which it is exposed. Also, the fire ring increases the gasket thickness around the cylinder bore so that it uses up to 75 percent of the clamping force to form a tight seal against combustion pressure losses.

A **fire ring** is a metal ring surrounding the cylinder opening in a head gasket.

Modern gasket designs generally do not require retorquing after initial installation; however, some head gaskets will require retorquing of the head bolts. This process is required on these type of gaskets because the heating and cooling of the engine causes the gasket to set, then relax, loosening the head bolt torque. When installing or replacing cylinder head gaskets, refer to the instructions provided by the supplier to determine if retorquing is required. Cylinder heads with gaskets that require retorquing are usually retorqued in three steps. First, the cylinder head bolts are torqued to the specified value when the engine is being assembled. Next, the engine is started and allowed to reach normal operating temperatures and cooled again. After the engine cools, the bolts are torqued a second time. A third retorquing may be required after the engine is driven about 500 miles.

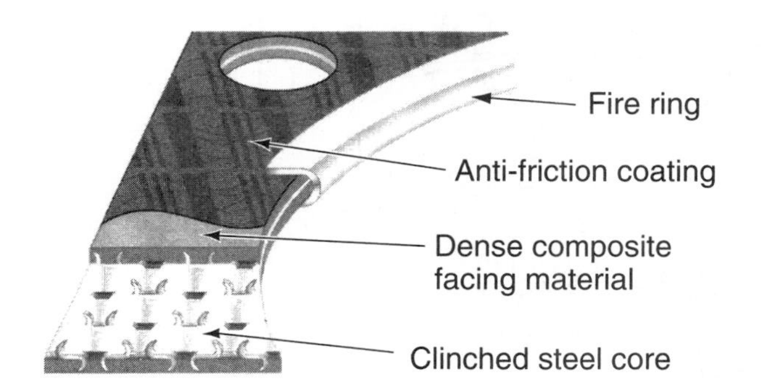


**Figure 9-3** Multilayers of a cylinder head gasket. (Courtesy of Fel-Pro, Inc.)

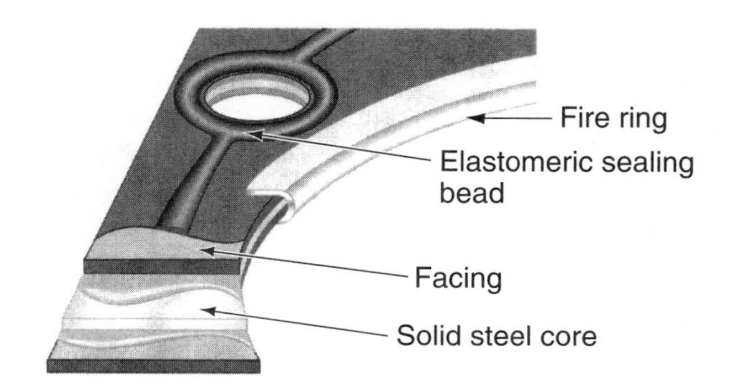


**Figure 9-4** Sealing beads work to prevent fluid leakage. (Courtesy of Fel-Pro, Inc.)

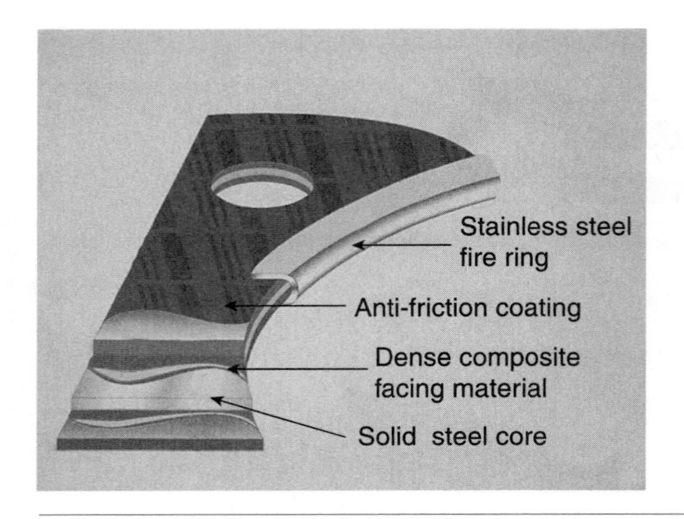


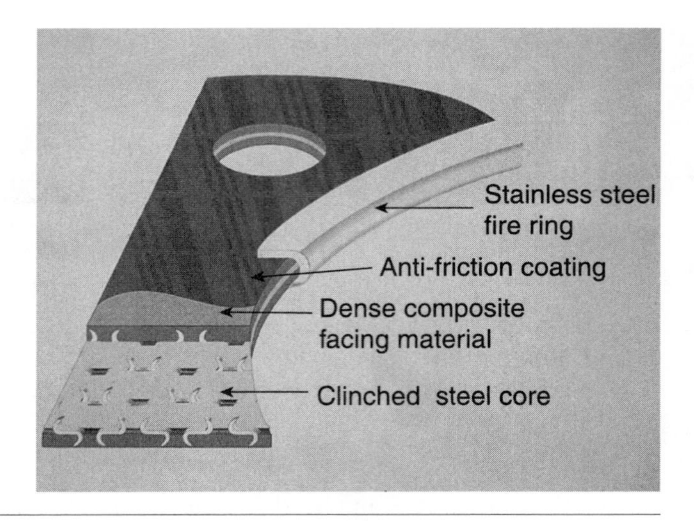


**Figure 9-5** The multilayer gasket uses sheets of steel with a rubber coating between the sheets. (Courtesy of Fel-Pro, Inc.)

The **intake manifold gasket** fits between the manifold and cylinder head to seal the air/fuel mixture or intake air.

**Valley pans** prevent the formation of deposits on the underside of the intake manifold.

A **strip seal** provides a seal between the flat surfaces on the front and back of the engine block and the intake manifold.

**Intake Manifold Gaskets.** A leak in the connection of the intake manifold and cylinder head can cause the mixture entering the combustion chamber to be too lean, resulting in rough idle and/or detonation. The **intake manifold gasket** is designed to provide a good seal under the changing temperatures it will be subject to (Figure 9-6).

Many intake manifold gaskets are constructed of a solid or perforated steel core with a fiber facing (Figure 9-7). Some manufacturers use gaskets made from a rubber silicone bonded to a steel or high-temperature plastic. On this style, the silicone provides the actual sealing. Some V-type engines use a **valley pan** style of intake manifold gasket (Figure 9-8). In addition, V-type engines use end strip seals to seal the connection between the ends of the manifold and the block (Figure 9-9). **Strip seals** are generally made from molded rubber or cork-rubber. When installing the end strips, use a silicone sealer on the corner joints where the strip meets the intake manifold gasket (Figure 9-10). Many fuel-injected engines use a plenum that requires a gasket between it and the intake manifold (Figure 9-11).

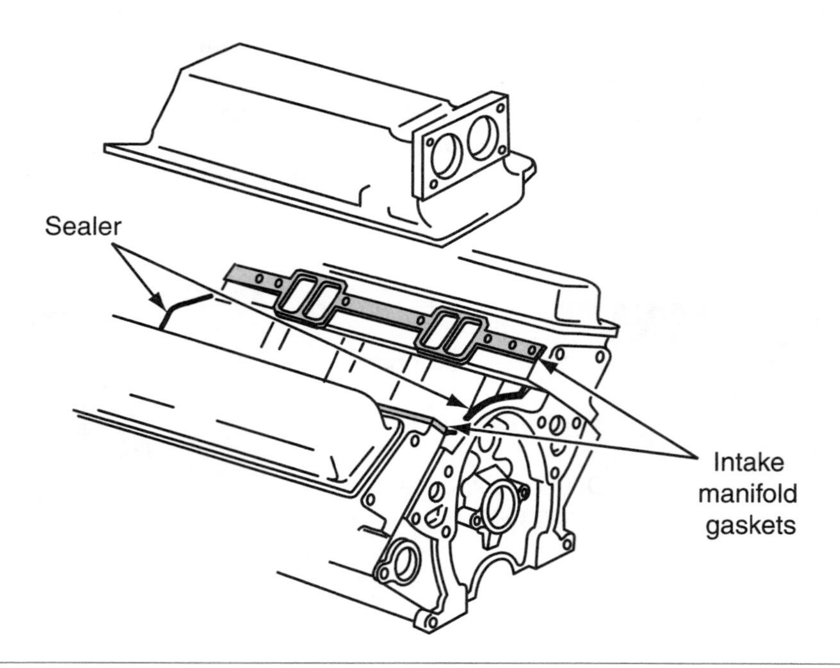


**Figure 9-6** Intake manifold gasket installation location.

Figure 9-7 Intake manifold gasket types. (Courtesy of Fel-Pro, Inc.)

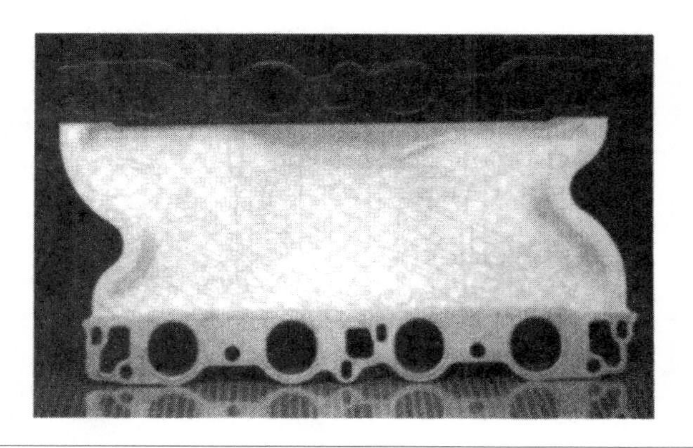

Figure 9-8 Valley pan. (Courtesy of Fel-Pro, Inc.)

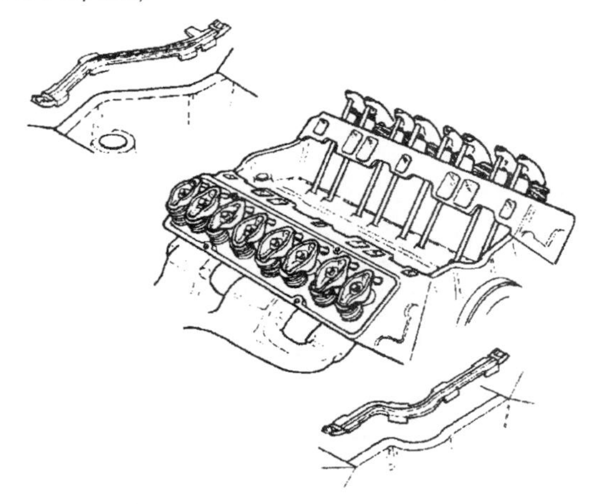

Figure 9-9 End strip seals used on V-type engines. (Courtesy of Fel-Pro, Inc.)

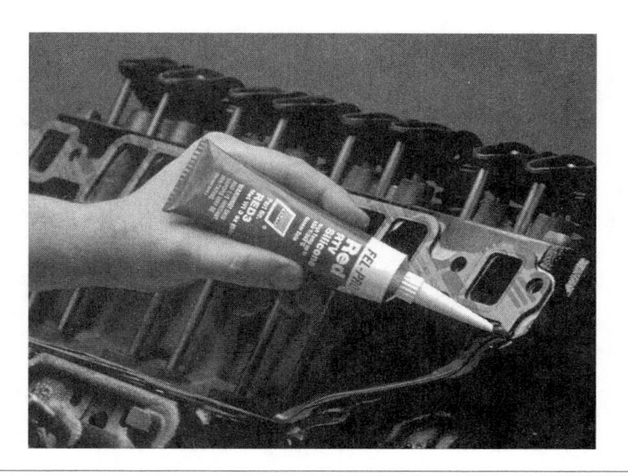

Figure 9-10 Use silicone sealer at the mating location of the seal strips and intake manifold. (Courtesy of Fel-Pro, Inc.)

With the advancement of plastics and composites, some manufacturers are designing intake manifolds from these products. Some of these do not use the typical intake manifold gasket; instead, they use a series of O-rings to seal each port (Figure 9-12).

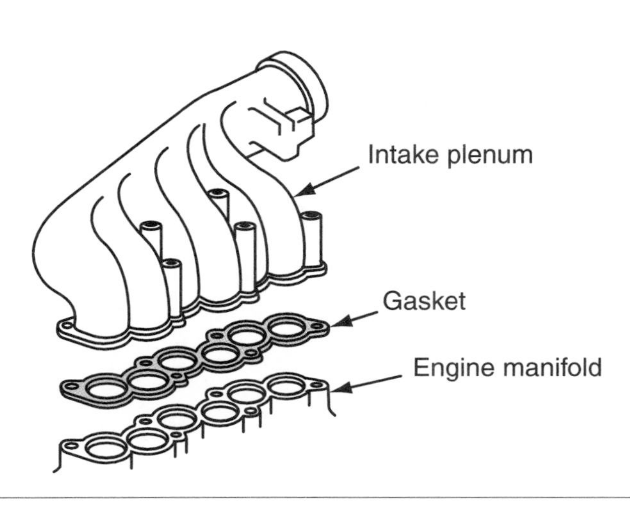


Figure 9-11 Intake plenum gasket location.

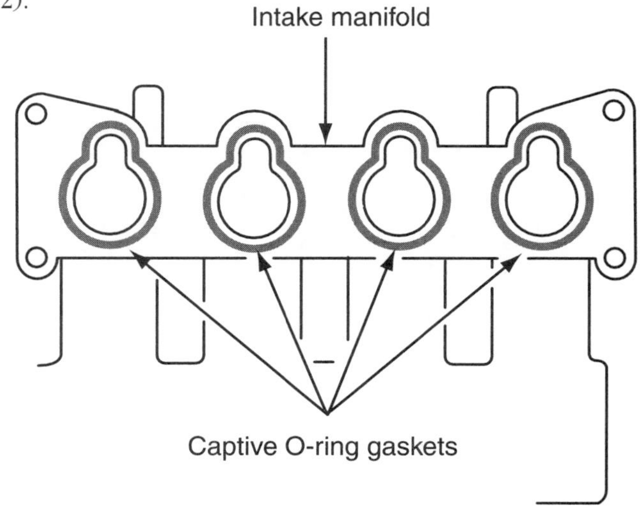


Figure 9-12 Some intake manifolds use captive O-rings to seal against the cylinder head.

When installing intake manifolds and gaskets on most in-line engines, the studs will hold the gasket in place while the manifold is installed. On V-type engines, the gasket may slip as the manifold is lowered into position. To prevent this, use an adhesive to hold the gasket in place.

**AUTHOR'S NOTE:** In my experience, the most common reasons for replacement head gasket failures are improper bolt service and torque procedures, and inadequate cleaning of the head and block mating surfaces. To prevent gasket failures, it is extremely important that the block and cylinder head mating surfaces are properly cleaned without damaging these surfaces. Replacement of torque-to-yield bolts and following the manufacturer's torque procedures are also essential in preventing gasket failures.

The **exhaust manifold gasket** seals the connection between the cylinder head and the exhaust manifold.

**Exhaust Manifold Gaskets.** The **exhaust manifold gasket** must prevent leakage under extreme temperatures (Figure 9-13). Leaks in this connection disrupt the flow of exhaust gases and can result in poor engine performance. In addition, exhaust leaks can lead to burned valves and objectionable noises. The use of exhaust manifold gaskets by engine manufacturers has declined in recent years with improved machining processes and better materials; however, when

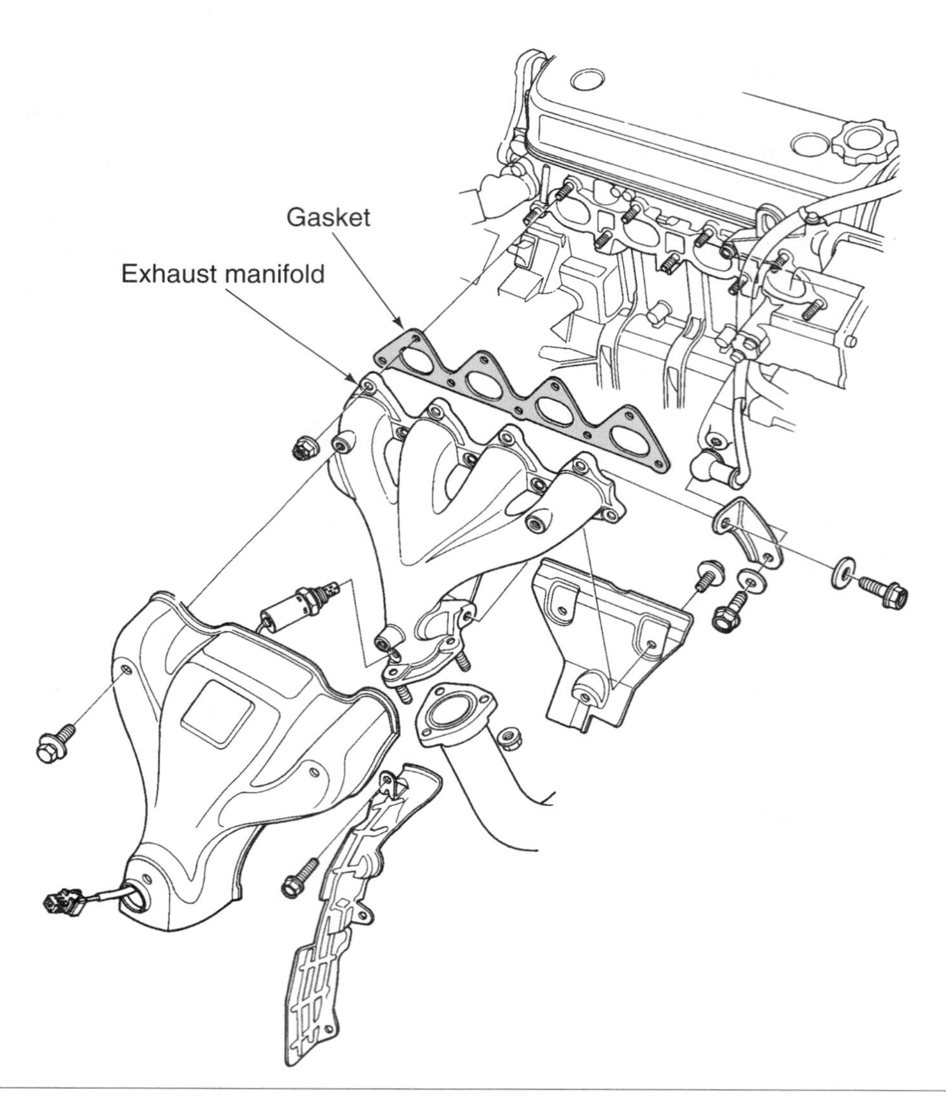


**Figure 9-13** Exhaust manifold gasket location. (Courtesy of American Honda Motor Co., Inc.)

the exhaust manifold is removed, it may become warped and require the use of a gasket. Most technicians opt to install an exhaust manifold gasket even if the engine was not originally equipped with one.

**Shop Manual**
Chapter 9, page 437

**Valve Cover Gaskets. Valve cover gaskets** are common locations for external oil leakage. This is largely due to the wide spacing between the attaching bolts. The wide spacing allows the stamped steel cover to distort easily and provides less clamping forces (Figure 9-14).

**Valve cover gaskets** seal the connection between the valve cover and cylinder head. The gasket is not subject to pressures, but must be able to seal hot, thinning oil.

To seal properly, the valve cover gasket must be highly compressible yet have good torque retention. To perform these tasks, there are a variety of materials used to construct valve cover gaskets. Some of the most common are synthetic rubber, cork-rubber, and molded rubber. Synthetic rubber gaskets seal by deforming instead of compressing. The synthetic rubber has a tendency to "remember" its original shape and will attempt to return to it. When the valve cover is tightened against the gasket, the gasket deforms. The gasket attempts to return to its original shape and pushes back against the valve cover, creating a tight seal. One drawback to this gasket is it can be difficult to install. Cork-rubber gaskets compress very well and provide good sealing (Figure 9-15). Molded rubber gaskets are the easiest to install and provide the best sealing. Sealers or adhesives should not be used on molded rubber gaskets. In addition to the valve cover gasket, many overhead camshaft engines use molded rubber semicircular plugs (Figure 9-16).

Valve covers may be called rocker arm covers.

Before installing a stamped steel valve cover, use a ball-peen hammer and a block of wood to remove deformities around the bolt holes (Figure 9-17). If a plastic or cast aluminum cover is

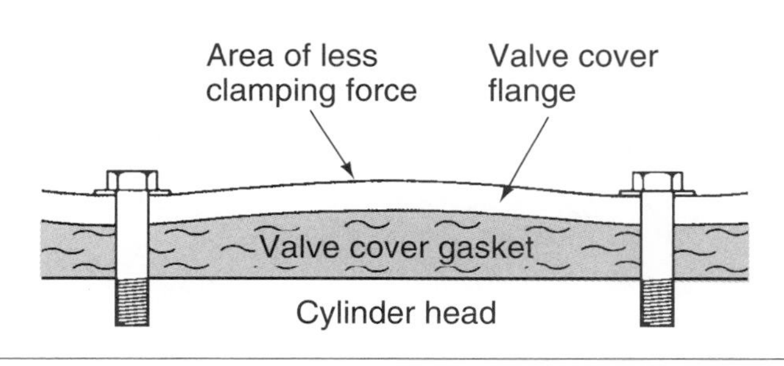


**Figure 9-14** The valve cover gasket can be easily deformed. (Courtesy of Fel-Pro, Inc.)

**Figure 9-15** Cork-rubber gasket composition. (Courtesy of Fel-Pro, Inc.)

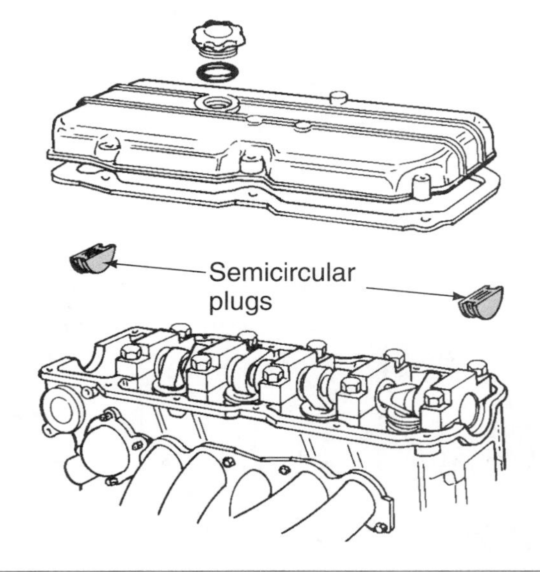


**Figure 9-16** Camshaft plugs work with the cover gasket to seal oil. (Courtesy of Fel-Pro, Inc.)

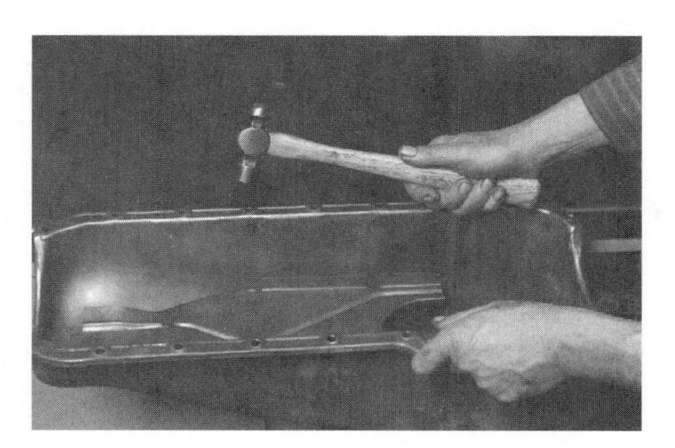

**Figure 9-17** Remove the deformation around the bolt holes of the valve cover before installing it.

used, it will require replacement if it is distorted. When the attaching bolts are torqued, use an X pattern, starting in the center and working outward (Figure 9-18).

**Oil pan gaskets** are used to prevent leakage from the crankcase at the connection between the oil pan and the engine block.

**Oil Pan Gaskets.** **Oil pan gaskets** can be multipiece or a single-unit molded gasket, depending on the crankcase design. The most common type of multipiece gasket uses two side pieces made of cork and rubber and two end pieces made of synthetic rubber (Figure 9-19). Many modern engines use a single-unit molded gasket (Figure 9-20).

Before installing the oil pan, check the mating surface for deforming around the bolt holes. If multisegment gaskets are used, place a drop of silicone sealant at all joints (Figure 9-21).

**Miscellaneous Gaskets.** Many additional gaskets are used on the engine. Each of these must be replaced during the assembling process. Additional gasket applications include:

- Exhaust gas recirculation (EGR) valve
- Water pump
- Fuel pump
- Carburetor or throttle body

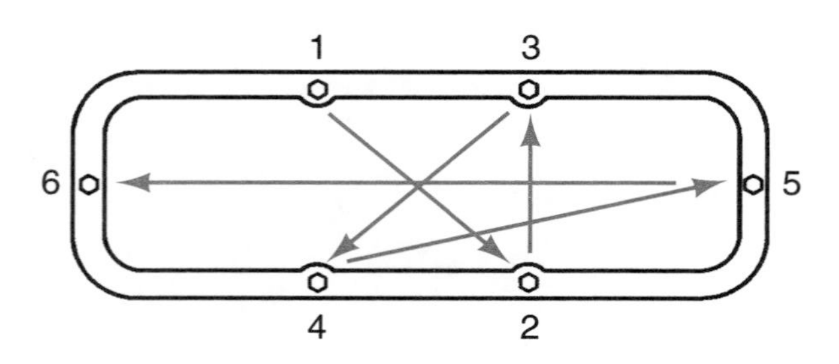


Figure 9-18 Typical valve cover tightening sequence. (Courtesy of Fel-Pro, Inc.)

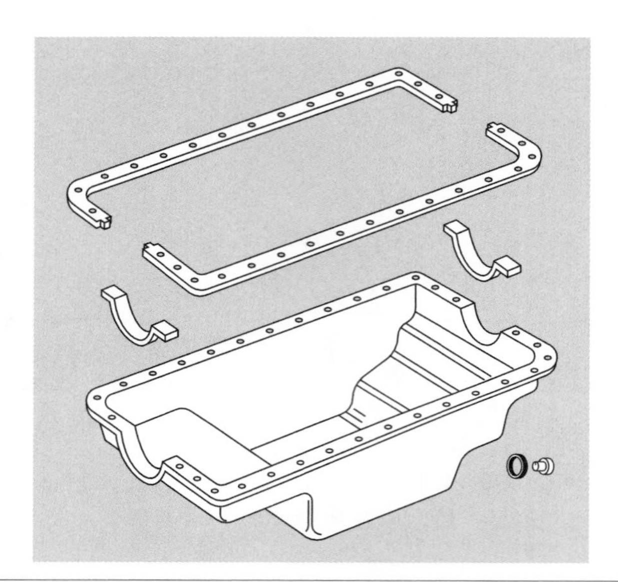

Figure 9-19 Multipiece oil pan gasket. (Courtesy of Fel-Pro, Inc.)

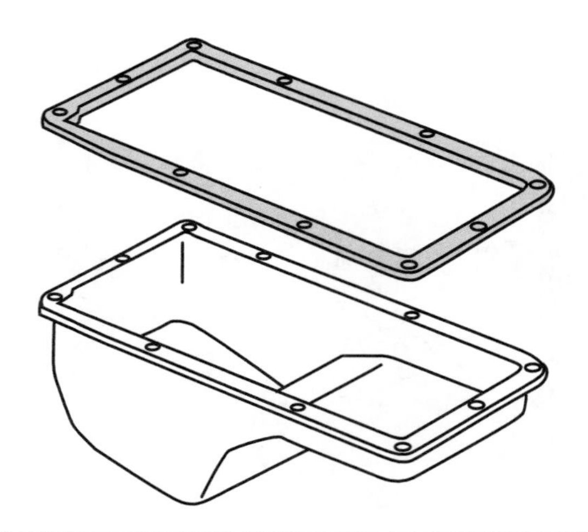

Figure 9-20 Single-unit molded oil pan gasket.

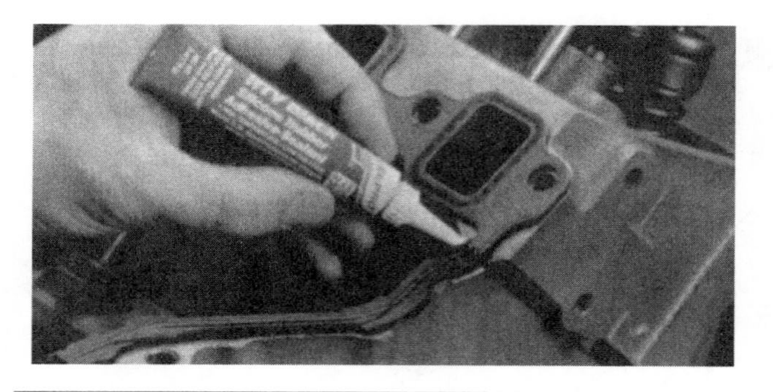

Figure 9-21 Using silicone sealer at the joints of the multi-piece oil pan gasket. (Courtesy of Fel-Pro, Inc.)

- Air cleaner
- Fuel injector mounting
- Timing cover
- Exhaust pipe

Each of these gaskets is designed for a particular purpose. The materials used must perform many of the tasks previously discussed, but under unique circumstances. Always refer to the service manual for the proper torque and sequence when installing these gaskets.

## Seals

**Shop Manual**
Chapter 9, page 423

A **timing cover seal** is a lip-type seal that prevents leaks between the timing gear cover and the front of the crankshaft.

The **rear main seal** is a seal installed behind the rear main bearing on the rear main bearing journal to prevent oil leaks.

**Lip seals** are formed from synthetic rubber and have a slight raise (lip) that is the actual sealing point. The lip provides a positive seal while allowing for some lateral movement of the shaft.

Service sleeves (sometimes referred to as speedy-sleeves) press over the damaged sealing area to provide a new, smooth surface for the seal lip.

The most common seals in the engine are the **timing cover seal** and the **rear main seal.** The timing cover seal seals around the harmonic balancer to prevent oil leakage from the front of the crankshaft, while the rear main seal prevents leakage from the rear of the crankshaft. Most modern engines use a type of **lip seal** to perform these functions. The lip seal is generally constructed of butyl rubber or neoprene. The rubber seal is attached to a metal case that is driven into the bore (Figure 9-22). The seal lip will use the pressure of the fluid between the seal and the case to force the lip tight against the shaft. To assist in providing a tighter seal, some lip seals use a garter spring behind the lip. If the seal is installed in a location where the front of the seal may be exposed to dirt, a dust lip may be formed into the seal to deflect dirt away from the outside of the seal. When installing a new lip seal, the lip always faces the hydraulic pressure side.

The timing cover and oil pump housing oil seals are generally one-piece lip seals with a steel outer ring (Figure 9-23). To aid in sealing during installation, apply silicone sealer to the outer diameter of the metal shell. Neoprene seals should always be lubricated prior to installation. This will prevent seal damage from overheating during initial startup. If the surface of the harmonic balancer is damaged, a service sleeve can be installed to provide a new sealing surface (Figure 9-24).

There are three common seal designs for rear main seals: rope-type, two-piece molded synthetic rubber, and one-piece rubber with steel ring. The rope-type rear main seal is found on many older engines. These are difficult to replace without removing the crankshaft. They were replaced with molded two-piece seals. Rope-type seals are also two pieces. The lower piece is installed into the bearing cap, while the upper piece is installed into the block. An installation tool is used to properly seat the seal (Figure 9-25). After the seal is properly seated, the excess rope is cut off.

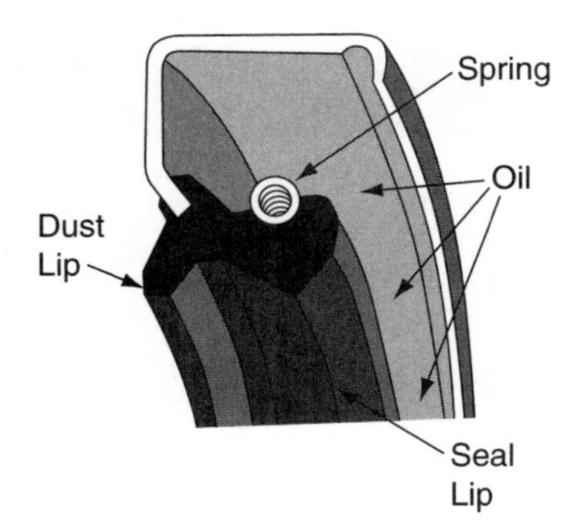


**Figure 9-22** The common lip seal has an outer case, rubber sealing lips, and a gator spring. Hydraulic pressure works to cause the lip seal to ride tight against the shaft.

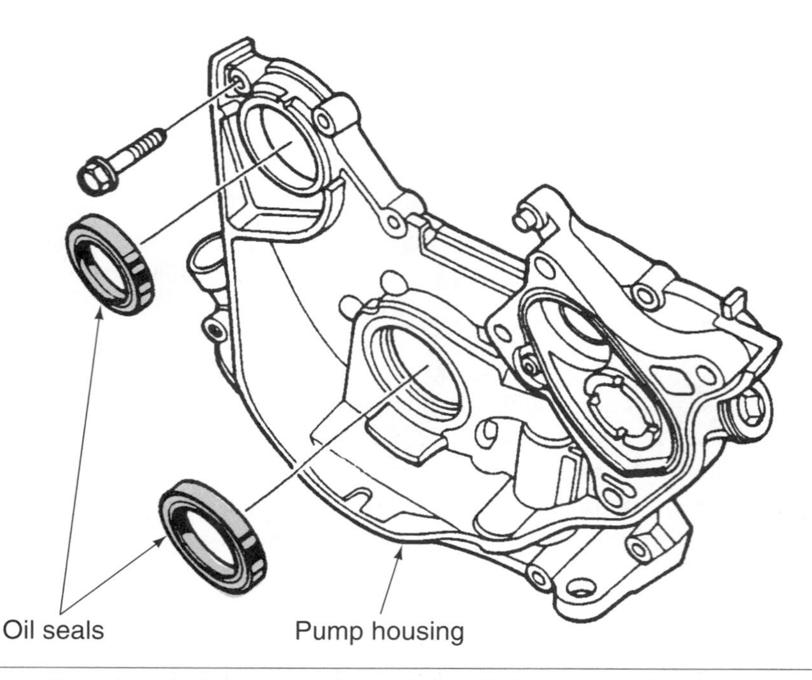


**Figure 9-23** One-piece seals are common on timing chain covers and oil pump housings. (Courtesy of American Honda Motor Co., Inc.)

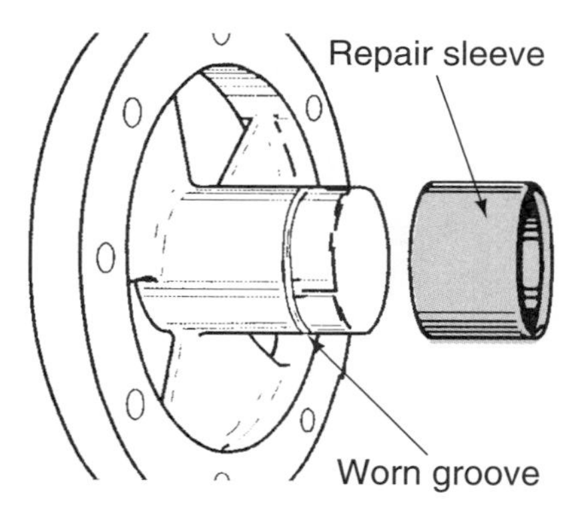


**Figure 9-24** A sleeve can be installed over damaged sealing surfaces to restore the surface. (Courtesy of Fel-Pro, Inc.)

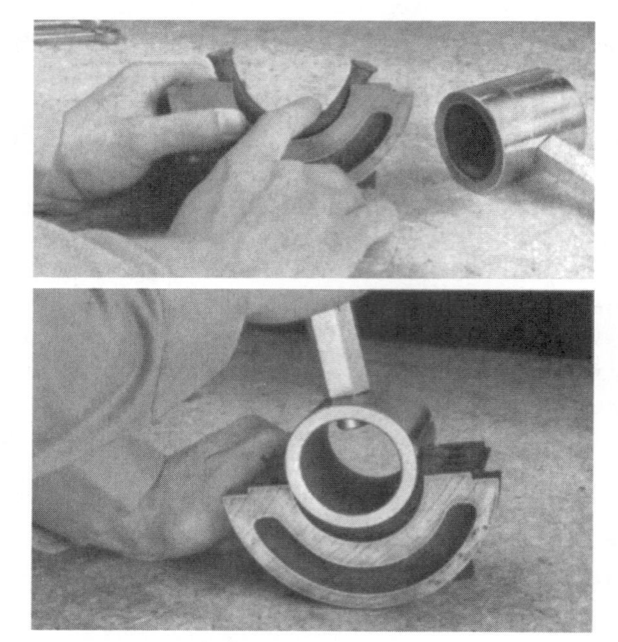

**Figure 9-25** Installing rope-type rear main seals. (Courtesy of Fel-Pro, Inc.)

When installing formed two-piece seals, the lip of the seal must face into the engine. Single-piece seals are installed in the same manner as the timing cover seal. Follow the service manual procedure since some manufacturers require the seal to be installed onto the crankshaft prior to installing the crankshaft into the engine, while others install the seal after the crankshaft is installed into the block. Some one-piece rear main seals use a silicone coating on the outer shell. Do not apply oil to the shell for aid during installation; use soapy water. The oil may not be compatible with the silicone. If the crankshaft flange is damaged or has a groove worn into it, it may be repaired using a service sleeve.

## Adhesives, Sealants, and Form-in-Place Gaskets

**Shop Manual**
Chapter 9, page 434

There are a number of chemical sealing materials available designed to reduce labor and increase the probability of a good seal. Proper use of these chemical materials is required to assure good results.

To assist in gasket material removal, spray or brush-on gasket remover solvent is available. Adhesives are designed to bond different types of gasket materials in place prior to installation. These would be used to hold the gasket in place during installation of parts that may require a lot of wiggling into place. Some gaskets must be coated with a sealer prior to installation. This is true of most MLS cylinder head gaskets, but not all. Some sealants can damage gasket coatings; be sure to check the instructions for the gasket you are installing.

In the 1980s, many manufacturers began switching to **sealants** and form-in-place gaskets instead of molded gaskets at many joint connections throughout the engine. When applying these materials, care must be taken to assure proper application. Bead size, continuity, and location are factors affecting proper sealing. If the bead is too thin, a leak can result. A bead that is too wide can result in spillover and clog the oil pump or oil galleries. Also, the manufacturer may require different types of sealers at specified locations in the engine. For this reason, always refer to the service manual for the correct type and usage of sealers.

**Sealants** are commonly used to fill irregularities between the gasket and its mating surface. Some sealants are designed to be used in place of a gasket.

There are two basic types of sealants:

- **Aerobic**
- **Anaerobic**

**Aerobic sealants** require the presence of oxygen to cure. **Anaerobic sealants** will only cure in the absence of oxygen.

Most aerobic sealants are silicone compounds and work well with metals and plastics. The most common type of silicone sealant is **room-temperature vulcanizing (RTV)** compound. This type of sealant forms a rubber seal by absorbing moisture from the air. Before attempting to use RTV, the sealing surfaces must be thoroughly cleaned. RTV begins to set in about 10 minutes, requiring the components to be assembled quickly. Some RTV suppliers use different colors to denote the temperature range the RTV is capable of withstanding. Use the correct RTV for the expected temperature. RTV cannot be used on the exhaust system components.

**Room-temperature vulcanizing (RTV)** is a type of engine sealer that may be used in place of a gasket on some applications.

Anaerobic sealants cure after the components have been assembled. These sealants are capable of filling gaps up to 0.030 in. (0.8 mm) and withstand temperatures up to 350°F (177°C). Some engines using a one-piece main bearing require the use of special anaerobic sealants that do not cure until a specified temperature is reached. This assures proper maintenance of oil clearances as the main bearing bolts are torqued, without fear of the sealant setting too soon.

### A BIT OF HISTORY

In the year 2000, sales of light trucks surpassed car sales. The light truck market consists of these vehicles:

1. Minivans—17.9 percent
2. Full-sized vans—5.5 percent
3. Full-sized pickups—26.6 percent
4. Compact pickups—14.7 percent
5. Full-sized sport utility vehicles (SUVs)—8 percent
6. Compact SUVs—23.3 percent
7. Small SUVs—3.9 percent

SUVs of all sizes make up over 35 percent of the light truck market. Surveys indicate that in American households with over two vehicles, only 28 percent use a car as their primary vehicle. In these households, 17 percent use an SUV as their primary vehicle, and 52 percent use a pickup as their primary vehicle.

Source: Automotive Service Association (ASA), Auto Inc., July 1999, p. 10.

Terms to Know

Aerobic
Anaerobic
Cylinder head gasket
Exhaust manifold gasket
Fire ring
Gaskets
Intake manifold gasket
Lip seal
Oil pan gaskets
Rear main seal
Room-temperature vulcanizing (RTV)
Sealants
Seals
Strip seals
Timing cover seal
Valley pan
Valve cover gaskets

## Summary


❒ Gaskets are used to prevent gas, coolant, oil, or pressures from escaping between two mating parts.

❒ Seals are used to seal between a stationary part and a moving one.

❒ Hard gaskets are made from metals or metals covered by a layer of clay/fiber compound. Hard gaskets include cylinder head, exhaust manifold, intake manifold, and EGR valve gaskets.

❒ Soft gaskets are made from rubber, cork, paper, and rubber-covered metal. Soft gaskets include valve cover, oil pan, thermostat housing, water pump, and timing cover gaskets.

❒ Sealants are a liquid material used to form gaskets. Examples of sealants are anaerobics and silicones.

❒ Cylinder head gaskets must be capable of sealing cylinder pressures up to 2,700 psi (18,616 kPa).

❒ Cylinder head gaskets must withstand temperatures over 2,000°F (1,100°C).

❒ Cylinder head gaskets must be able to withstand the scrubbing stress caused by an aluminum cylinder head expanding more than the cast-iron cylinder block.

❒ Conventional cylinder head gaskets contain a core made from solid or clinched steel, a facing constructed of graphite or rubber fiber, and a Teflon or silicone-based coating.

❒ A rubber-coated embossed (RCE) steel shim head gasket contains a steel shim gasket with a rubber coating.

❒ Most cylinder head gaskets have a fire ring surrounding the combustion chamber opening to protect the gasket material from high combustion temperatures.

❒ Many intake manifold gaskets contain a solid or perforated steel core with a fiber facing. On OHV (pushrod) engines, end strips are used to seal the ends of the intake manifold to the block.

❒ Some manufacturers depend on precision machining of the exhaust manifold and cylinder head mating surfaces to provide a seal and prevent exhaust leaks without the use of a gasket.

❒ Valve cover gaskets are made from synthetic rubber, cork-rubber, or molded rubber.

❒ There are three common designs used for rear main seals: rope-type, two-piece molded synthetic rubber, and one-piece rubber with steel ring.

❒ Sealants are commonly used to fill irregularities between the gasket and its mating surface. Some sealants are designed to be used in place of a gasket.

❒ Aerobic sealants require the presence of oxygen to cure. Anaerobic sealants will only cure in the absence of oxygen.

# Review Questions

## Short Answer Essays

1. Explain the purpose of gaskets.
2. What is the function of a seal?
3. Explain the difference between aerobic and anaerobic sealants.
4. Describe the demands to which the head gasket is subjected.
5. Explain how the head gasket is subjected to lateral scrubbing stress.
6. Explain the location and purpose of a fire ring in a head gasket.
7. Describe the construction of a conventional head gasket.
8. Explain the advantages of rubber-coated embossed (RCE) steel shim head gaskets.
9. Explain the purpose of strip seals on an intake manifold in an OHV engine.
10. List three different types of valve cover gaskets.

## Fill-in-the-Blanks

1. ________________ sealants require the presence of oxygen to cure.
2. When installing a lip seal, the lip always faces the ________________ ________________ side.
3. Neoprene seals should always be ______________ before installation.
4. Head gaskets must be able to seal __________ and ___________ under pressure.
5. A conventional head gasket contains a ___________, _____________, and a _____________ .
6. Some head gaskets have a ________________ ______________ _____________ to increase the clamping force around fluid passages.
7. Some intake manifold gaskets are made from a solid or perforated steel core with a ______________ _____________ .
8. Some engine manufacturers use ______________ ________________ on the exhaust manifold and cylinder head mating surfaces rather than an exhaust manifold gasket.
9. The most common locations for lip seals on a typical engine are the ______________ ____________ ______________ and the ______________ ______________ ____________ .
10. Three common types of rear main bearing seals are:

    a. ______________________________________

    b. ______________________________________

    c. ______________________________________

## Multiple Choice

1. Lip seals are being discussed.
   *Technician A* says to install the seal with the lip facing the hydraulic pressure.
   *Technician B* says some lip seals use a garter spring behind the lip to assist in better sealing.
   Who is correct?
   **A.** A only **C.** Both A and B
   **B.** B only **D.** Neither A nor B

2. *Technician A* says seals are used to seal surfaces of stationary parts.
   *Technician B* says gaskets are used to seal between a stationary part and a moving one.
   Who is correct?
   **A.** A only **C.** Both A and B
   **B.** B only **D.** Neither A nor B

3. Cylinder head gaskets are being discussed.
   *Technician A* says cylinder head gaskets may require retorquing after the engine has been run.
   *Technician B* says cylinder head gaskets are designed to be reused.
   Who is correct?
   **A.** A only **C.** Both A and B
   **B.** B only **D.** Neither A nor B

4. Exhaust manifold gaskets are being discussed.
   *Technician A* says all engines use this gasket.
   *Technician B* says a damaged gasket can result in poor engine performance.
   Who is correct?
   **A.** A only **C.** Both A and B
   **B.** B only **D.** Neither A nor B

5. *Technician A* says aerobic sealants cure in the absence of oxygen.
   *Technician B* says anaerobic sealants cure in the presence of oxygen.
   Who is correct?
   **A.** A only **C.** Both A and B
   **B.** B only **D.** Neither A nor B

6. All of these statements about head gaskets are true EXCEPT:
   **A.** A conventional head gasket contains a core, facing, and coating.
   **B.** The core is usually made from silicone.
   **C.** The facing is usually made from graphite or rubber fiber.
   **D.** The coating may be made from Teflon.

7. The fire ring in a head gasket:
   **A.** Compensates for movement between metals with different expansion rates.
   **B.** Increases heat transfer to the cooling system.
   **C.** Protects the gasket material from high combustion temperatures.
   **D.** Seals coolant passages through the head gasket.

8. While discussing head gaskets and fasteners:
   *Technician A* says head gaskets may be installed with either side facing upward.
   *Technician B* says torque-to-yield head bolts should be replaced each time the head is removed.
   Who is correct?
   **A.** A only **C.** Both A and B
   **B.** B only **D.** Neither A nor B

9. When installing intake manifold gaskets:
   **A.** End strips are used to seal the ends OHV intake manifolds to the block.
   **B.** A valley pan–type intake gasket is used on an in-line four-cylinder engine.
   **C.** A leaking intake manifold gasket may cause a rich air/fuel mixture.
   **D.** Silicone sealer should be placed on both surfaces of intake manifold gaskets before installation.

10. When removing and replacing valve cover gaskets:
    **A.** Valve cover gaskets may be made from cork-rubber, molded rubber, or synthetic rubber.
    **B.** The flange on a plastic or cast aluminum valve cover may be straightened with a hammer and a block of wood.
    **C.** A sealant should be used on molded rubber valve cover gaskets.
    **D.** Valve cover fasteners should be tightened in sequence in a clockwise direction around the valve cover.

# High-Performance Engines

Upon completion and review of this chapter, you should be able to:

- ❑ Describe intake manifold modifications that will improve airflow.
- ❑ Explain the operation of a dual runner intake manifold with butterfly valves in the secondary runners.
- ❑ Explain the operation of a fuel injection system with dual injectors in each intake port.
- ❑ Describe combustion chamber design modifications that will improve volumetric efficiency and engine power.
- ❑ Explain camshaft modifications that improve volumetric efficiency and engine power.
- ❑ Explain valve float, and describe how this problem can be reduced.
- ❑ Describe how exhaust manifolds may be designed to improve exhaust flow.
- ❑ Explain basic turbocharger operation.
- ❑ Describe how the turbocharger boost pressure is controlled.
- ❑ List three items that cause premature turbocharger failure.
- ❑ Explain basic supercharger operation.
- ❑ Describe the difference in compression ratio in a turbocharged or supercharged engine compared to a normally aspirated engine.

## Introduction

Engines that are produced and sold in most passenger cars and light trucks are compromises. To some extent they must sacrifice high performance for fuel economy, driveability, and low emissions. Ever since the automobile was first introduced to the public, some people have been trying to figure out how to make it go faster. This chapter discusses a few of the many things that can be done to an engine to increase its horsepower and internal strength. Keep in mind that when an engine is built for increased performance, the components and alternations added to the engine must complement each other; for example, installing a camshaft that has an effective rpm above 6,500 and an intake manifold that has an effective rpm of 4,000 would result in an engine that would fight against itself. Throughout the process of building a performance engine, the intended purpose must be the focal point. An engine that is built for top fuel racing is built different from an engine that will be used in mud bog racing. One racing engine builder's advertising logo is, "From Mild to Wild." This racing engine builder will build an engine with the amount of performance required by the customer, and this performance depends on what the customer wants to do with the engine. The amount of engine performance in a street rod is much different from the performance required in a dragster. A formula one race car requires much different engine performance from the dragster or street rod. The racing engine builder will use all the methods of improving engine performance to design an engine to meet the customer's requirements.

Some of the major vehicle manufacturers are involved in auto racing. The advantages of vehicle manufacturers being involved in auto racing are technology transfer and engineer training. When discussing engineers' involvement in racing, one executive representing a vehicle manufacturer said, "You bring them (engineers) in, you infect them with this disease called racing, then you send them back to infect other people." Technology is sometimes developed and used on the race tracks before it is made available to the car buyer. This technology may be related to engines or chassis; for example, the General Motors Stabilitrak system was used in racing prior to production vehicles. Stabiltrak is a computer-controlled system that maintains vehicle stability when the vehicle begins to swerve sideways. The Night Vision system now available on some models of Cadillac was first used in the Le Mans 24-hour race.

The engine is extremely important to a race car's success, but other factors such as chassis systems and aerodynamics are equally important. Since this book discusses automotive engines, the information related to high performance engines is limited to this subject.

Throughout this book we have introduced concepts that will carry over to high-performance engine building. The basics are the same: we discussed compression ratios and how they can be modified, we discussed piston head configurations, cylinder head combustion chamber designs, crankshaft design, and camshaft lobe characteristics. We also discussed some machining operations (such as peening) that can be performed to increase the strength of a component. In this chapter, we will discuss how to improve the engine's rpm, power, and torque.

# Methods of Improving Engine RPM, Power, and Torque

## Improving Volumetric Efficiency

Before any further discussion, let us ask the question, What limits the engine torque and rpm on a typical car or light truck engine? There are several answers to this question. Probably the most important factor in determining the torque is the breathing capacity of the engine. After a specific engine rpm, the design of the air intake system, valve train, combustion chambers, and exhaust system limits the airflow into the engine to the point at which the pistons are moving so fast the air just does not have time to get into the cylinder. Under this condition, compression pressure and engine torque begin to decrease.

In a **normally aspirated** engine, atmospheric pressure is used to force air into cylinders.

The horsepower output of the engine is directly related to the amount of air that is compressed in the cylinders. In a **normally aspirated** engine, atmospheric pressure is used to cause airflow into the combustion chambers. Atmospheric pressure at sea level is 14.7 psi (101 kPa). The engine's ability to produce power is limited by its volumetric efficiency. As we discussed earlier, pumping losses reduce the ability of the engine to breath. Without the ability to breath, all other performance enhancements done to the engine are irrelevant.

Volumetric efficiency is the amount of air taken into an engine compared to the amount of air that could be theoretically taken into the engine.

In order to increase horsepower output, the amount of air compressed in the cylinder must be increased. Airflow is the main focus over fuel delivery because airflow is the limiting factor. Usually any amount of air delivered to the cylinders can be matched with the correct amount of fuel; however, the inverse is not true.

Therefore, if engine torque is to be improved and maintained to a higher rpm, the air intake system must be improved to increase the airflow into the cylinders at high rpm. Increasing airflow into the cylinders at higher rpm may be accomplished in many different ways, including:

1. Intake manifold design
2. Valve train design
3. Combustion chamber design
4. Exhaust manifold design
5. Increasing the number of valves per cylinder and/or the valve diameter
6. Increasing valve lift

## Intake Manifold Design

The 2000 Special Vehicle Team (SVT) Mustang Cobra R is equipped with a modified 4.6L engine. Engineers achieved a 25 percent increase in airflow into this engine by reshaping the intake and exhaust ports, redesigning the intake and exhaust valves, and increasing exhaust valve diameter by 0.080 in. (2 mm). The modified 4.6L engine produces 385 hp at 6,250 rpm and 385 ft. lb. of torque at 4,250 rpm.

**Intake manifold tuning** is the process of designing the intake manifold to increase the airflow into the cylinders.

Intake manifolds can be designed to improve the volumetric efficiency of the engine. This process is called **intake manifold tuning.** As the air rushes through the intake manifold into the cylinders, the opening and closing of the intake valves causes the airflow to pulse. If the intake manifold is divided into individual runners for each cylinder, the pulsing effect of the airflow pushes or rams more air into the cylinder. By adjusting or tuning the length of the individual intake manifold runners, the manufacturer can design an intake manifold to supply the amount of air required by the engine. Smaller diameter, longer intake manifold runners ram more air into the cylinders at lower engine rpm. To ram more air into the cylinders at high rpm when air pulsing is faster, the intake manifold runners should be larger in diameter and shorter. Intake manifold runners must be curved to avoid sharp bends that offer more airflow restriction. Some engines have dual runners into each intake port (Figure 10-1). In this type of intake manifold, butterfly valves may be positioned in the high-speed runners. These butterfly valves are normally closed at lower speeds up to approximately 3,500 rpm. Above this engine speed, the powertrain control module (PCM) operates an electric/vacuum solenoid that opens and supplies vacuum to an actuator connected to the butterfly valves (Figure 10-2). Under this condition, the butterfly valves open the high speed runners in the intake manifold to increase airflow into the intake

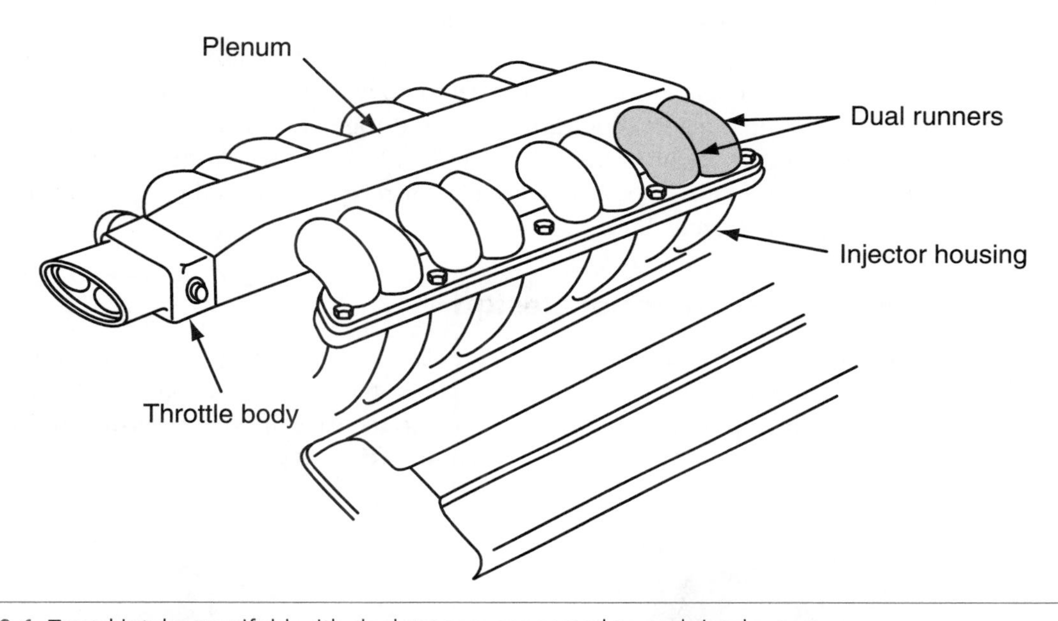


Figure 10-1 Tuned intake manifold with dual runners connected to each intake port.

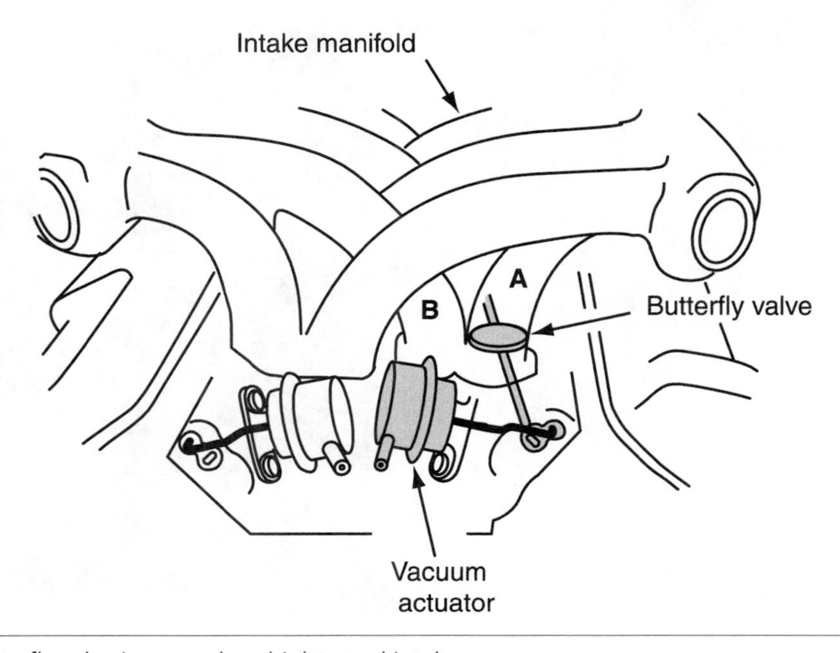


Figure 10-2 Butterfly valve in secondary, high speed intake runners.

manifold. In a maximum performance racing engine that operates continually at high rpm, the main concern is power and torque at high rpm. Therefore, the intake manifold is designed with only short, large diameter intake runners. The legendary Offenhauser engine dominated the Indy 500 racing scene for many years in the 1960s and 1970s. One of these 4-cylinder in-line 159 cubic inch displacement (CID) engines produced over 1,200 horsepower. Figure 10-3 indicates the short, large diameter intake runners on this engine. To increase engine rpm, power, and torque, the first requirement is an intake manifold that moves the required amount of air into the engine.

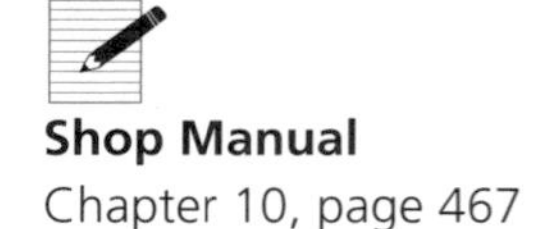

**Shop Manual**
Chapter 10, page 467

The fuel system must be designed to deliver the amount of fuel required to mix with the air entering the engine so the proper air/fuel ratio is maintained.

## Fuel System

Some high-performance engines with dual intake runners and butterfly valves in the secondary runners have injectors in each intake runner near the intake ports. Therefore, these engines have two injectors per cylinder (Figure 10-4). When the PCM operates the vacuum/electric solenoid to open the butterfly valves in the secondary runners, the PCM also begins operating the injectors in the secondary runners to supply more fuel to go with the additional air entering the engine. The dual injectors in each cylinder are connected together on the ground side and operated by the same injector driver in the PCM (Figure 10-5). To switch the secondary injectors on and off, the PCM operates two solid-state injector relays connected on the power side of the secondary injectors (Figure 10-6).

The **compression ratio** is a comparison between the volume above the piston at bottom dead center and the volume above the piston at top dead center.

**Thermal efficiency** is a measurement comparing the amount of energy present in a fuel and the actual energy output of the engine.

## Combustion Chamber Design

Various combustion chamber designs were discussed previously in Chapter 6. The hemispherical or pent roof combustion chamber, or a variation thereof, is commonly used in high-performance engines. Domed pistons are usually installed to increase the **compression ratio.** Increasing the compression ratio improves the **thermal efficiency** of the engine. When the thermal

**Figure 10-3** Offenhauser engine.

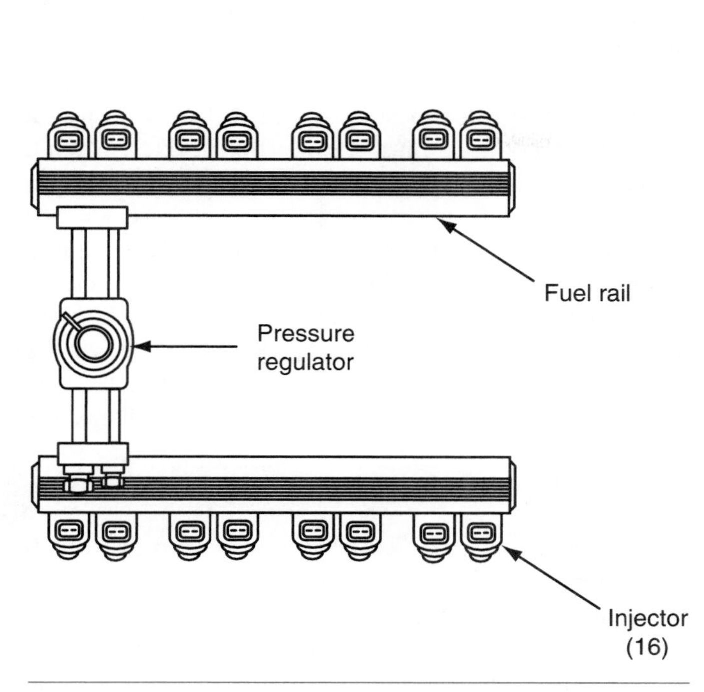


**Figure 10-4** Fuel system with two injectors for each cylinder.

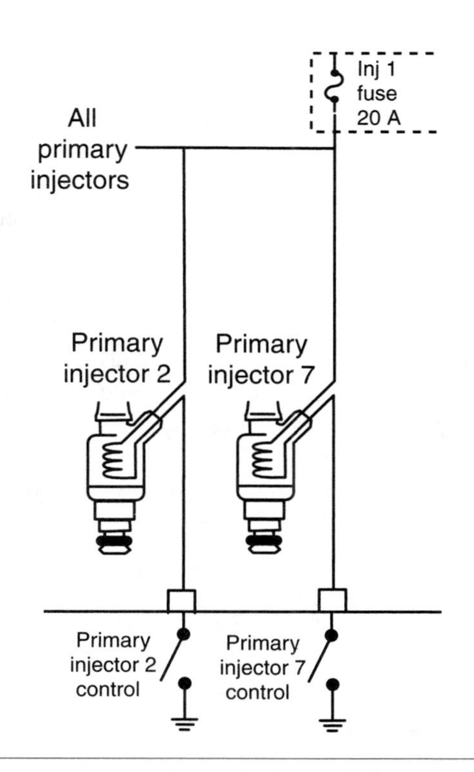


**Figure 10-5** Primary injector electrical circuit.

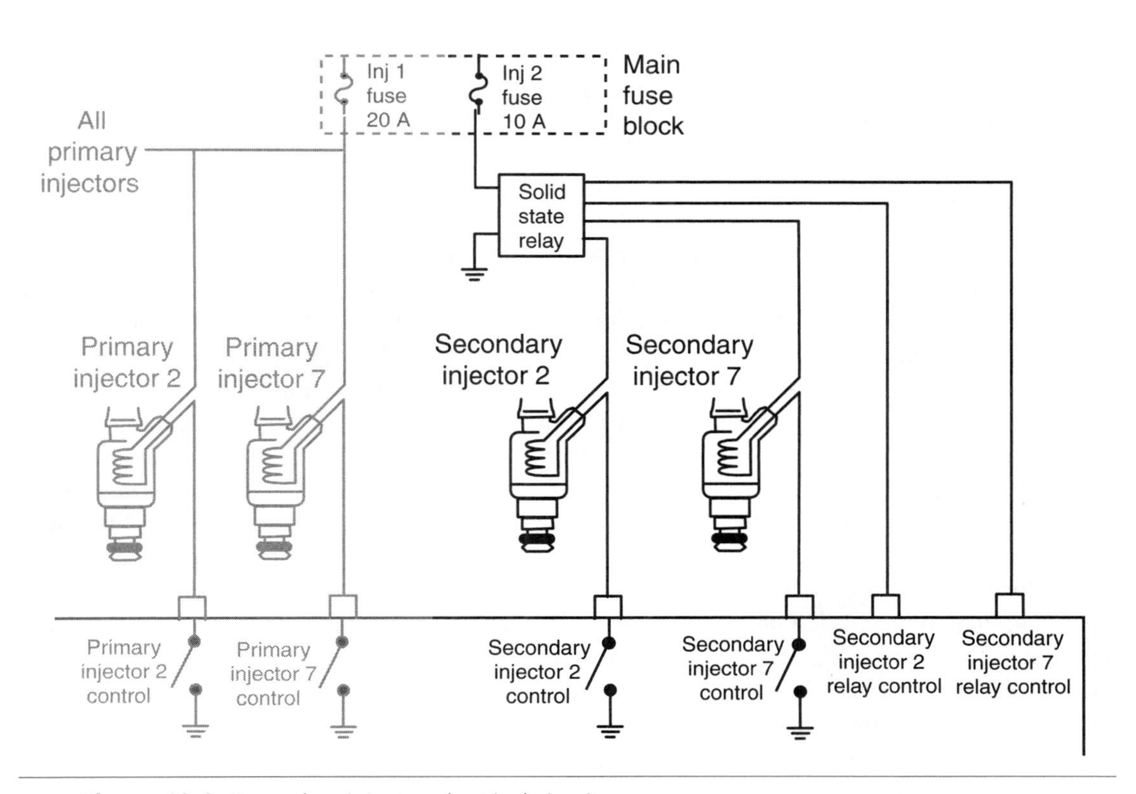


**Figure 10-6** Secondary injector electrical circuit.

efficiency is improved, more of the energy in the fuel is converted into engine power and torque. The domed pistons may have valve recesses in the tops of the pistons. The increased compression ratio increases detonation and temperature in the combustion chambers. Racing engines with considerably higher compression ratios use different fuel to reduce detonation. These fuels may include octane boosters mixed with gasoline or an alcohol such as a methanol blend with a higher octane rating. Compared to gasoline, methanol has a must greater cooling effect when it changes from a liquid to a vapor. This increased cooling effect provides a denser air/fuel mixture and reduced combustion temperatures.

The hemispherical or pent roof combustion chamber allows the installation of larger valves and four valves per cylinder. Using larger valves or more valves increases the volumetric efficiency of the engine. Intake ports in the cylinder head must be large enough so they match the intake manifold runner size and do not restrict airflow. To increase engine rpm, power, and torque, the cylinder head must be designed to flow more air into the combustion chambers, and the combustion chambers must be designed to obtain as much power as possible from the burning of the air/fuel mixture.

## Camshaft

To improve volumetric efficiency and engine performance, camshafts must be designed with higher valve lift, longer duration of valve opening, and more valve overlap. Increasing the valve lift opens the valves wider and allows more air into and out of the combustion chamber. When an intake valve opens, the air does not begin entering the combustion chamber instantly. Therefore, when an intake valve opens, some of the valve open time is actually wasted. Designing the camshaft lobes so the intake valves stay open longer allows more air/fuel mixture to be moved into the combustion chambers.

**Valve overlap** is the length of time, measured in degrees of crankshaft revolution, that the intake and exhaust valves of the same combustion chamber are open simultaneously.

**Valve overlap** is the number of degrees the crankshaft rotates when both valves are open at the same time when the piston is near top dead center (TDC) on the exhaust stroke. When the piston is a few crankshaft degrees before TDC, the intake valve is just beginning to open, and a few degrees after TDC, the exhaust valve closes. If the intake valves open sooner, this action compensates for the delay in airflow mixture movement when the valves first open; however, increasing the valve overlap causes some air/fuel mixture to be swept out through the exhaust valve at low rpm, and this causes rough engine operation at low speeds. However, this is not a concern on a racecar engine that is not driven on the street.

## Valve Train

The valve train plays a significant role in improving engine performance. Overhead (OHC) engines eliminate the use of pushrods. If the camshaft lobes directly contact the valve lash adjusters or valves, rocker arms are also eliminated. Most engines with four valves per cylinder have dual overhead camshafts.

**Valve lift** is the amount of valve movement from the closed to the open position.

Engines with higher **valve lift** and longer duration of valve opening usually have high-tension nickel-chrome valve springs. When valves have a longer duration of valve opening, there is less time between valve closing and valve opening. If valve springs with normal tension are used, the valve may "float" and never really close at high rpm. Valve float causes a loss of compression pressure when the piston is on the compression stroke. High-performance engines are fitted with high-tension valve springs to prevent valve float. The elimination of pushrods and rocker arms also reduces the possibility of valve float.

If the engine has valve lifters, roller-type lifters reduce friction between the camshaft lobes and the lifters. If the high-performance engine has rocker arms, rollers are usually mounted in the rocker arms, and the camshaft lobes contact these rollers. Anything that reduces internal engine friction increases engine power and performance. Some performance engines have solid lifters in place of hydraulic lifters, because hydraulic lifters tend to pump up and hold the valves open too long at high rpm, much like a valve float condition.

## Exhaust Manifolds

A high-performance engine must be able to supply more air/fuel mixture to the combustion chambers and utilize more of the energy that is in the air/fuel mixture. These engines must also be able to deliver a greater volume of exhaust gas through the exhaust system. If the exhaust system does not have the capability to handle a greater volume of exhaust flow, the additional air/fuel mixture supplied to the combustion chambers is wasted. Most high-performance engines have fabricated steel exhaust manifolds. This type of manifold is easier to design into tubes for tuning purposes. Exhaust manifolds may be tuned much like intake manifolds. Proper tuning of an exhaust manifold can actually create a vacuum in the exhaust manifold between pressure pulses, and this vacuum helps to move the exhaust gases out of the system to improve volumetric efficiency. In a **tuned exhaust manifold** a separate tuned pipe is connected to each exhaust port in the cylinder head. These individual pipes are then joined together some distance from the engine (Figure 10-7). A tuned exhaust manifold with individual pipes is designed so the exhaust flow from one cylinder does not interfere with the exhaust flow from the other cylinders, and this improves volumetric efficiency.

A **tuned exhaust manifold** is designed with equal length passages to improve exhaust flow.

## Engine Features

In a high-performance engine, many components have increased durability to withstand the additional pressures created by higher compression ratio. Crankshafts are forged steel and may have additional main bearings. Pistons are stronger, and timing chains are used in place of the timing belt found in many car engines. Connecting rods may be made from powdered iron compounds, which provide a denser, stronger, and lighter component compared to cast or forged rods. Internal friction must be minimized. This may be accomplished through such features as low tension rings and even narrower rings.

Many engine developments that became available in car engines in the 1990s were used in the race car industry for decades. These developments include four valves per cylinder and dual overhead camshafts. In a car engine, the use of four valves per cylinder provides more power

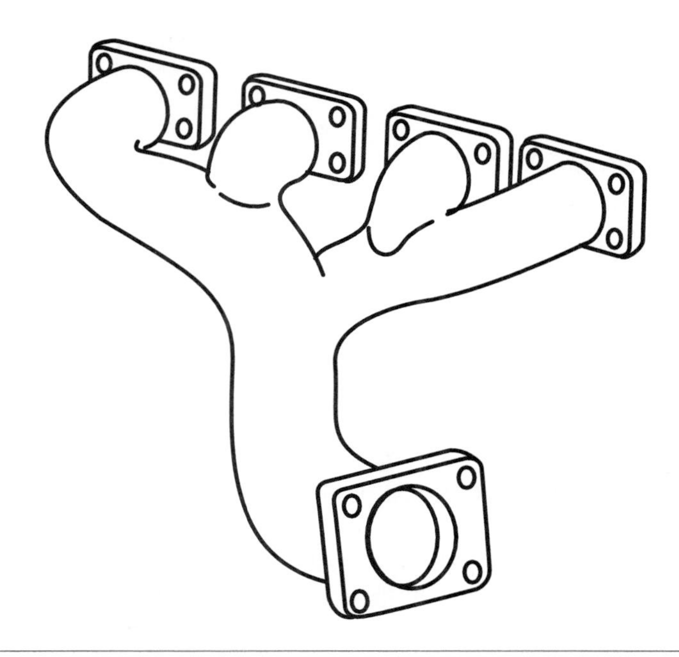

**Figure 10-7** Tuned exhaust manifolds.

and performance from a smaller CID engine. In the 1960s and 1970s, many V8 car engines were between 400 and 500 CID (6.6 to 8.3L) These engines were rated between 200 and 300 HP. The 2000 Chrysler 3.5 V6 four-valve engine in LHS and 300M cars is rated at 251 HP.

Following its debut in 1997, the Oldsmobile Aurora V8 racing engine became the dominant powerplant in the Indy Racing League (IRL). The winning car in each of the nine races in the IRL was powered by an Aurora V8. Race cars powered by the Aurora V8 have won the Indy 500 race for three consecutive years.

The average passenger car or light truck engine may encounter 4,000 to 5,000 rpm during wide-open throttle acceleration; however, some high-performance engines in race cars are capable of speeds up to 14,000 rpm. At this rpm, if the engine is not perfectly balanced, imbalanced components will cause the engine to disintegrate. Therefore, all components in high-performance engines must be balanced.

A race car engine is vastly different from a car engine. Even the lubrication and cooling systems are modified. In a race car engine, an external sump may be used with one or two external pumps to circulate the oil through the engine. When cornering at very high speeds, the oil in a conventional oil pan sloshes to one side, and this may cause the oil pump in the oil pan to pull in air. This action causes lack of lubrication and destroys engine components. The external oil sump and pumps eliminate this problem. The cooling system may be modified with dual water pumps that force coolant in each side of the block.

## Turbochargers

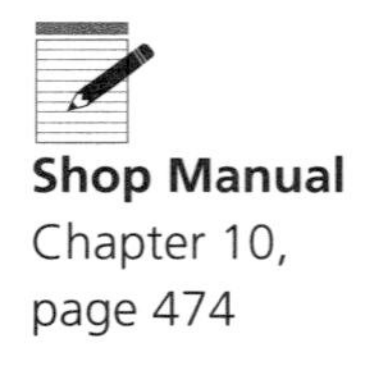

**Shop Manual**
Chapter 10,
page 474

Another method of improving volumetric efficiency and engine performance is to compress the intake air before it enters the combustion chamber. This can be accomplished through the use of a turbocharger or a supercharger.

A turbocharger or supercharger changes the effective compression ratio of an engine, simply by packing in air at a pressure that is greater than atmospheric; for example, an engine that has a compression ratio of 8:1 and receives 10 pounds (69 kPa) of boost will have an effective compression ratio of 10.5:1.

A **turbocharger** uses the expansion of exhaust gases to rotate a fan type wheel which increases the pressure inside of the intake manifold.

A **turbocharger** is a blower or special fan assembly that utilizes the expansion of hot exhaust gases to turn a turbine and compress incoming air. In a typical car engine, a turbocharger may increase the horsepower approximately 20 percent. A typical turbocharger consists of the following components (Figure 10-8).

- Turbine wheel
- Shaft
- Compressor wheel
- Wastegate valve
- Actuator
- Center housing and rotating assembly (CHRA)

### Basic Operation

A **turbine wheel** is a vaned wheel in a turbocharger to which exhaust pressure is supplied to provide shaft rotation.

The **turbine wheel** and the compressor wheel are mounted on a common shaft. Both wheels have fins or blades, and each wheel is encased in its own spiral-shaped housing. The shape of the housing works to control and direct the flow of the gases. The shaft is supported on bearings in the turbocharger housing. The expelled exhaust gases from the cylinders are directed through a nozzle against the blades of the turbine wheel. When engine load is high enough, there is enough exhaust gas flow to cause the turbine wheel and shaft to rotate at a high speed. This action creates a vortex flow. Since the compressor wheel is positioned on the opposite end of the shaft, the compressor wheel must rotate with the turbine wheel (Figure 10-9).

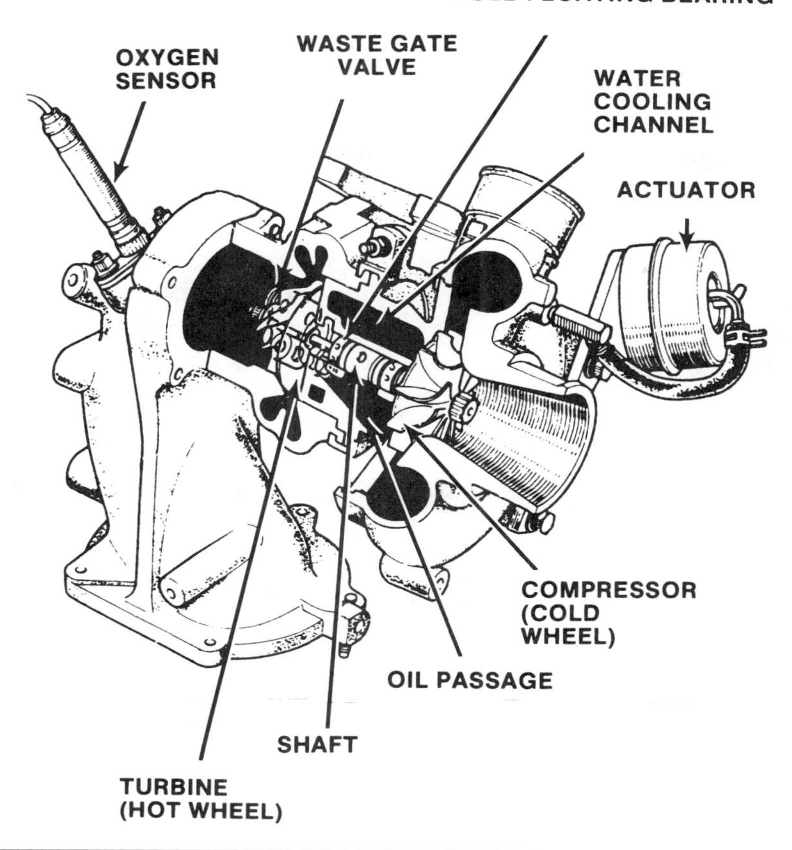


**Figure 10-8** Cross section of a typical turbocharger. (Reprinted with permission)

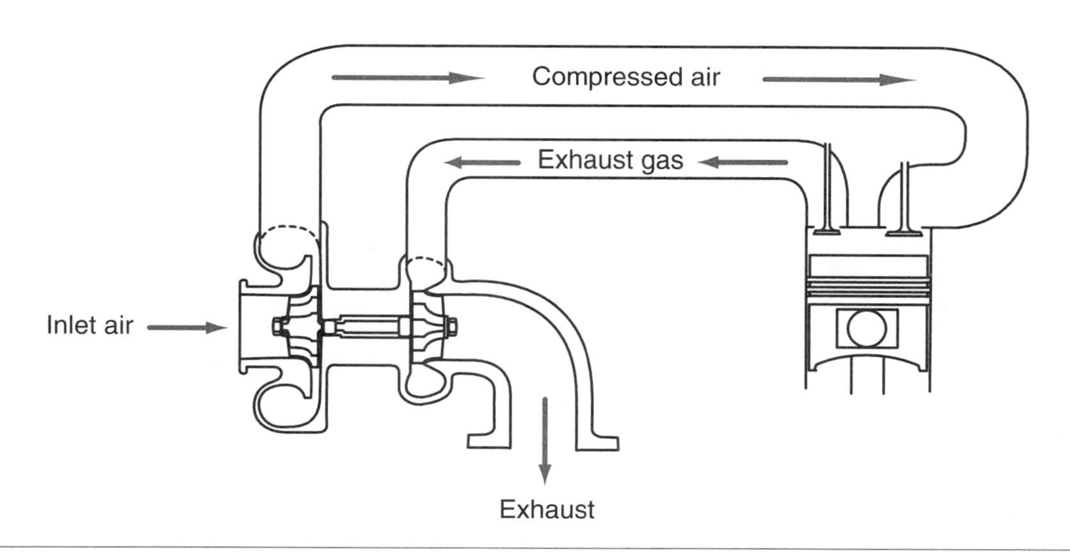


**Figure 10-9** Basic turbocharger operation.

A **compressor wheel** is a vaned wheel mounted on the turbocharger shaft that forces air into the intake manifold.

The **compressor wheel** is mounted in the air intake system. As the compressor wheel rotates, air is forced into the center of the wheel, where it is caught by the spinning blades and thrown outward by centrifugal force. The air leaves the turbocharger housing and enters into the intake manifold. Since most turbocharged engines are port injected, the fuel is injected into the intake ports. The rotation of the compressor wheel compresses the air and fuel in the intake manifold,

Turbo boost is a term used to describe the amount of positive pressure increase of the turbocharger; for example, 10 psi (69 kPa) of boost means the air is being induced into the engine at 24.7 psi (170 kPa). Normal atmospheric pressure is 14.7 psi (101 kPa); add the 10 psi (69 kPa) to get 24.7 psi (170 kPa).

creating a denser air/fuel mixture. This increased intake manifold pressure forces more air/fuel mixture into the cylinders to provide increased engine power. The turbocharger shaft must reach a certain rpm before it begins to pressurize the intake manifold. Some turbochargers begin to pressurize the intake manifold at 1,250 engine rpm, and reach full boost pressure in the intake manifold at 2,250 rpm.

In a normally aspirated engine, air is drawn into the cylinders by the difference in pressure between the atmosphere and engine vacuum. A turbocharger pressurizes the intake charge to a point above normal atmospheric pressure. **Turbo boost** is the positive pressure increase created by the turbocharger.

Turbocharger wheels rotate at very high speeds, in excess of 100,000 rpm. Engineers design and balance the turbocharger to run in excess of 150,000 rpm, about twenty-five times the maximum rpm of most engines. In comparison, the typical automotive air-conditioning generator will run at about 20,000 rpm. Due to this high speed operation, turbocharger wheel balance and bearing lubrication are very important.

**A BIT OF HISTORY**

Turbochargers have been common in heavy-duty applications for many years, but they were not widely used in the automotive industry until the 1980s. Turbochargers had two traditional problems that prevented their wide acceptance in automotive applications. Older turbochargers had a lag, or hesitation, on low-speed acceleration, and there was the problem of bearing cooling. Engineers greatly reduced the low-speed lag by designing lighter turbine and compressor wheels with improved blade design. Water cooling combined with oil cooling provided improved bearing life. These changes made the turbocharger more suitable for automotive applications.

A wastegate limits the maximum amount of turbocharger boost by directing the exhaust gases away from the turbine wheel. The wastegate valve is also referred to as a bypass valve.

**Boost Pressure Control.** If the turbocharger boost pressure is not limited, excessive intake manifold and combustion pressure can destroy engine components. Also, if the amount of boost is too high, detonation knock can occur and decrease engine output. To control the amount of boost developed in the turbocharger, most turbochargers have a **wastegate** diaphragm mounted on the turbocharger. A linkage is connected from this diaphragm to a wastegate valve in the turbine wheel housing (Figure 10-10).

Under low to partial load conditions, the diaphragm spring holds the wastegate valve closed. This routes all of the exhaust gases through the turbine housing. Boost pressure from the intake manifold is also supplied to the wastegate diaphragm (Figure 10-11).

Under full load, when the boost pressure in the intake manifold reaches the maximum safe limit, the boost pressure pushes the wastegate diaphragm and opens the wastegate valve. This action allows some exhaust to bypass the turbine wheel, which limits turbocharger shaft rpm and boost pressure (Figure 10-12)

On some engines, the boost pressure supplied to the wastegate diaphragm is controlled by a computer-operated solenoid. In many systems, the powertrain control module (PCM) pulses the wastegate solenoid on and off to control boost pressure. Some computers are programmed to momentarily allow a higher boost pressure on sudden acceleration to improve engine performance.

Turbo lag is a short delay period before the turbocharger develops sufficient boost pressures.

## Turbo Lag

**Turbo lag** occurs when the turbocharger compressor and turbine wheels are not spinning fast enough to create boost. It takes time to get the exhaust gases to bring the turbocharger wheels up to operating speed. The size and weight of the turbine and compression wheels, along with

Figure 10-10 Wastegate diagram used to control boost pressure. (Courtesy of JTG Associates)

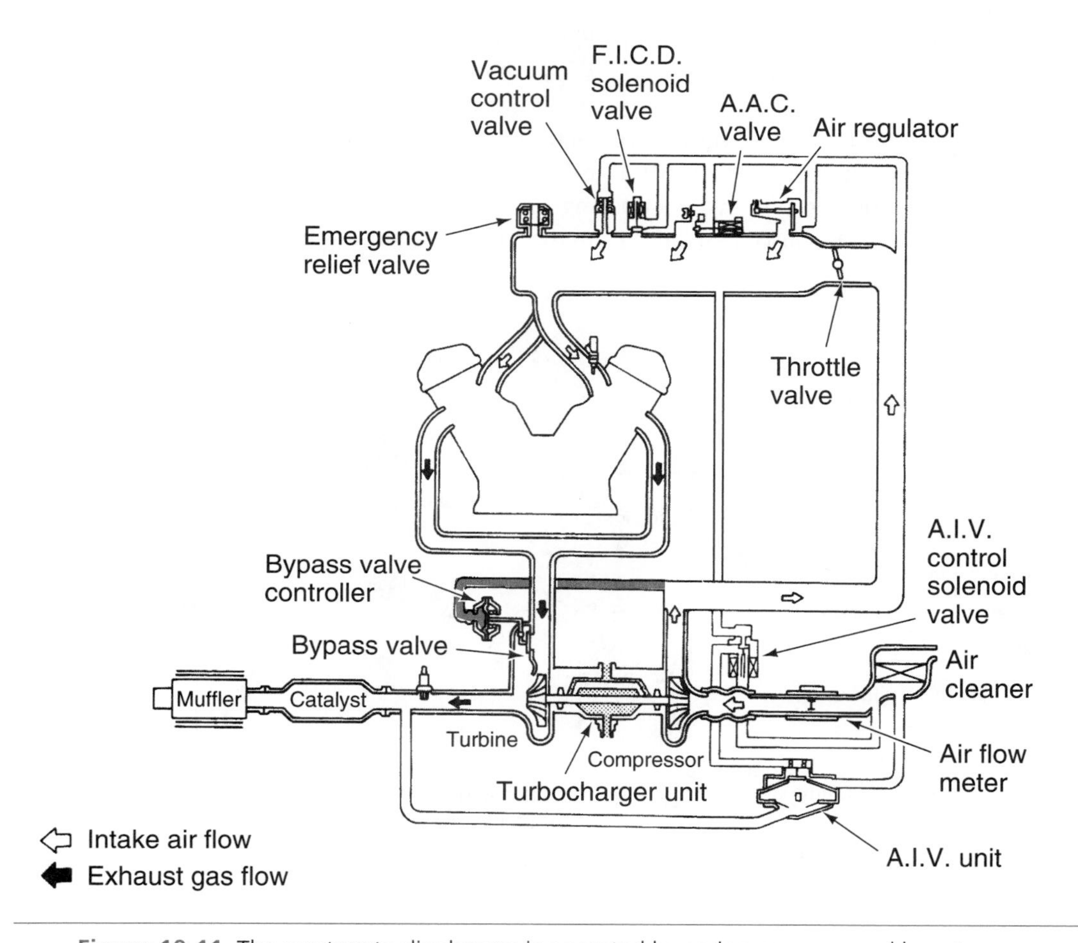


Figure 10-11 The wastegate diaphragm is operated by spring pressure and boost pressure. (Used with permission from Nissan North America, Inc.)

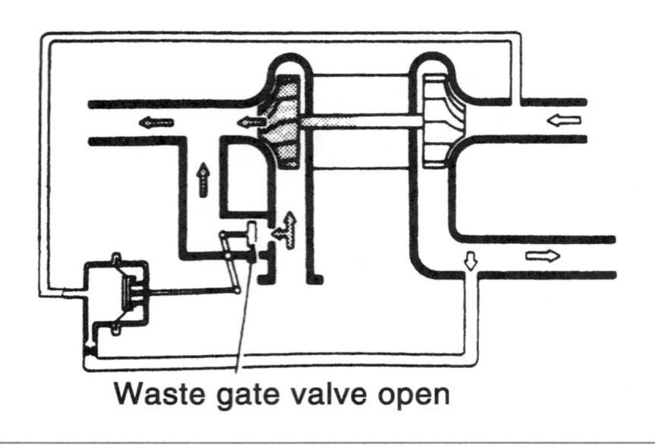


Figure 10-12 As boost pressure increases, the waste-gate valve opens, diverting exhaust gases away from the turbine wheel.

Figure 10-13 Variable nozzle turbine-type turbocharger.

housing design, are factors that affect the amount of turbo lag. Lighter, smaller wheels will get up to speed faster and reduce the effects of turbo lag. Also, variable nozzle turbine (VNT) turbocharger systems have been developed to reduce the lag period (Figure 10-13).

## Turbocharger Cooling

Exhaust flow past the turbine wheel creates very high turbocharger temperature, especially under high engine load conditions. Many turbochargers have coolant lines connected from the turbocharger housing to the cooling system (Figure 10-14). Coolant circulation through the turbocharger housing helps to cool the bearings and shaft. Full oil pressure is supplied from the main oil gallery to the turbocharger bearings and shaft to lubricate and cool the bearings. This oil is drained from the turbocharger housing back into the crankcase. Seals on the turbocharger shaft prevent oil leaks into the compressor or turbine wheel housings. Worn turbocharger seals allow oil into the compressor or turbine wheel housings, resulting in blue smoke in the exhaust and oil consumption. Some heat is also dissipated from the turbocharger to the surrounding air.

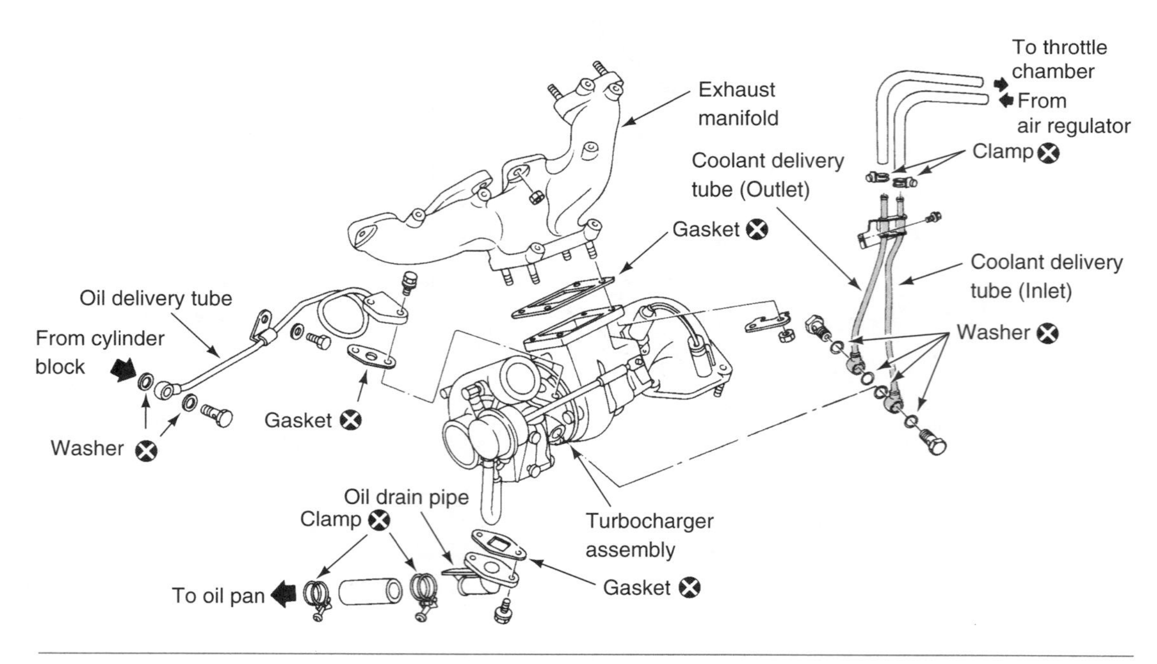


Figure 10-14 Some turbochargers use coolant to control the temperatures of the bearings and the shaft. (Used with permission from Nissan North America, Inc.)

Some turbochargers do not have coolant lines connected to the turbocharger housing. These units depend on oil and air cooling. On these units, if the engine is shut off immediately after heavy-load or high-speed operation, the oil may burn to some extent in the turbocharger bearings. When this action occurs, hard carbon particles, which destroy the turbocharger bearings, are created. The coolant circulation through the turbocharger housing lowers the bearing temperature to help prevent this problem. When turbochargers that depend on oil and air cooling have been operating at heavy load or high speed, idle the engine for at least one minute before shutting it off. This action will help to prevent turbocharger bearing failure.

## Intercoolers

The **intercooler** cools the intake air temperature to increase the density of the air entering the cylinders. It is also referred to as a charge air cooler or an aftercooler.

A disadvantage of the turbocharger (and supercharger) is that it heats the incoming air. The hotter the air, the less dense it is. As the air gets hotter, fewer air molecules can enter the cylinder on each intake stroke. Also, hotter intake air leads to detonation problems. To combat these effects, many turbocharger systems use an **intercooler.** The intercooler is like a radiator in that it removes heat from the turbocharger system by dissipating it to the atmosphere. The intercooler can be either air cooled or water cooled. Cooling the air makes it denser, increasing the amount of oxygen content with each intake stroke. The intercooler cools the air that leaves the turbocharger at about 100°F (38°C) before it enters the cylinders (Figure 10-15). For every 10°F (5.5°C) that the air is cooled, a power gain of about 1 percent is obtained. If the intercooler is capable of cooling the air by 100°F (38°C), then a 10 percent power increase is obtained.

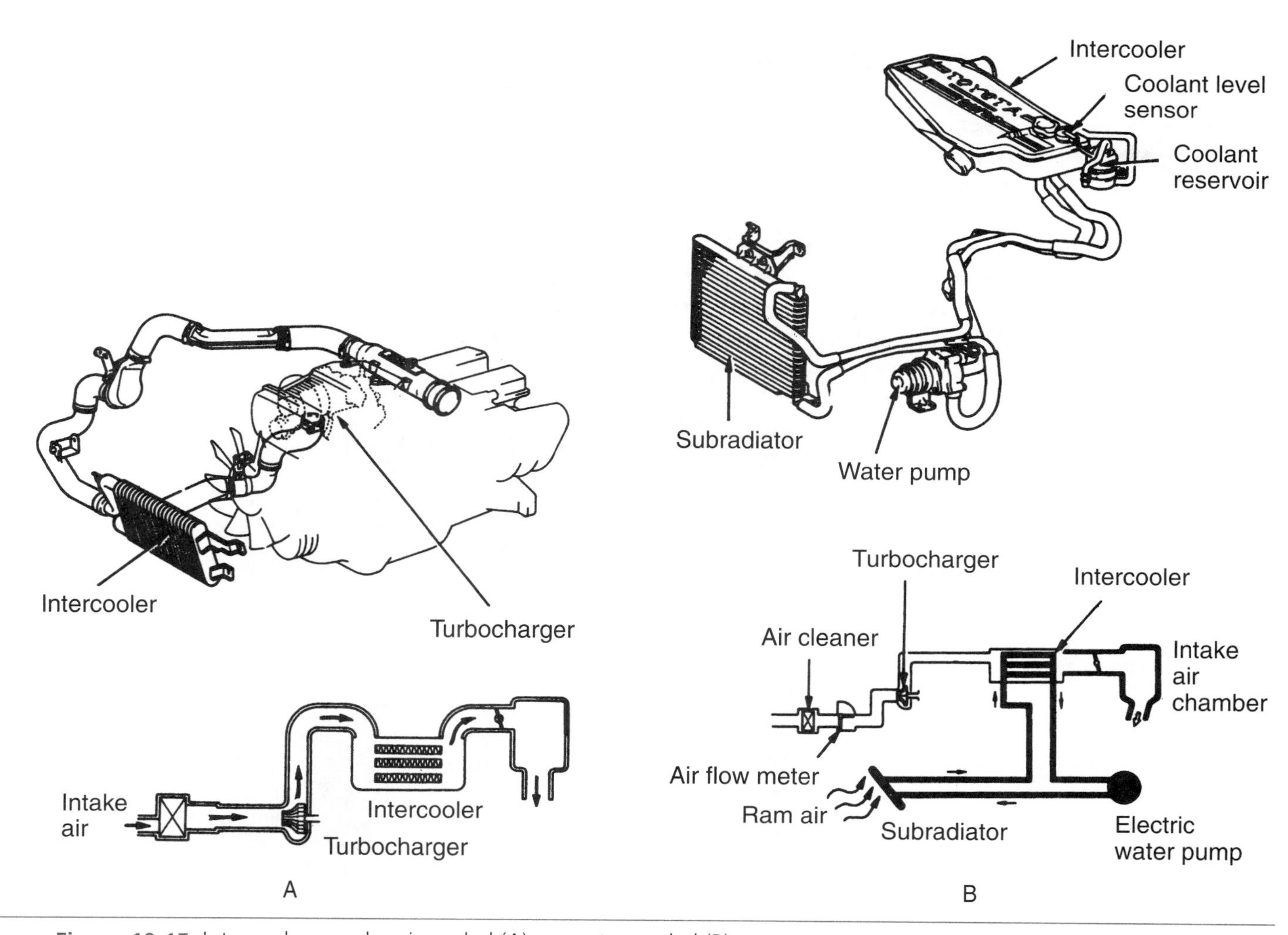


**Figure 10-15** Intercoolers can be air cooled (A), or water cooled (B).

## Scheduled Maintenance

There are three main things that will reduce the life of a turbocharger: lack of oil, contaminants in the oil, and ingestion of foreign material through the air intake. To prevent premature turbocharger failure, engine oil and filters should be changed at the vehicle manufacturer's recommended intervals. The engine oil level must be maintained at the specified level on the dipstick. The air cleaner element and the air intake system must be maintained in satisfactory condition. Dirt entering the engine through an air cleaner will damage the compressor wheel blades. When coolant lines are connected to the turbocharger housing, the cooling system must be maintained according to the vehicle manufacturer's maintenance schedule to provide normal turbocharger life.

Turbocharged engines have a lower compression ratio than a normally aspirated engine, and many parts are strengthened in a turbocharged engine because of the higher cylinder pressure. Therefore, many components in a turbocharged engine are not interchangeable with the parts in a normally aspirated engine.

**AUTHOR'S NOTE:** During my experience in the automotive service industry, I encountered a significant number of turbocharged engines with low boost pressure. This low boost pressure was often caused by low engine compression. Because low engine compression reduces the amount of air taken into the engine, this problem also reduces turbocharger speed and boost pressure. Therefore, when diagnosing low boost pressure, always verify the engine condition before servicing or replacing the turbocharger.

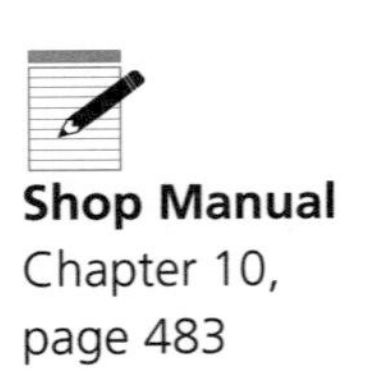

**Shop Manual**
Chapter 10,
page 483

# Superchargers

The **supercharger** is a belt-driven air pump used to increase the compression pressures in the cylinders

The supercharger may be called a blower.

After falling into disuse for a number of years (except for racing applications), the **supercharger** started to reappear on automotive engines as original equipment manufacturer (OEM) equipment in 1989 (Figure 10-16). The supercharger is belt driven from the crankshaft by a ribbed belt (Figure 10-17). A shaft is connected from the crankshaft pulley to one of the drive gears in the front supercharger housing, and the driven gear is meshed with the drive gear. The rotors inside the supercharger are attached to the two drive gears (Figure 10-18).

The drive gear design prevents the rotors from touching; however, there is a very small clearance between the drive and driven gears. In some superchargers, the rotor shafts are sup-

Figure 10-16 A factory-installed supercharger.

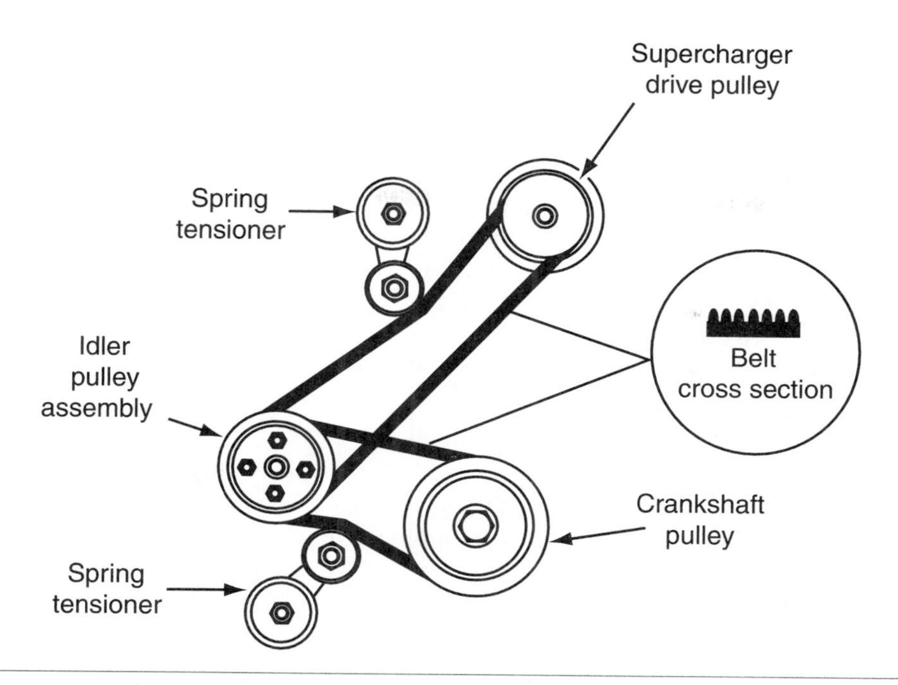


**Figure 10-17** The supercharger is driven by a belt off of the crankshaft.

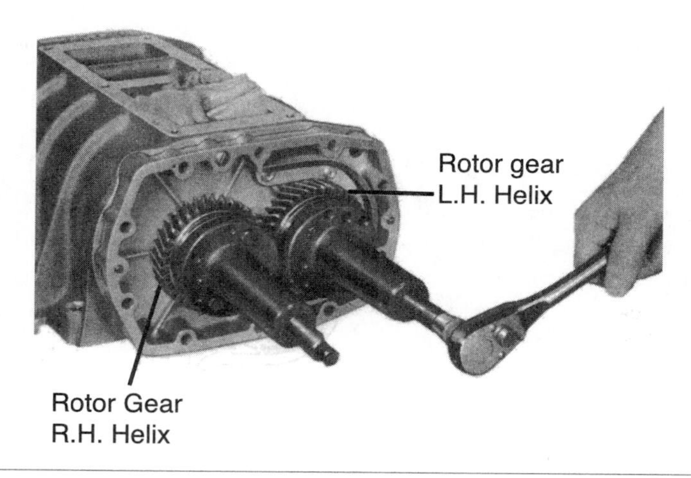


**Figure 10-18** The rotors are driven by gears in the front of the supercharger. (Courtesy of Detroit Diesel Allison)

ported by roller bearings on the front and needle bearings on the back. During the manufacturing process, the needle bearings are permanently lubricated. The ball bearings are lubricated by a synthetic base high-speed gear oil. A plug is provided for periodic checks of the front bearing lubricant. Front bearing seals prevent lubricant loss into the supercharger housing.

In a typical car engine, a supercharger increases horsepower approximately 20 percent compared to a normally aspirated engine with the same CID; for example, the General Motors normally aspirated 1997 3800 V6 engine is rated at 200 HP. The supercharged version of this engine is rated at 240 HP. The supercharger will operate around 10,000 to 15,000 rpm. Unlike the turbocharger, the amount of boost the supercharger produces is a function of engine rpm (not load). The advantage of the supercharger is that it will produce more torque at lower engine speeds than a turbocharger. Also, there is no lag time associated with a supercharger. The disadvantage of the supercharger is that it consumes horsepower as it is driven.

## Supercharger Operation

The **Roots**-type supercharger is a positive displacement supercharger that uses a pair of lobed vanes to pump and compress the air.

While there have been a number of supercharger designs on the market over the years, the most popular is the **Roots** type. The pair of three-lobed rotor vanes (Figure 10-19) in the Roots supercharger is driven by the crankshaft. The lobes force air into the intake manifold. The helical design evens out the pressure pulses in the blower and reduces noise. It was found that a 60-degree helical twist works best for equalizing the inlet and outlet volumes. Another benefit of the helical rotor design is it reduces carryback volumes—air that is carried back to the inlet side of the supercharger because of the unavoidable spaces between the meshing rotors which represents a loss of efficiency.

To handle the higher operating temperatures imposed by supercharging, an engine oil cooler is usually built into the engine lubrication system. This water-to-oil cooler is generally mounted between the engine front cover and oil filter.

Intake air enters the inlet plenum at the back of the supercharger, and the rotating blades pick up the air and force it out the top of the supercharger. The blades rotate in opposite directions, acting as a pump as they rotate. This pumping action pulls air through the supercharger inlet and forces the air from the outlet. There is a very small clearance between the meshed rotor lobes and between the rotor lobes and the housing (Figure 10-20).

Air flows through the supercharger system components in the following order:

1. Air flows through the air cleaner and mass airflow sensor into the throttle body.
2. Airflow enters the supercharger intake plenum.
3. From the intake plenum the air flows into the rear of the supercharger housing (Figure 10-21).
4. The compressed air flows from the supercharger to the intercooler inlet.
5. Air leaves the intercooler and flows into the intercooler outlet tube.
6. Air flows from the intercooler outlet tube into the intake manifold adapter.
7. Compressed, cooled air flows through the intake manifold into the engine cylinders.
8. If the engine is operating at idle or very low speeds, the supercharger is not required. Under this condition, airflow is bypassed from the intake manifold adapter through a butterfly valve to the supercharger inlet plenum (Figure 10-22).

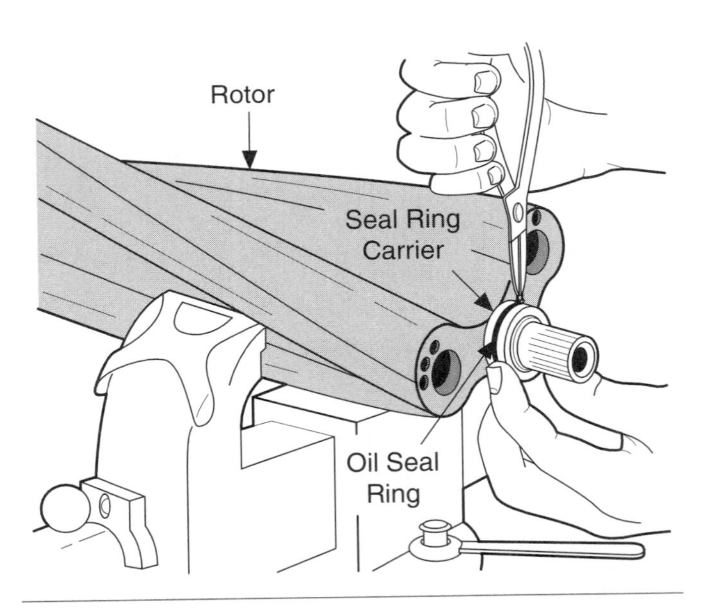


**Figure 10-19** Common design of the rotor. (Courtesy of Detroit Diesel Allison)

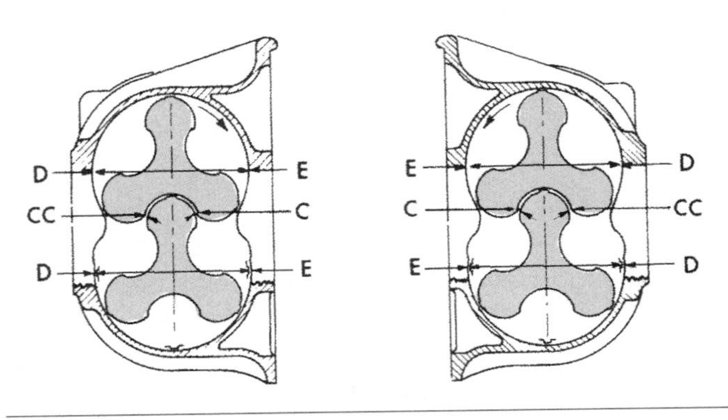


**Figure 10-20** Pumping action of the rotors. (Courtesy of Detroit Diesel Allison)

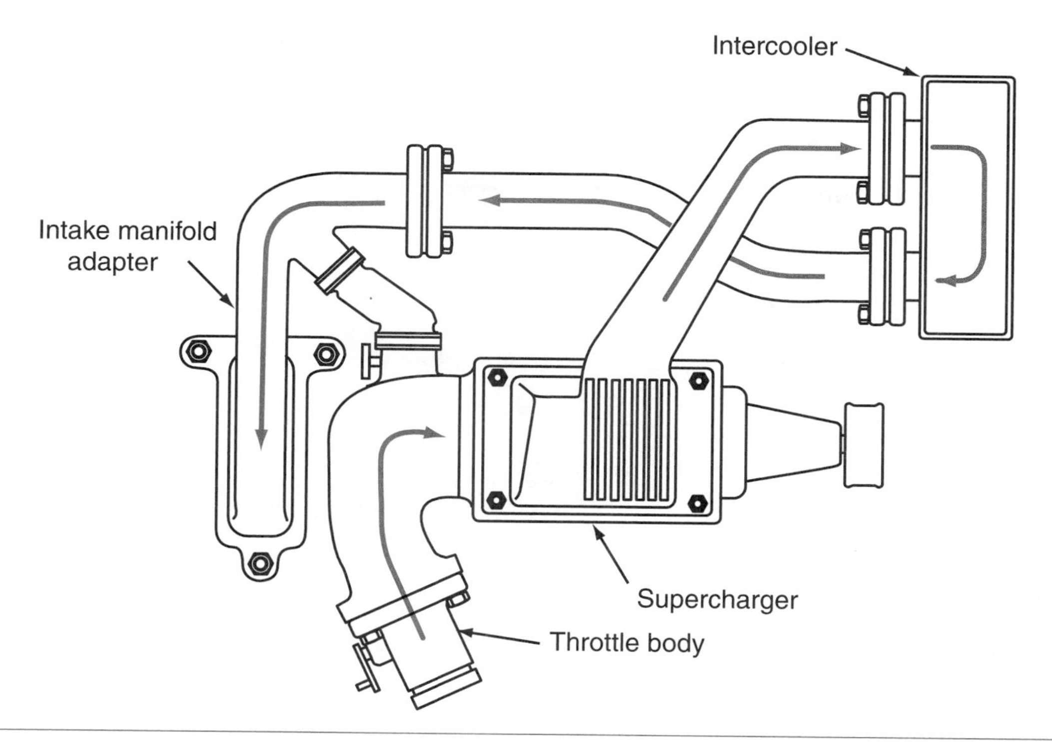


**Figure 10-21** Air flow through a supercharger.

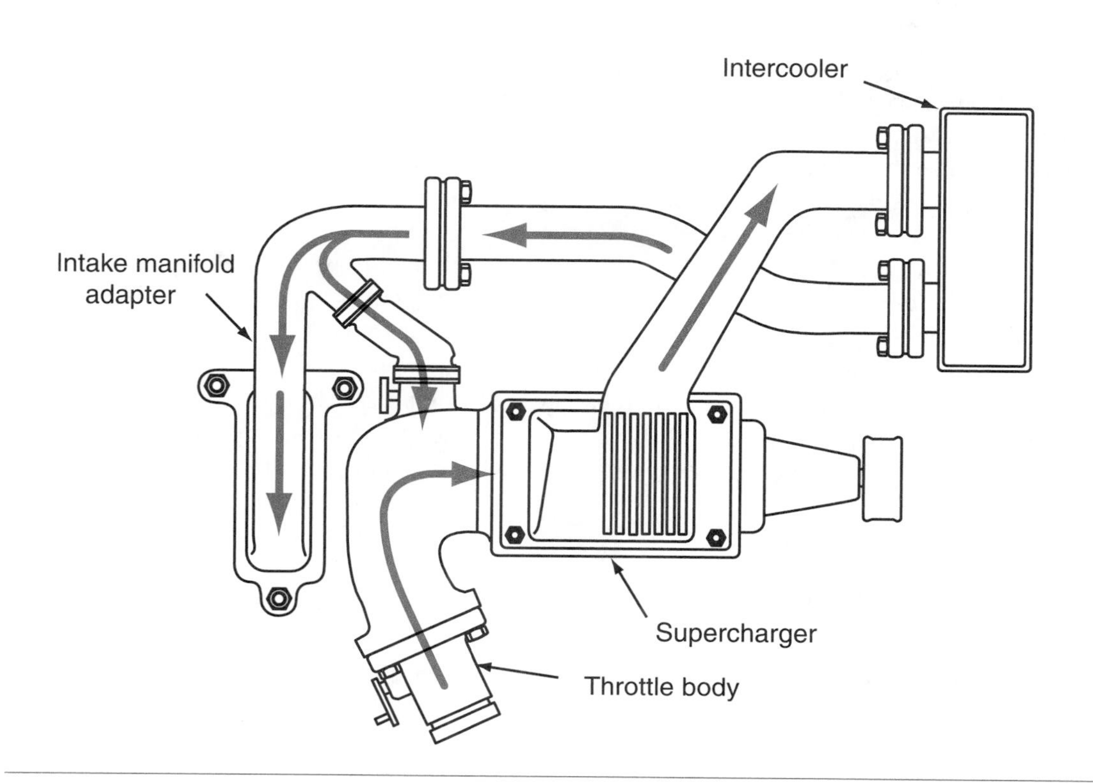


**Figure 10-22** Supercharger air bypass actuator.

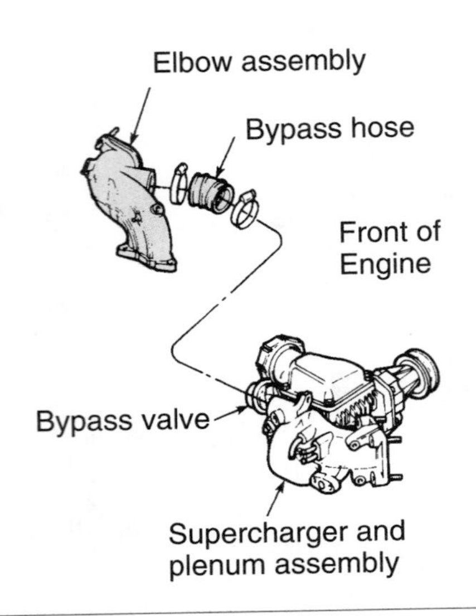


**Figure 10-23** Supercharger bypass hose and intake elbow.

The bypass butterfly valve (Figure 10-23) is operated by an air bypass actuator diaphragm as follows:

1. When manifold vacuum is 7 in. Hg (23.6 kPa) or higher, the bypass butterfly valve is completely open and a high percentage of the supercharger air is bypassed to the supercharger inlet.
2. If the manifold vacuum is 3 to 7 in. Hg (10 to 23.6 kPa), the bypass butterfly valve is partially open and some supercharger air is bypassed to the supercharger inlet, while the remaining airflow is forced into the engine cylinders.
3. When the vacuum is less than 3 in. Hg (10 kPa), the bypass butterfly valve is closed and all the supercharger airflow is forced into the engine cylinders.

On some superchargers, the pulley size causes the rotors to turn at 2.6 times the engine speed. Since supercharger speed is limited by engine speed, a supercharger wastegate is not required. Belt-driven superchargers provide instant low-speed action compared to exhaust-driven turbochargers, which may have a low-speed lag because of the brief time interval required to accelerate the turbocharger shaft. Compared to a turbocharger, a supercharger turns at much lower speeds.

Friction between the air and the rotors heats the air as it flows through the supercharger. The intercooler dissipates heat from the air in the supercharger system to the atmosphere, creating a denser air charge. When the supercharger and the intercooler supply cooled, compressed air to the cylinders, engine power and performance are improved.

The compression ratio in the supercharged 3.8L engine is 8.2:1, compared to a normally aspirated 3.8L engine, which has a 9.0:1 compression ratio. The following components are reinforced in the supercharged engine because of the higher cylinder pressure:

- Engine block
- Main bearings
- Crankshaft bearing caps
- Crankshaft
- Steel crankshaft sprocket
- Timing chain
- Cylinder head
- Head bolts
- Rocker arms

Superchargers can be enhanced with electrically operated clutches and bypass valves. These allow the same computer that controls fuel and ignition to kick the boost on and off precisely as needed. This results in far greater efficiency than a full-time supercharger.

**Shop Manual**
Chapter 10, page 484

## Summary

- ❑ Intake manifold tuning may be described as designing individual intake runners that provide the maximum airflow at the desired engine rpm.
- ❑ Some tuned intake manifolds have dual runners into each intake port.
- ❑ The primary low-speed intake runners may be smaller in diameter and longer. The shorter, larger diameter secondary intake runners may have butterfly valves that are opened by the PCM at a specific rpm.
- ❑ Some engines with dual intake runners have dual injectors in each intake runner. The injectors in the secondary runners are activated by the PCM when the butterfly valves open the secondary runners.
- ❑ Increasing the compression ratio improves thermal efficiency and allows more energy in the fuel to be converted to engine power and torque.
- ❑ Using larger valves or more valves in the cylinder head increases volumetric efficiency.
- ❑ A camshaft with longer duration of valve opening, increased overlap, and more valve lift improves volumetric efficiency.
- ❑ Valve float may be reduced by installing high-tension valve springs.
- ❑ Tuned exhaust manifolds contain a separate tuned pipe for each cylinder.
- ❑ In a high-performance engine, many components such as the crankshaft, camshaft, pistons, connecting rods, main bearings, and engine block have increased strength.
- ❑ In a turbocharger, the exhaust gas from the cylinders flows past the turbine wheel causing it to rotate at high speed.
- ❑ The turbine wheel and compressor wheel in a turbocharger are mounted on a common shaft.
- ❑ The compressor wheel in a turbocharger forces more air into the intake manifold. Since the intake manifold is pressurized, more air flows into the engine cylinders, resulting in increased engine power.
- ❑ The wastegate valve in a turbocharger is controlled by a wastegate diaphragm, and this diaphragm is operated by boost pressure.
- ❑ When the wastegate diaphragm opens the wastegate valve, some exhaust bypasses the turbine wheel, which limits boost pressure.
- ❑ A supercharger is belt-driven from the engine.
- ❑ Supercharger rotors usually contain three lobes, which force air into the intake manifold.
- ❑ The compression ratio is lower in a supercharged or turbocharged engine compared to a normally aspirated engine.
- ❑ Many components are strengthened in a supercharged or turbocharged engine because of the higher cylinder pressures.

**Terms to Know**

Compression ratio
Compressor wheel
Intake manifold tuning
Intercooler
Normally aspirated
Roots
Supercharger
Thermal efficiency
Tuned exhaust manifold
Turbine wheel
Turbo boost
Turbocharger
Turbo lag
Valve lift
Valve overlap
Wastegate

# Review Questions

## Short Answer Essays

1. Explain how the intake manifold may be designed to improve airflow.
2. Explain the operation of a fuel system with dual intake runners and dual injectors for each cylinder.
3. Describe combustion chamber modifications that increase volumetric efficiency and engine power.
4. Explain camshaft modifications that increase volumetric efficiency.
5. Explain valve float and describe how it may be reduced.
6. Describe the design of a tuned exhaust manifold.
7. Explain how a turbocharger or supercharger supplies more engine power.
8. Describe basic turbocharger operation.
9. What is the purpose of the intercooler?
10. Explain basic supercharger operation.

## Fill-in-the-Blanks

1. At low engine speeds, ____________ ____________ - ____________ intake runners improve volumetric efficiency.
2. A tuned exhaust manifold has ____________ ____________ ____________ connected to each exhaust port in the cylinder head.
3. At high engine speeds ____________ ____________ - ____________ intake manifold runners improve volumetric efficiency.
4. The exhaust flows past the ____________ wheel in a turbocharger.
5. The intake airflow is forced into the intake manifold by the ____________ wheel in a turbocharger.
6. The wastegate diaphragm is moved by ____________ pressure from the intake manifold.
7. Turbocharged or supercharged engines have ____________ compression ratios compared to normally aspirated engines.
8. A supercharger is ____________-driven from the engine.
9. The air flows through the intercooler ____________ it flows through the supercharger.
10. Supercharger rotor speed is limited by ____________ ____________ .

## Multiple Choice

1. A tuned intake manifold may have these features:
   **A.** Low-speed runners that are larger in diameter than the high-speed runners.
   **B.** Runners with sharp bends.
   **C.** High-speed runners that are longer than the low-speed runners
   **D.** Dual intake runners with secondary runners that are open only at high engine speeds.

2. All of these statements about combustion chamber design are true EXCEPT:
   **A.** Hemispherical or pent roof combustion chambers can accommodate four valves per cylinder.
   **B.** Increasing the compression ratio improves thermal efficiency and engine power.
   **C.** The compression ratio may be increased by installing pistons with dished tops.
   **D.** A higher compression ratio may require a higher octane fuel to reduce detonation.

3. Increasing the valve overlap by opening the intake valves sooner results in:
   **A.** Improved low-speed engine power.
   **B.** Improved engine idle quality.
   **C.** Improved engine power at high rpm.
   **D.** Decreased volumetric efficiency.

4. A tuned exhaust manifold:
   **A.** Has dual pipes connected to each exhaust port.
   **B.** Creates a vacuum in the exhaust manifold between pressure pulses.
   **C.** Is made from cast iron or nodular iron.
   **D.** Allows the exhaust flow from one cylinder to interfere with the exhaust flow from another cylinder.

5. While discussing turbocharger operation:
   *Technician A* says the turbocharger shaft may turn at speeds in excess of 100,000 rpm.
   *Technician B* says the exhaust is routed past the compressor wheel.
   Who is correct?
   **A.** A only **C.** Both A and B
   **B.** B only **D.** Neither A nor B

6. When discussing turbocharger boost pressure control:
   *Technician A* says when the wastegate valve is open, intake air is by-passed around the compressor wheel.
   *Technician B* says a hose is connected from the intake manifold to the wastegate diaphragm.
   Who is correct?
   **A.** A only **C.** Both A and B
   **B.** B only **D.** Neither A nor B

7. While discussing turbocharger bearing cooling:
   *Technician A* says inadequate bearing cooling may result in the formation of hard carbon particles in the bearings.
   *Technician B* says if turbocharger bearing cooling is inadequate, the oil may burn to some extent in the bearings after a hot engine is shut off.
   Who is correct?
   **A.** A only **C.** Both A and B
   **B.** B only **D.** Neither A nor B

8. While discussing turbocharged engines:
   *Technician A* says the parts are interchangeable between a turbocharged engine and the same size normally aspirated engine.
   *Technician B* says a turbocharged engine has a higher compression ratio than a normally aspirated engine.
   Who is correct?
   **A.** A only **C.** Both A and B
   **B.** B only **D.** Neither A nor B

9. While discussing supercharger operation:
   *Technician A* says a supercharger requires a wastegate to prevent excessive intake manifold pressure at high engine speeds.
   *Technician B* says a typical supercharger speed would be 2.6 times the engine speed.
   Who is correct?
   **A.** A only **C.** Both A and B
   **B.** B only **D.** Neither A nor B

10. While discussing supercharger control:
    *Technician A* says the vacuum diaphragm opens the bypass butterfly valve at low engine speeds, and this valve bypasses a high percentage of the supercharger airflow to the supercharger inlet.
    *Technician B* says the vacuum diaphragm opens the bypass butterfly valve under hard acceleration, and this valve bypasses some of the supercharger airflow to the atmosphere.
    Who is correct?
    **A.** A only **C.** Both A and B
    **B.** B only **D.** Neither A nor B

# GLOSSARY

*Note: Terms are highlighted in color, followed by Spanish translation in bold.*

A

Abrasion Wearing or rubbing that damages the surface area of a part.
**Abrasión** El desgaste o frotamiento de una parte que daña la superficie de una área.

Additive A material added to the engine oil to provide additional properties not originally found in the oil.
**Aditivo** Una materia añadida al aceite de motor para proporcionar unas propriedades adicionales que no se encuentran originalmente en el aceite.

Adhesion The property of oils to cling to surfaces.
**Adhesión** La propriedad de los aceites que permite que se adhieren a las superficies.

Advanced camshaft Camshaft design that has the intake valves opened more than the exhaust valves at TDC. This type of camshaft is used to increase output at low rpm.
**Arbol de levas de avance** Un diseño del árbol de levas que tiene las válvulas de admisión abiertas más que las válvulas de escape en PMS. Este tipo de árbol de levas se emplea para aumentar la producción del par en bajas rpm.

AERA Automotive Engine Rebuilder's Association.
**AERA** La Asociación de Reacondicionadores Automotivos.

Aerobic sealants Sealants that require the presence of oxygen to cure.
**Sellantes aeróbicos** Los sellantes que requieren la presencia del oxígeno para curarse.

Aftermarket Equipment and parts sold to consumers after the production of the vehicle.
**Repuestos no originales** Los accesorios y partes vendidos al consumidor después de la producción de un vehículo.

Align bore The process of boring, reaming, or honing the main bearing bores to center all bores onto a true centerline.
**Calibrado en serie** El proceso de taladrar, escariar o bruñir los taladros de los muñones del cigüeñal para alinearlos todos en una linea recta central.

Align honing Machining process of the main bearing journals used to remove in excess of 0.050 inch, but this will result in changing the location of the crankshaft in the block. Align honing restores original bore size by removing metal from the entire circumference of the bore.
**Rectificado en serie** Un proceso de rectificar en máquina los muñones del cigüeñal para quitar más de 0.050 de una pulgada, pero ésto resulta en que cambia de lugar el cigüeñal en el monoblock. El rectificado en serie restaura el tamaño original del taladro removiendo el metal de toda su circunferencia.

Aligning bars Tools used to determine the proper alignment of the crankshaft saddle bores.
**Barras para alinear** Las herramientas que sirven para determinar el alineamiento correcto de los taladros del asiento del cigüeñal.

Alkaline A chemical class that is the opposite of acidic, but is just as corrosive to certain materials.
**Alcalino** Un tipo de químico que es lo contrario del ácido, pero igualmente corrosivo para algunas materiales.

Alloys Mixtures of two or more metals; for example, brass is an alloy of copper and zinc.
**Aleación** La mezcla de dos o más metales; por ejemplo, el latón es una aleación de cobre con zinc.

Anaerobic sealants Sealants that will only cure in the absence of oxygen.
**Sellantes anaeróbicos** Los sellantes que sólo se curan en la ausencia del oxígeno.

Antifriction bearing A bearing constructed with balls or rollers between the journal and the bearing surface.
**Cojinete antifricción** Un cojinete construido con bolas o rodillos entre las superficies de apoyo.

Asymmetrical pattern camshaft Camshaft design in which the lobes on the opening ramps are ground differently than on the closing ramps.
**Arbol de levas modelo asimétrico** Un diseño de árbol de levas en el cual el lado del lóbulo de apertura se muela de una manera distinta que el lado de cerradura.

Atmospheric pressure The weight of the air.
**Presión atmosférica** El peso del aire.

Austenitic steel A corrosion resistant steel made with carbide or carbon alloys.
**Acero austenítico** Un acero resistente a la corrosión hecho con el carburo o las aleaciones de carbono.

B

Backlash The clearance between two parts.
**Culateo/juego** La holgura entre dos partes.

Backpressure The pressure created within the engine cylinder as a result of a restricted exhaust system. With the creation of backpressure, vacuum cannot be produced as efficiently.
**Contrapresión** La presión creada dentro del cilindro del motor causada por una restricción en el sistema de escape. Al crear esta presión, el vacío no se puede producir eficazmente.

Balance Weight introduced to prevent vibration of moving parts.
**Equilibrio** Un peso metido para prevenir la vibración de las partes en movimiento.

**Balancing** The process of removing vibrations that are caused by the entire engine's reciprocating and rotational mass.
**Equilibrar** El proceso de quitar las vibraciones causadas por la masa recíproca y rotativa del motor.

**Bead blasters** Parts cleaners that use beads of abrasive media carried by air pressure to clean the parts. The abrasives knock off the contaminants.
**Chorro de perlitas** Una limpiadora de partes que usa una materia abrasiva impulsada por el aire bajo presión para limpiar las partes. Los abrasivos desprenden los contaminantes.

**Bearing** Soft metallic shells used to reduce friction created by rotational forces.
**Cojinete** Una pieza hueca de metal blanda que sirve para reducir la fricción creada por las fuerzas giratorias.

**Bearing cap** The removable half of the saddle that holds the bearing in place.
**Tapa del cojinete** La mitad removable del asiento que sostiene al cojinete en su lugar.

**Bearing crush** The extension of the bearing half beyond the seat which is crushed into place when the seat is tightened.
**Aplastamiento del cojinete** La parte de extensión del cojinete debajo del asiento que se aplasta en su lugar al apretar el asiento.

**Bearing lining** A layer of alloy that is adhered to the bearing back and forms the bearing surface.
**Revestamiento del cojinete** Una capa de aleación que se adhiere a la parte exterior de un cojinete y forma la superficie del cojinete.

**Bearing spread** The distance between the outside parting edges is larger than the diameter of the bore.
**Envergadura del cojinete** La distancia al través de los rebordes exteriores es más grande que el diámetro del taladro.

**Big end** The end of the connecting rod that attaches to the crankshaft.
**Extremo grande** Refiere a la extremidad de la biela que se conecta al cigüeñal.

**Blowby** The unburned fuel and combustion products that leak past the piston rings and enter the crankcase.
**Soplado** El combustible no consumido y los productos de la combustión que escapen por los anillos de pistón y entran al cárter.

**Blueprinting** A technique of building an engine using stricter tolerances than those used by most manufacturers. This results in a smoother running, longer lasting, and higher output engine.
**Especificado por el plan detallado** Una técnica de construcción del motor usando las tolerancias más exactas de las que usan la mayoría de los fabricantes. Esto resulta en un motor que marcha mejor, dura más, y de mejor rendimiento.

**Bob weights** Devices that are attached to the throws of the crankshaft to simulate the rotating and reciprocating mass of the piston assembly.
**Pesos de contra-balanzón** Los dispositivos que se conectan al brazo excéntrico del cigüeñal para simular la masa giratoria y recíproca del conjunto de los pistones.

**Bore** The diameter of a hole.
**Taladro** El diámetro de un agujero.

**Borescope** Special tool that uses fiber optics to allow the technician to see the internal condition of the engine without having to disassemble it.
**Calibrescopio** Una herramienta especial que usa la tecnología fibro-óptica para permitir que el técnico vea las condiciones internas del motor sín tener que desarmarlo.

**Boring** The process of enlarging a hole.
**Escariar** El proceso de agrandar un agujero.

**Boss** The cast or forged part of a piston that can be machined for balance.
**Resalto** La parte colada o forjada de un pistón que se puede rebajar por máquina para el equilibrio.

**Bottom Dead Center (BDC)** Term used to indicate the piston is at the very bottom of its stroke.
**Punto Muerto Inferior (PMI)** El término uqe indica que el pistón esta en el punto más inferior de su carrera.

**Broach** To finish the inside surface of a bore by forcing a multiple-edge cutting tool through it.
**Fresar con barrena** Acabar una superficie interior de un taladro al atravesarla con una herramienta que tiene múltiples hojas cortantes.

**Burn time** The amount of time from the instant the mixture is ignited until the combustion is complete.
**Tiempo de encendido** La cantidad del tiempo desde el instante que se enciende la mezcla hasta que se haya completado la combustión.

**Burned valves** Valves that have warped and melted, leaving a groove across the valve head.
**Válvulas quemadas** Las válvulas que se han deformado y fundido, dejándo una ranura al través de la cabeza de la válvula.

**Burnish** To smooth or polish with a sliding tool under pressure.
**Bruñir** Alisar o pulir con una herramienta deslizandola bajo presión.

**Bushing** A removable liner for a bearing.
**Casquillo** Un forro removable para un cojinete.

**Bypass valve** A safety feature to prevent engine failure. The valve opens when there is a pressure differential of 5 to 15 psi between the outside and inside of the filter element.
**Válvula de paso** Un componente de seguridad para prevenir los fallos del motor. La válvula se abre cuando hay una diferencial de presión de entre 5 a 15 libras por pulgada cuadrada entre el exterior y el interior del elemento de filtro.

## C

**Calibrate** To determine the scale of an instrument giving quantitative measurements.
**Calibrar** Determinar la escala de un instrumento dando medidas cuantitativas.

**Caliper** Measuring tool capable of taking readings of inside, outside, depth, and step measurements in 0.001-inch increments.
**Calibre** Una herramienta de medir capaz de tomar las medidas interiores, exteriores, de profundidad y de paso en incrementos de 0.001 de una pulgada.

**Cam ground pistons** Pistons cast or forged into a slight oval or cam shape to allow for expansion. As the piston warms, it will become round.
**Pistones rectificados por leva** Los pistones colados o forjados en una forma lijeramente ovulada o en forma de la leva para permitir la expansión. Al calentarse el pistón se redondea.

**Camshaft** The shaft containing lobes to operate the engine valves.
**Arbol de levas** Un eje que tiene lóbulos que operan las válvulas del motor.

**Camshaft degreeing** The altering of the point where the camshaft activates the valves in relation to the crankshaft.
**Graduación del árbol de levas** Cambiar el punto en donde el árbol de levas acciona las válvulas con respeto al cigüeñal.

**Carbon** A nonmetallic element that forms inside of the combustion chamber as a product of burning fuel.
**Carbono** Un elemento nometálico que forma en el interior de la cámara de combustión como un producto del combustible al quemarse.

**Carbon monoxide** An odorless, colorless, and toxic gas that is produced as a result of incomplete combustion.
**Monóxido de carbono** Un gas sin olor, sin color y tóxico que se produce como resultado de una combustión incompleta.

**Carbonize** The process of carbon formation.
**Carbonizar** El proceso de formar el carbono.

**Carburetor** A fuel delivery device that mixes fuel and air to the proper ratio to produce a combustible gas.
**Carburador** Un dispositivo para entregar el combustible que mezcla el combustible y el agua en proporciones correctas para producir un gas combustible.

**Caustic solutions** Cleaning solutions usually consisting of a mixture of water, sodium hydroxide, and sodium carbonate. This solution is extremely alkaline, with a pH rating of 10 or more.
**Soluciones cáusticas** Las soluciones de limpieza normalmente compuestas de una mezcla del agua, el óxido de sodio, y el carbonato sódico. Esta solución es extremadamente alcalino con una clasificación pH de 10 o más.

**Cc-ing** A method of measuring the volume of the combustion chamber by measuring the amount of oil the chamber can hold.
**Midiendo en centímetros cúbicos** Un método para medir el volumen de la cámara de combustión al medir la cantidad de aceite que puede contener la cámara.

**Centrifugal force** A force that tends to move a body away from its center of rotation.
**Fuerza centrífuga** Una fuerza cuya tendencia es de mudar un cuerpo fuera de su centro de rotación.

**Chamfer** A bevel or taper at the edge of a bore.
**Chaflán** Un bisel o chaflán en el borde de un taladro.

**Channeling** Local leakage around a valve head caused by extreme temperatures developing at isolated locations on the valve face and head.
**Acanalado** Una fuga local alrededor de una cabeza de válvula causada por las temperaturas severas que ocurren en puntos aislados en la cara y en la cabeza de la válvula.

**Chemical cleaning** The process of using chemical action to remove soil contaminants from the engine components.
**Limpieza química** El proceso de usar la acción química para desprender los contaminantes sucios de los componentes del motor.

**Clearance** The space allowed between two parts.
**Holgura** El espacio permitido entre dos partes.

**Clearance volume** The volume of the combustion chamber when the piston is at TDC. The size of the clearance volume is a factor in determining compression ratio.
**Volumen de la holgura** El volumen de la cámara de combustión cuando el pistón esta en PMS. El tamaño del volumen de la holgura es un factor en determinar el índice de compresión.

**Combustion** The process of burning.
**Combustión** El proceso de quemar.

**Combustion chamber** The volume of the cylinder above the piston with the piston at TDC.
**Cámara de combustión** El volumen del cilindro arriba del pistón cuando el pistón está en PMS.

**Composites** Man-made materials using two or more different components tightly bound together. The result is a material consisting of characteristics that neither component possesses on its own.
**Compuestas** Las materiales artificiales que usan dos o más componentes distinctos combinados estructuralmente. El resultado es una material que tiene las características que ninguno de los componentes posee individualmente.

**Compound** A mixture of two or more ingredients.
**Compuesta** Una mezcla de dos o más ingredientes.

**Compression** The reduction in volume of a gas.
**Compresión** La reducción del volumen de un gas.

**Compression ratio** A comparison between the volume above the piston at BDC and the volume above the piston at TDC.
**Indice de compresión** Una comparación entre el volumen arriba del pistón en el PMI y el volumen arriba del pistón en el PMS.

**Compression rings** The upper rings of the piston designed to hold the compression in the cylinder.
**Anillos (aros) de compresión** Los anillos superiores de un pistón diseñados a mantener la compresión dentro del cilindro.

**Compression testing** A diagnostic test to determine the engine cylinder's ability to seal and to maintain pressure.
**Prueba de compresión** Una prueba diagnóstica que determina la habilidad del cilindro del motor de sellar y mantener la presión.

**Concentric** Two or more circles having a common center.
**Cencéntrico** Dos o más círculos que comparten un centro común.

**Conductor** A material capable of supporting the flow of electricity.
**Conductor** Una materia que permite el flujo de la electricidad.

**Conformability** The ability of the bearing material to conform itself to slight irregularities of a rotating shaft.
**Conformidad** La habilidad de una material de un cojinete de conformarse a las pequeñas irregularidades de un eje giratorio.

**Connecting rod** The link between the piston and crankshaft.
**Biela** La conexión entre el pistón y la cigüeñal.

**Contraction** A reduction of mass.
**Contracción** Una reducción de la masa.

**Convection** A transfer of heat by circulating heated air.
**Convección** Una transferencia del calor por medio de la circulación del aire calentado.

**Coolant** The liquid that is circulated through the engine to absorb heat and transfer it to the atmosphere.
**Fluido refrigerante** El líquido circulado por el motor que absorba el calor y lo transfere a la atmósfera.

**Core plugs** Metal plugs screwed or pressed into the block or cylinder head at locations where drilling was required or sand cores were removed during casting.
**Tapones del núcleo** Los tapones metálicos fileteados o prensados en el monoblock o en la cabeza de los cilindros ubicados en donde se había requerido taladrar o quitar los machos de arena durante la fundición.

**Counterbore** To enlarge a bore to a given depth.
**Ensanchar** Extender un taladro a una profundidad designada.

**Counterweight** Weight cast or forged into the crankshaft to reduce rotational vibration.
**Contrapeso** Un peso colado o forjado en un cigüeñal para reducir las vibraciones giratorias.

**Crankcase** The area of the lower engine block that contains the oil and fumes from the combustion process.
**Cárter** El área inferior del monoblock que contiene el aceite y los vapores del proceso de combustión.

**Cranking vacuum test** Used to compare results with the running vacuum test. Low vacuum during cranking can indicate external leaks such as broken or disconnected vacuum hoses that may not be indicated on running engine vacuum tests.
**Prueba de arranque en vacío** Se efectúa para comprar los resultados con los resultados de la prueba de marcha en vacío. Un nivel bajo de vacío en el arranque puede indicar la presencia de unas fugas externas tal como las mangueras de vacío desconectadas o quebradas que no se pueden descubrir en la prueba de marcha en vacío.

**Crankpins** Another term for connecting rod bearing journals.
**Codo de cigüeñal** Otro término para indicar los muñones de la biela.

**Crankshaft** A mechanical device that converts the reciprocating motions of the pistons into rotary motion.
**Cigüeñal** Un dispositivo mecánico que convierte los movimientos recíprocos de los pistones a un movimiento giratorio.

**Crankshaft end play** The measure of how far the crankshaft can move lengthwise in the block.
**Juego en el extremo del cigüeñal** La medida de cuánto puede moverse el cigüeñal a lo largo del monoblock.

**Crankshaft throw** The measured distance between the centerline of the rod bearing journal and the centerline of the crankshaft.
**Codo de cigüeñal** Una distancia medida entre el centro del muñon del cojinete de biela y el centro del cigüeñal.

**Crown** The center area of a bearing half.
**Centrado de la punta** El área central de un mitad del cojinete.

**Crush** The press-fit allowance required to maintain bearing position in the bore.
**Aplastamiento** El margen requirido en un ajuste prensado para mantener al cojinete en su posición en el taladro.

**Cycling** A systematic driving method that varies the load on the engine.
**Ciclar** Un método sistemático de conducir que diversifica la carga en el motor.

**Cylinder** A hole bored into the engine block that the piston travels in, and that makes up part of the combustion chamber where the air/fuel mixture is compressed.
**Cilindro** Un agujero taladrado en el monoblock en el cual viaja el pistón, y que forma parte de la cámara de combustión en donde se comprime la mezcla de aire/combustible.

**Cylinder block** The main structure of the engine. Most of the other engine components are attached to the block.
**Monoblock** La estructura principal del motor. La mayoría de los otros componentes del motor se conectan al monoblock.

**Cylinder bore dial gauge** An instrument used to measure the cylinder bore for wear, taper, and out-of-round.
**Calibre carátula del taladro del cilindro** Un instrumento de medida que sirve para averiguar si el taladro del cilindro está gastado, cónico, o ovulado.

**Cylinder head** On most engines, the cylinder head contains the valves, valve seats, valve guides, valve springs, and the upper portion of the combustion chamber.
**Cabeza del cilindro** En la mayoría de los motores la cabeza del cilindro contiene las válvulas con sus asientros, sus guías, sus resortes y la parte superior de la cámara de combustión.

**Cylinder leakage test** A test that determines the condition of the piston ring, intake or exhaust valve, and head gasket. It uses a controlled amount of air pressure to determine the amount of leakage.
**Prueba de fugas en el cilindro** Una prueba para determinar la condición del anillo de pistón, la válvula de entrada o escape, y la junta de la cabeza. Se emplea una cantidad controlada de presión de aire para determinar la cantidad de la fuga.

**Cylinder ridge** An area of no wear resulting from the piston ring not traveling the full height of the cylinder.
**Cilindro con reborde** Una área sin desgaste que resulta cuando el anillo no suba la altura completa del cilindro.

## D

**Deceleration** A reduction of speed.
**Deceleración** Una reducción de la velocidad.

**Deck** The top of the engine block where the cylinder head is attached.
**Cubierta** La parte superior del monoblock en donde se conecta la cabeza del cilindro.

**Deck clearance** A measure of the distance the top of the piston is below, or above, the deck of the engine block (when the piston is at TDC).
**Holgura de la cubierta** La distancia que queda la parte superior del pistón abajo, o arriba, de la cubierta del monoblock (cuando el pistón está en PMS).

**Deflection** The bending or movement away from normal due to loading.
**Desviación** El abarquillamiento o movimiento fuera de lo normal debido a la carga.

**Deglazing** The process of roughening the cylinder wall without changing its diameter.
**Lijar** El proceso de poner áspero el muro del cilindro sin cambiar su diámetro.

**Density** The compactness or relative mass of matter in a given volume.
**Densidad** Lo compacto o la masa relativa de la materia en un volumen designado.

**Depth micrometers** An instrument designed to measure the depth of a bore.
**Micrómetro de profundidad** Una herramiento diseñada para medir la profundidad de un taladro.

**Detonation** A defect that occurs if the air/fuel mixture in the cylinder is burned too fast.
**Detonación** Un defecto que ocurre si la mezcla aire/combustible en el cilindro se quema demasiado rápido.

**Diagnosis** The use of instruments, service manual, and experience to determine the action of parts or systems to determine the cause of the failure.
**Diagnosis** El uso de los instrumentos, el manual de servicio, y la experiencia para determinar el acción de los partes o los sistemas y descubrir la causa de un fallo.

**Dial caliper** Measuring instrument similar to vernier caliper except the dial performs the function of the vernier scale.
**Calibre de carátula** Un instrumento de medida parecido al pie de rey menos que la carátula funciona como una regla vernier.

**Dial indicator** An instrument used to measure the travel of a plunger in contact with a moving component.
**Calibre de cáratula** Un instrumento que sirve para medir el viaje de un émbolo que está en contacto con un componente en movimiento.

**Displacement** A measure of engine volume.
**Deplazamiento** Una medida del volumen del motor.

**Distortion** A warpage or change in form from original.
**Distorción** Un abarquillamiento o un cambio en la forma de la original.

**Distributor** The mechanism within the ignition system that controls the primary circuit and directs the secondary voltage to the correct spark plug.
**Distribuidor** El mecanismo dentro del sistema de arranque que controla al circuito primario y dirige el voltaje secundario o la bujía correcta.

**Dry compression testing** A compression test performed with no additional oil added to the cylinders.
**Prueba de compresión en seco** Una prueba de compresión que se afectúa sin añadir aceite adicional a los cilindros.

**Dry sleeves** Replacement cylinder sleeves that do not come into contact with engine coolant. They are surrounded by the cylinder bore.
**Camisas secas** Las camisas de repuesto del cilindro que no se ponen en contacto con el fluido refrigerante del motor. Se ajusten en el taladro del cilindro.

**Duration** The length of time, expressed in degrees of crankshaft rotation, the valve is open.
**Duración** La cantidad del tiempo, representado por los grados de rotación del cigüeñal, que esta abierta la válvula.

**Dynamometer** A device used to brake the power produced by the engine for testing purposes.
**Dinamómetro** Un dispositivo que sirve en frenar la fuerza producida por el motor para que se puede efectuar las pruebas.

## E

**Eccentric** One circle within another circle with neither having the same center.
**Excéntrico** Un círculo que queda dentro de otro círculo sin compartir el mismo centro.

**Eccentricity** The physical characteristics designed into some bearings calling for an inside assembled vertical diameter that is slightly smaller than the horizontal diameter.
**Exentricidad** Las características físicas diseñadas en algunos cojinetes que requieren que le diámetro interior asemblado sea un poco más pequeño que el diámetro horizontal.

**Eddy** A current that runs against the main current.
**Interrupción** Un corriente que fluye contra el corriente principal.

**Efficiency** A ratio of the amount of energy put into an engine as compared to the amount of energy produced by the engine.
**Rendimiento** Un índice de la cantidad de energía introducido en el motor comparado a la cantidad de energía que produce el motor.

**Electrolysis** The result of two different metals in contact with each other. The lesser of the two metals is eaten way.
**Electrólisis** El resultado de dos metales distinctos que se ponen en contacto. El menor de los dos metales se consume.

**Emission system** Helps to reduce the harmful emissions resulting from the combustion process.
**Sistema de emisión** Ayuda en reducir las emisones nocivos que resultan del proceso de combustión.

**Energy** The ability to do work.
**Energía** La habilidad de hacer un trabajo.

**Engine** The power plant that propels the vehicle.
**Motor** El central de energía que propulsa el vehículo.

**Engine block** The main structure of the engine that houses the pistons and crankshaft. Most other engine components attach to the engine block.
**Monoblock** La estructura principal del motor que contiene los pistones y el cigüeñal. La mayoría de los otros componentes del motor se conectan al monoblock.

**Engine hoist** A special lifting tool designed to remove the engine through the hood opening. Most are portable or fold up for easy storage. The lifting of the boom is performed by a special long reach hydraulic jack.
**Grúa** Una herramienta especial de izar diseñada para remover el motor por la apertura del cofre. Suelen ser portátiles o se doblan fácilmente para guardarse. El brazo de la grúa se levanta con un gato hidráulico de larga extensión.

**Engine stand** A special holding fixture that attaches to the back of the engine, supporting it at a comfortable working height. In addition, most stands allow the engine to be rotated for easier disassembly and assembly.
**Bancada para motor** Un accesorio de apoyo especial que se conecta a la parte trasera del motor. Permite que se soporta el motor en una altura ideal para trabajar. Además, la mayoría de las bancadas permiten la rotación del motor para facilitar el desmontaje y montaje.

**Exhaust manifold** A component that collects and then directs engine exhaust gases from the cylinders.
**Múltiple de escape** Un componente que colecciona y luego dirige los gases de escape del motor desde los cilindros.

**Exhaust manifold gasket** A part that seals the connection between the cylinder head and the exhaust manifold.
**Junta del múltiple de escape** Una parte que sella la conexión entre la cabeza del cilindro y el múltiple de escape.

**Exhaust system** A system that removes the byproducts of the combustion process from the cylinders.
**Sistema de escape** Un sistema que remueva los subproductos del proceso de combustión de los cilindros.

**Exhaust valve** An engine part that controls the expulsion of spent gases and emissions out of the cylinder.
**Válvula de escape** Una parte del motor que controla la expulsión de los gases consumidos y los emisiones del cilindro.

**Expansion** An increase in size.
**Dilatación** Un incremento en el tamaño.

**Externally balanced engine** Engines balanced by using counterweights on the flywheel and vibration damper in conjunction with the crankshaft counterweights.
**Motor de equilibración externo** Los motores que se equilibran empleando los contrapesos en el volante y un amortiguador de vibraciones junto con los contrapesos del cigüeñal.

## F

**Face shield** A clear plastic shield that protects the entire face.
**Careta** Un escudo transparente de plástico que proteja la cara entera.

**Fast burn combustion chamber** A chamber designed to increase the speed of combustion by creating a turbulence as the air/fuel mixture enters the chamber.
**Cámara de combustión rápido** Una cámara diseñada para aumentar la rapidez de la combustión creando una turbulencia mientras que la mezcla de aire/combustible entra a la cámara.

**Fatigue** The deterioration of metal under excessive loads.
**Fatiga** El deterioro de un metal bajo cargas excesivas.

**Feeler gauge** Thin metallic strip of a known thickness.
**Galga calibrada** Una hoja metálica delgada de un espesor conocido.

**Ferrous metal** Metal containing iron or steel.
**Metal férreo** Un metal que contiene el hierro o el acero.

**Fillets** Small, rounded corners machined on the edges of journals to increase strength.
**Cantos redondeados** Las esquinas pequeñas, redondeadas labradas por máquina en los bordes de un muñon para reinforzarla.

**Flap** A stip of emery cloth, or other abrasive material, wound around a slotted mandrel. The loose end of the flap is allowed to slap against the surface being machined.
**Aleta** Un tiro de tela de esmeril, o otra material abrasiva, envuelto alrededor de un mandrino hendido. Se permite que la extremidad suelta de la aleta golpea la superficie que esta trabajando por máquina.

**Flex plate** A stamped steel coupler bolted to the rear of the crankshaft. The flex plate provides a mounting for the torque convertor.
**Placa flexible** Un acoplador de acero embutido empernado a la parte trasera del cigüeñal. La placa flexible provee el asisento del convertidor de par.

**Floating piston pin** A piston pin that is not locked in the connecting rod or the piston, allowing it to turn or oscillate in both the connecting rod and piston.
**Perno flotante del pistón** Un perno del pistón que no esta clavado en la biela o en el pistón, permitiendo girar o osilar en la biela y el pistón.

**Floor jack** A portable hydraulic tool used to raise and lower a vehicle.
**Gato** Una herramienta portátil hidráulica que sirve para levantar y bajar un vehículo.

**Flywheel** A heavy circular component located on the rear of the crankshaft that keeps the crankshaft rotating during nonproductive strokes.
**Volante** Un componente pesado circular ubicado en la parte trasera del cigüeñal que procure que el cigüeñal sigue girando durante las carreras no productivas.

**Followers** Similar to rocker arms, followers are used on many OHC engines. Followers run directly off of the camshaft.
**Seguidor** Parecidos a los balancines, los seguidores se emplean en muchos motores OHC. Los seguidores se accionan directamente por el árbol de levas.

**Foot-pound (ft.-lb.)** A measure of the amount of energy required to lift 1 pound 1 foot.
**Pie-libra (pies-lb.)** Una medida de la cantidad de energía o fuerza que requiere mover una libra la distancia de un pie.

**Four gas engine analyzer** A device that measures that exhaust of the engine. It enables the technician to look at the effects of the combustion process by measuring hydrocarbons, carbon monoxides, carbon dioxide, and oxygen levels in the exhaust.
**Analizador de cuatro gases del motor** Un dispositivo que mide los vapores de escape del motor. Permite que el técnico vea los efectos del proceso de combustión midiendo los niveles de los hidrocarburos, el monóxido de carbono, el bióxido de carbono y el oxígeno en los vapores del escape.

**Free-wheeling engine** An engine in which valve lift and angle prevent valve-to-piston contact if the timing belt or chain breaks.
**Motor de piñón libre** Un motor en el cual el levantamiento y el ángulo de las válvulas previenen el contacto entre la válvula y el pistón si se quiebra la correa o la cadena de sincronización.

**Fuel system** A system that includes the intake system which brings air into the engine and the components that deliver the fuel to the engine.
**Sistema de combustible** Un sistema que incluye al sistema de admisión que introduce el aire dentro del motor y a los componentes que entregan el combustible al motor.

## G

**Galling** A displacement of metal.
**Raspar** El desplazamiento del metal.

**Gasket** A rubber, felt, cork, or metallic material used to seal surfaces of stationary parts.
**Empaque** Una material de caucho, fieltro, corcho o metal que sirve para sellar las superficies de las partes fijas.

**Glaze** Polishing of the cylinder wall resulting from piston ring travel in the cylinder in conjunction with combustion heat and engine oil.
**Porcelana** El pulido del pared del cilindro que resulta del viaje del anillo del pistón en el cilindro en combinación con el calor de combustión y el aceite del motor.

**Glow plugs** Electrical heating elements threaded into the combustion chamber of a diesel engine to assist in cold temperature starts by heating the intake air.
**Bujías del precalentamiento** Los elementos eléctricos incandescentes fileteados dentro de la cámara de combustión de un motor diesel que calientan al aire de admisión para asistir en los encendidos fríos.

**Grade marks** Radial lines on the bold head that indicate the strength of the bolt.
**Marcos de grado** Las lineas radiales en la cabeza del perno que indican su fuerza.

**Grinding** A machining process of removing metal from the crankshaft journals by use of a special machine and stones.
**Rectificado a esmeril** Un proceso maquinario de rebajar el metal de los muñones del cigüeñal usando una máquina especial y las piedras de afiler.

**Grit** Cleaning sand that is angular in shape and is used for aggressive cleaning.
**Grano** La arena de limpieza cuyos partículos son de forma angular y que se emplea en la limpieza agresiva.

**Gross valve lift** The theoretical amount of opening of the valve. Gross valve lift is the distance the valve would lift from its seat if there were no deflections or clearances in the valve train. It can be determined by multiplying the lobe lift by the rocker arm ratio.

**Total bruto del levantamiento de la válvula** La cantidad teórica de apertura de la válvula. El total bruto del levantamiento de la válvula es la distancia que se levantaría una válvula de su asientos si no hubieran deflecciones ni holguras en el tren de las válvulas. Se determine al multiplicar el levantamiento de los lóbulos por el índice del balancín.

## H

**Hand tools** Tools that use only the force generated from the body to operate. They multiply the force though leverage to accomplish the work.

**Herramienta de mano** Las herramientas que sólo emplean la fuerza proporcionado por el cuerpo. Utilizan el acción de palanca para multiplicar la fuerza recibido e efectuar el trabajo.

**Harmonic balancer (vibration dampener)** A component attached to the front of the crankshaft used to reduce the torsional or twisting vibration that occurs along the length of the crankshaft.

**Equilibrador armónico (amortiguador de vibraciones)** Un componente conectado a la parte delantera de un cigüeñal que sirve para reducir las vibraciones torsionales que ocurren por la longitud del cigüeñal.

**Hazardous material** A material that could cause injury or death to a person, or could damage or pollute land, air, or water.

**Materiales peligrosas** Una material que podría causar daños o la muerte a una persona o podría causar los daños a la contaminación de la tierra, el aire o el agua.

**HC** The abbreviation for hydrocarbons. Hydrocarbons are present in particles of unburned gasoline.

**HC** La abreviación de hidrocarburos. Los hidrocarburos se encuentran en los partículos no consumidos de la gasolina.

**Head gasket** Gasket used to prevent compression pressures, gases, and fluids from leaking. It is located on the connection between the cylinder head and engine block.

**Junta de la cabeza** Una junta que se emplea en prevenir que se escapen las presiones, los gases y los fluidos de compresión. Se ubica en la conexión entre la cabeza de los cilindros y el monoblock.

**Heat treated** Metal hardened by a process of heating it to a high temperature, then quenching it in a cool bath.

**Tratado térmico (cementado)** El metal endurecido por el proceso de calentándolo a una temperatura muy elevada, luego amortiguándolo en un baño frio.

**Heli-coil** A spiral thread used to restore damaged threads to original size.

**Heli-coil** Un hilo espiral que se emplea para restaurar las roscas dañadas a su tamaño original.

**Hemispherical chamber** A nonturbulence, half circle–shaped combustion chamber used on many older high-performance engines. The valves are on either side with the spark plug in the center.

**Cámara hemisférica** Una cámara de combustión sin turbulencia, en forma de una semiesfera que se empleaban en los viejos modelos de motores de alta producción. Las válvulas se ubican da cada lado con la bujía de chispa en el centro.

**Hoist** A lift used to raise the entire vehicle.

**Elevador** Una grúa que se emplea en levantar el vehículo completo.

**Horsepower** The measure of the rate of work.

**Caballo de fuerzas** La medida del tiempo en que se realiza el trabajo.

**Hot spray tank** Clearing tank that sprays caustic solutions onto the components along with soaking them.

**Tanque de rocío caliente** Un tanque de limpieza que rocía las soluciones cáusticas sobre los componentes mientras que éstos remojan.

**Hydraulic lifters** Lifters that use oil to absorb the resultant shock of valve train operation.

**Elevadores hidráulicas** Los elevadores que usan el aceite para absorber los golpes causados por la operación del tren de válvulas.

**Hydrocarbon** A chemical composition, made up of hydrogen and carbon.

**Hidrocarburo** Una composición quimica, compuesta del hidrógeno y el carbono.

## I

**Ignition** The process of igniting the air/fuel mixture in the combustion chamber.

**Encendido** El proceso de encender la mezcla de aire/combustible en la cámara de combustión.

**Ignition system** System that delivers the spark used to ignite the compressed air/fuel mixture.

**Sistema de encendido** Un sistema que entrega la chispa que sirve para encender la mezcla comprimida de aire/combustible.

**Indexing** The process of offsetting the crankshaft so the rod journals are centered in the grinding machine. The offset from the main bearing journal is the same distance as from the center of the main journal to the center of the crank throw.

**Rectificación graduada** El proceso de descentrar el cigüeñal para que los muñones de las bielas sean centrales en la rectificadora. La desviación del muñon del cojinete principal queda la misma distancia que la distancia del centro del muñon del cojinete principal al centro del codo del cigüeñal.

**Induction hardening** A process that uses an electromagnet to heat the seat through induction. The seat is heated to a temperature of about 1700°F (930°C). The harden depth is about 0.060 in. (1.5 mm).

**Endurecimiento por inducción** Un proceso que usa un electroimán para calentar el asiento por medio de la inducción. El asiento se calienta a una temperatura de aproximadamente 1700°F (930°C). La profundidad del endurecimiento es aproximadamente de un 0.060 pulgada (1.5 mm).

**Inertia** The tendency of objects in motion to remain in motion and of objects at rest to remain at rest.

**Inercia** La tendencia de que los objetos en movimiento quedan en movimiento y los objetos en reposo permanezcan en reposo.

**Inlet check valve** Keeps the oil filter filled at all times so when the engine is started an instantaneous supply of oil is available.

**Válvula de retención** Mantiene siempre lleno al fitro de aceite para que cuando se arranque el motor dispone de un suministro instantáneo de aceite.

**Insert bearings** An interchangeable type of bearing. The bearing is a self-contained part that is inserted into the bearing housing.

**Cojinetes de inserción** Un cojinete te tipo intercambiable. El cojinete es de una parte entera que se puede insertar en la cubierta del cojinete.

**Insert guides** Removable valve guides that are pressed into the cylinder head.

**Guías para inserción** Las guías desmontables de las válvulas que han sido prensadas en la cabeza de los cilindros.

**Inside micrometer** Precision instrument designed to measure the inside diameter of a hole.

**Micrómetro interior** Una herramienta de medir diseñada para medir el diámetro interior de un agujero.

**Intake manifold** Component that delivers the air or air/fuel mixture to each engine cylinder.

**Múltiple de admisión** Un componente que entrega el aire o la mezcla del aire/combustible a cada cilindro del motor.

**Intake manifold gasket** Gasket that fits between the manifold and cylinder head to seal the air/fuel mixture or intake air.

**Junta del múltiple de admisión** Una junta que queda entre el múltiple y la cabeza del cilindro para sellar la mezcla de aire/combustible o el aire de admisión.

**Intake valve** The control passage of the air/fuel mixture entering the cylinder.

**Válvula de entrada** El pasaje de control para la mezcla de aire/combustible entrando al cilindro.

**Integral guides** Valve guides that are manufactured and machined as part of the cylinder head.

**Guías integrales** Las guías de las válvulas fabricadas e incorporadas a máquina como parte de la cabeza del cilindro.

**Interference angle** Valve design in which the seat angle is 1 degree greater than the valve face angle to provide a more positive seal.

**Angulo de interferencia** Un diseño de la válvula en el cual el ángulo del asiento de la válvula es un grado además del ángulo de la cara de la válvula para proveer un sello más positivo.

**Interference engine** An engine design in which, if the timing belt or chain breaks or is out of phase, the valves will contact the pistons. In addition, in multivalve interference engines, valve-to-valve contact is possible if the belt or chain is out of phase.

**Interferencia del motor** Un diseño del motor en el cual, si se quiebra la correa o la cadena de sincronización, o si esta fuera de fase, las válvulas se pondrán en contacto con los pistones. Además, en los motores de interferencia de múltiples válvulas, es posible el contacto de válvula a válvula si la correa o la cadena esta fuera de fase.

**Internal combustion engine** An engine that burns its fuels within the engine.

**Motor de combustión interna** Un motor que quema su combustible dentro del motor.

**Internally balanced engine** Engine balanced by the counterweights only.

**Motor de equilibración interno** El motor que se equilibra solamente por medio de los contrapesos.

## J

**Jack stands (safety stands)** Support devices used to hold the vehicle off the floor after is has been raised by the floor jack.

**Torres (soportes de seguridad)** Los dispositivos de soporte que se emplean para mantener el vehículo levantado del piso después de que haya sido levantado del piso por un gato hidráulico.

**Journal** An inner bearing operated by a shaft.

**Muñón** Un cojinete interior operado por una flecha.

## K

**Kinetic energy** Energy that is working.

**Energía cinética** La energía trabajnado.

**Knock** A descriptive term used to identify various noises occurring in an engine.

**Golpe** Un término descriptivo que sirve para identificar los varios ruidos que ocurren en un motor.

**Knurl** A special bit that rolls a thread into the guide and causes the metal to rise.

**Moleta** Una broca especial que enreda un hilo en la guía y causa que se levanta el metal.

**Knurling** A machining process that decreases the size of a bore by forcing a bit that swells the metal much like a tap does when it cuts threads.

**Moletear** Un proceso de rebajar a máquina el tamaño de un taladro forzando una broca que hincha al metal de una manera muy parecida un macho cortando las roscas.

## L

**Lapping** The process of fitting one surface to another by rubbing them together with an abrasive material between the surfaces.
**Pulido** El proceso de ajustar una superficie a otra rozándolas juntas con una material adhesiva entre las superficies.

**Leakdown** The relative movement of the lifter's plunger in respect to the lifter body.
**Tiempo de fuga** El movimiento relativo del émbolo de levantaválvulas con relación al cuerpo del levantaválvulas.

**Leakdown testing** Testing that determines the lifter's ability to hold hydraulic pressure and maintain zero lash.
**Prueba de fuga** La prueba que verifica la habilidad del levantaválvulas de mantener la presión y un juego cero.

**Lean-burn miss** The result of incomplete combustion due to the lack of oxygen.
**Marcha irregular de mezcla pobre** El resultado de la combustión incompleta debido a la falta de oxígeno.

**Lift** The maximum distance the valve is lifted from its seat. This distance is determined by multiplying cam lobe lift by rocker arm ratio.
**Levantamiento** La distancia máxima que se puede levantar una válvula de su asiento. Esta distancia se calcula multiplicando el levantamiento del lóbulo del árbol de levas por el índice del balancín.

**Lifters** Mechanical (solid) or hydraulic connections between the camshaft and the valves. Lifters follow the contour of the camshaft lobes to lift the valve off its seat.
**Levantaválvulas** Las conexiones mecánicas (sólidas) o hidráulicas entre el árbol de levas y las válvulas. Los levantaválvulas siguen el contorno de los lóbulos del árbol de levas para levantar la válvula de su asiento.

**Line boring** A machining process of the main bearing journals that restores the original bore size by cutting metal using cutting bits.
**Taladrar en serie** Un proceso de rectificación a máquina de los muñones del cigüeñal que restaura el tamaño original del taladro cortando el metal con brocas para cortar.

**Liners** Thin tubes placed between two parts.
**Manguitos** Los tubos delgados puestos entre dos partes.

**Lip seal** Molded synthetic rubber seal with a slight raise (lip) that is the actual sealing point. The lip provides a positive seal while allowing for some lateral movement of the shaft.
**Junta con reborde** Las juntas de caucho sintético moldeado que tienen un reborde ligero (un labio) que es el punto efectivo del sello. El reborde provee un sello positivo mientras que permite algún movimiento lateral de la flecha.

**Lobe** The part of the camshaft that raises the lifter.
**Lóbulo** La parte del árbol de levas que alza al levantaválvulas.

**Lobe center** The angle between a line drawn from the center of the base to the highest point on the lobe and the base circle position with the crankshaft at TDC.
**Centro del lóbulo** EL ángulo entre una linea que se dibuja del centro del base al punto más alto del lóbulo y la posición del círculo del base con el cigüeñal en PMS.

**Lower end** Refers to the main bearing journals of the cylinder block.
**Extremo inferior** Refiera a los muñones del cigüeñal del monoblock.

**Lubrication system** System that supplies oil to high friction and wear locations.
**Sistema de lubricación** Un sistema que proporciona el aceite a las áreas da alta fricción y desgaste.

## M

**Machinist's rule** A multiple scale ruler used to measure distances or components that do not require precise measurement.
**Regla de acero** Una regla con graduaciones múltiples para medir las distancias o los componentes que no requieren una medida precisa.

**Main bearing** A bearing used as a crankshaft support.
**Cojinete del cigüeñal** Un cojinete que se emplea como un soporte para el cigüeñal.

**Main bearing clearance** The distance between the main bearing journal and the main bearings.
**Holgura del cojinete principal** La distancia entre el muñón del cigüeñal y los cojinetes principales del cigüeñal.

**Main bearing journal** The crankshaft journal that is supported by the main bearing.
**Muñón del cojinete principal** El muñón del cigüeñal apoyado por el cojinete del cigüeñal.

**Main bearing saddle bore** The housing that is machined to receive a main bearing.
**Taladro del asiento del cojinete principal** Un cárter se ha labrado a máquina para acceptar un cojinete de cigüeñal.

**Major thrust surface** The side of the piston skirt that pushes against the cylinder wall during the power stroke.
**Superficie de empuje principal** El lado de la faldilla que empuja contra el muro del cilíndro durante la carrera de fuerza es la superficie de empuje principal.

**Manifolds** Tubular channels used to direct gases into or out of the engine.
**Múltiples** Los canales tubulares que sirven para dirigir los gases dentro o fuera del motor.

**Mechanical efficiency** A comparison of the power actually delivered by the crankshaft to the power developed within the cylinders at the same rpm.
**Rendimiento mecánica** Una comparación entre la energía que actualmente entrega el cigüeñal y la energía desarrollado dentro de los cilindros en la misma rpm.

**Microinch** One millionth of an inch. The microinch is the standard measurement for surface finish in the American customary system.
**Micropulgada** La millonésima parte de una pulgada. Una micropulgada es una unidad común para el acabado de la superficie en el sistema de medida americana.

**Micrometer** One millionth of a meter. The micrometer is the standard measurement for surface finish in the metric system.
**Micrómetro** Una millonésima parte de un metro. La medida común para el acabado de la superficie en el sistema métrica.

**Micron** A thousandth of a millimeter or about .0008 inch.
**Micrón** La milésima parte de un milímetro o aproximadamente un .0008 de una pulgada.

**Minor thrust side** The side opposite the side of the rod that is stressed during the power stroke.
**Lado de empuje menor** El lado opuesto del lado cuyo biela recibe el esfuerzo durante la carrera de potencia.

**Minor thrust surface** The areas of the piston skirt that pushes against the cylinder wall during the compression stroke.
**Superficie de empuje menor** El faldón del pistón que empuja contra el muro del cilindro durante la carrera de compresión.

## N

**Necking** A valve stem defect in which the stem narrows near the head.
**Vástago ahusado** Un defecto del vástago de válvula en el cual el vástago se adelgaza cerca de la cabeza del cilindro.

**Net valve lift** The actual amount a valve lifts off its seat. It is found by subtracting the lash specification and amount of component deflection from the gross valve lift.
**Producto neto del levantamiento de la válvula** La cantidad actual que se levanta una válvula de su asiento. Para determinarlo se substrae la especificación del juego de las válvulas más la cantidad de desviación del componente del total bruto de la cantidad del levantamiento de la válvula.

**Nucleate boiling** The process of maintaining the overall temperature of a coolant to a level below its boiling point, but allowing the portions of the coolant actually contacting the surfaces (the nuclei) to boil into a gas.
**Ebullición nucleido** El proceso de mantener un nivel de temperatura general de un fluido refrigerante menos de la de su punto de ebullición, pero permitiendo que las porciones del fluido refrigerante que actualmente estan en contacto con las superficies (el nucleido) hiervan para formar un gas.

## O

**Occupational safety glasses** Eye protection device designed with special high-impact lenses and frames and side protection.
**Lentes de seguridad** Un dispositivo de protección para los ojos diseñado con los cristales y el armazón resistentes a los impactos fuertes, y que provee protección a los lados de los ojos.

**Off-square** When the seat and valve stem are not properly aligned, they are off-square. This condition causes the valve stem to flex as the valve face is forced into the seat by spring and combustion pressures.
**Fuera de escuadra** Refiere a que el asiento y el vástago de la válvula no estén alineadas correctamente. Esta condición causa que el vástago de la válvula se dobla cuando la cara de la válvula se asienta bajo la fuerza de las presiones del resorte y la combustión.

**Oil breakdown** Condition of oil that results from exposure to high temperatures for extended periods of time. The oil will combine with oxygen and can cause carbon deposits in the engine.
**Deterioro del aceite** Un resultado de las temperaturas elevadas por largos periódos de tiempo. El aceite combinará con el oxígeno y puede causar los depósitos del carbono en el motor.

**Oil clearance** The difference between the inside bearing diameter and the journal diameter.
**Holgura de aceite** La diferencia entre el diámetro interior de un cojinete y el diámetro del muñón.

**Oil gallery** The main oil supply line in the engine block.
**Canalización de aceite** La linea principal del suministro de aceite en el monoblock.

**Oil pan gaskets** Gaskets used to prevent leakage from the crankcase at the connection between the oil pan and the engine block.
**Empaque de la tapa del cárter** Los empaques que se emplean para prevenir las fugas del cárter en la conexión entre el colector de aceite y el monoblock.

**Oil pressure test** Test to determine the condition of the bearings and other internal engine components.
**Prueba de presión de aceite** Una prueba que se afectua para determinar la condición de los cojinetes u otros componentes interiores del motor.

**Oil pump** A device that pulls oil from the sump and pressurizes, then delivers it to throughout the engine by use of galleries.
**Bomba de aceite** Un dispositivo que toma el aceite del suministro, lo presurisa, y lo entrega por el motor por medio de las canalizaciones.

**One-piece valves** A valve stem design in which the head is not welded to the stem. In a two-piece valve, the head is welded to the stem.
**Válvulas de una pieza** Un diseño del vástago de válvula que no tiene una cabeza soldada al vástago. Una válvula de dos piezas tiene una cabeza soldada al vástago.

**Open pressure** Spring tension when the valve spring is compressed and the valve is fully open.

**Presión abierta** La tensión de un resorte de la válvula al estar comprimido con la válvula completamente abierta.

**Opposed cylinder engine** Engine block design in which the cylinders are across from each other. Also referred to as a horizontally opposed or "pancake" engine.

**Motor de cilindros opuestos** Un diseño de monoblock que coloca los cilindros en lados opuestos. Tambien conocido con el nombre motores de cilindros opuestos horizontalmente o motores achatados.

**Out-of-round** The condition when measurements of a diameter differ at different locations. The term out-of-round applies to inside or outside diameters.

**Ovulado** Una condición en la cual las medidas de un diámetro varían an lugares distinctos. El término ovulado aplica a los diámetros interiores o exteriores.

**Out-of-round gauges** Instruments used to measure the concentricity of connecting rod bores.

**Calibradores de ovulado** Los instrumentos para medir la concentricidad de los taladros de las bielas.

**Outside micrometer** Tool designed to measure the outside diameter or thickness of a component.

**Micrómetros exteriores** Una herramienta diseñada para medir los diámetros exteriores o el espesor de un componente.

**Overhead hoist** A lifting tool that uses a chain fall or electric motor to hoist the engine out of the hood opening. The hoist can be attached to a moveable A-frame or on an I-beam across the shop ceiling.

**Grúa (montacarga) en alto** Una herramienta de izar que utiliza una caída de cadena o un motor eléctrico para levantar el motor por la apertura del capó. El aparato de izar puede conectarse a un armazón en forma de A o a un hierro en T a través del techo del taller.

**Overhead valve engine** An engine with the camshaft located in the engine block and the valves in the cylinder head.

**Motor con válvulas en cabeza** Un motor que tiene el árbol de levas ubicado en el monoblock y las válvulas en la cabeza del cilindro.

**Oversize bearings** Bearings that are thicker than standard to increase the outside diameter of the bearing to fit an oversize bearing bore. The inside diameter is the same as standard bearings.

**Cojinete de medidas superiores** Los cojinetes que son de un espesor más grueso de lo normal para aumentar el diámetro exterior del cojinete para quedarse en un taladro que rebasa la medida. El diámetro interior es lo mismo del cojinete normal.

## P

**Parts washers** Parts washers generally use a mild solvent to soak the components. Some provide for agitation and spraying of the solvent.

**Lavadora de partes** Las lavadoras de partes generalmente remojan los componentes en un solvente debil. Algunos proveen la agitación y rocían el solvente.

**Peen** To stretch or clinch over by pounding.

**Martillazo** Estirar o remachar por machacado.

**Peening** The process of removing stress in a metal by striking it.

**Martillar** El proceso de quitar la fatiga de un metal golpeandolo.

**Pent roof chamber** A combustion chamber shaped as an inverted "V." This chamber design provides a low volume, resulting in high compression ratios with less chances of detonation.

**Cámara encerrada** Una cámara de combustión que tiene la forma de un "V" invertido. Este diseño de cámara provee un volumen bajo, lo que resulta en un índice de compresión con menos tendencias de la detonación.

**pH** A value expressing acidity or basicity in terms of the relative amounts of hydrogen ions (H+) and hydroxide ions (OH–) present in a solution.

**pH** Un valor que expresa la acidez o lo básico en términos de las cantidades relativas de los iones de hidrógeno (H+) y los iones hidróxidos (OH–) presentes.

**Pin boss** A bore machined into the piston that accepts the piston pin to attach the piston to the connecting rod.

**Mamelón** Un taladro tallado a máquina en el pistón que acomoda la espiga del pistón para conectarlo a la biela.

**Piston** An engine component in the form of a hollow cylinder that is enclosed at the top and open at the bottom. Combustion forces are applied to the top of the piston to force it down. The piston, when assembled to the connecting rod, is designed to transmit the power produced in the combustion chamber to the crankshaft.

**Pistón** Un componente del motor que consiste de un cilindro hueco cerrado en la parte de arriba y abierto en la parte de abajo. Las fuerzas de combustión se aplican en la parte superior del pistón para forzarlo hacia abajo. El pistón, al conectarse a la biela, es diseñado para transmitir la fuerza producida en la cámara de combustión al cigüeñal.

**Piston balance pads** Some manufacturers provide balance pads just below the pin boss. The piston can be balanced by removing material from this area.

**Placa (pastilla) de equilibración del pistón** Algunos fabricantes proveen las placas para equilibrar ubicadas justo abajo del mamelón. Se puede equilibrar al pistón quitando la material de esta área.

**Piston collapse** A condition describing a collapse or reduction in diameter of the piston skirt due to heat or stress.

**Caída del pistón** Una condición que describe el fallo o la reducción en el diámetro de la falda del pistón debido al calor o la fatiga.

**Piston dwell time** The length of time in crankshaft degrees the piston remains at top dead center without movng.

**Angulo de cierre del pistón** La cantidad del tiempo en los grados del ángulo del cigüeñal que el pistón se queda en la posición de punto muerto superior sin moverse.

**Piston head (crown)** The top of the piston that forms the bottom of the combustion chamber.

**Cabeza de pistón** La parte superior del pistón qu forma la parte inferior de la cámara de combustión.

**Piston land** Area used to confine and support the piston rings in their grooves.

**Meseta del pared del pistón** El área de un pistón que sirve para restringir y sostener los anillos del pistón en sus muescas.

**Piston offset** A piston design that offsets the pin bore to provide more effective downward force onto the crankshaft by increasing the leverage applied to the crankshaft.

**Desviación del pistón** Un diseño del pistón de desvía el taladro del eje para proveer una fuerza descendente en el cigüeñal más eficáz aumentando la acción de palanca que se aplica en el cigüeñal.

**Piston pin** Component that connects the piston to the connecting rod. There are three basic designs used: a piston pin anchored to the piston and floating in the connecting rod, a piston pin anchored to the connecting rod and floating in the piston, and a piston pin full floating in the piston and connecting rod.

**Eje del pistón** Un componente que conecta el pistón a la biela. Se usan tres diseños básicos: un eje de pistón fijo al pistón y libre en la biela, un eje de pistón fijo a la biela y libre en el pistón, y un eje de pistón completamente libre en el pistón y la biela.

**Piston pin knock** A noise caused by a worn piston pin or bushing, worn piston pin boss, and worn bearings.

**Golpe del eje del pistón** Un ruido causado por un eje o manguito de pistón desgastado, un mamelón desgastado, y los cojinetes desgastados.

**Piston rings** Components that seal the compression and expansion gases, and prevent oil from entering the combustion chamber.

**Anillos (aros) del pistón** Un componente que sella los gases de compresión y expansión, y previene que entra el aceite en la cámara de combustión.

**Piston skirt** A component that forms a bearing area in contact with the cylinder wall and helps to prevent the piston from rocking in the cylinder.

**Faldilla del pistón** Un componente que forma una área de apoyo en contacto con el muro del cilindro y ayuda en prevenir que el pistón oscila en el cilindro.

**Piston slap** A sound that results from the piston hitting the side of the cylinder wall.

**Golpeteo del pistón** Un ruido resultando del pistón golpeando contra el muro del cilindro.

**Piston stroke** The distance the piston travels from TDC to BDC.

**Carrera del pistón** La distancia que viaja el pistón del PMS al PMI.

**Plasma** A material containing positive ions and unbound electrons in which the total number of positive and negative charges are almost equal. The properties of plasma are sufficiently different from those of solids, liquids, and gases for it to be considered a fourth state of matter.

**Plasma** Una material comprendida de los iones positivos y los electrones libres en la cual en número total de cargas positivas y negativas son casi iguales. Las propriedades del plasma son bastante diferentes de las de los sólidos, los líquidos y los gases para que se considera un cuarto estado de materia.

**Plastigage** A string-like plastic that is available in different diameters used to measure the clearance between two components. The diameter of the plastigage is exact, thus any crush of the gage material will provide an accurate measurement of oil clearance.

**Plastigage** Un plástico en forma de hilo disponible en diámetros distinctos que se emplea en medir la holgura entre dos componentes. El diámetro del calibre plástico es preciso, asi cualquier aplastamiento del material del calibre provee una medida precisa de la holgura del aceite.

**Plateau honing** A cylinder honing process that uses a coarse stone to produce the finished cylinder size and a very fine stone to remove an immeasurable amount and plateau the cut.

**Esmerilado de nivelación** Un proceso de esmerilar el cilindro usando una piedra tosca para producir el tamaño del cilindro acabado y una piedra muy fina para quitar una cantidad inmensurable y nivelar el corte.

**Plunge grinding** A crankshaft grinding method that uses a dressed stone the exact shape of the new journal surface. The stone is fed straight into the journal.

**Rectificación empujado** Un metodo de afilar de cigüeñal usando una piedra de afilar en forma exacta de la superficie del muñon nuevo. La piedra de afilar se empuja directamente dentro del muñon.

**Pneumatic tools** Tools powered by compressed air.

**Herramientas neumáticas** Las herramientas que derivan su poder del aire bajo presión.

**Polishing** The process of removing light roughness from the journals by using a fine emery cloth. Polishing can be used to remove minor scoring of journals that do not require grinding.

**Bruñido** El proceso de quitar la aspereza ligera de los muñones usando una tela abrasiva muy fina. El bruñido puede emplearse en quitar las rayas superficiales de los muñones que no requieren rectificación.

**Poppet valve** A valve design consisting of a circular head with a stem attached in the center. Poppet valves are used to control the opening or closing of a passage by linear movement.
**Válvula champiñón** Un diseño de una válvula que consiste de una cabeza redonda con un vástago conectado en el centro. Las válvulas champiñones controlan la apertura o cerradura de un pasaje por medio de un movimiento linear.

**Positive crankcase ventilation (PCV) system** An emission control system that routes blowby gases and unburned oil/fuel vapors to the intake manifold to be added to the combustion process.
**Sistema (PCV) de ventilación positiva de la caja del cigüeñal** Un sistema de emisión que lleva los gases soplados y los vapores del aceite/combustible no quemados al múltiple de entrada para que se pueden añadir al proceso de combustión.

**Positive displacement pumps** Pumps that deliver the same amount of oil with every revolution, regardless of speed.
**Bombas de desplazamiento positivo** Las bombas que entregan la misma cantidad del aceite con cada revolución, sin que importa la velocidad.

**Power balance test** Test used to determine if all cylinders are producing the same amount of power output. In an ideal situation, all cylinders would produce the exact same amount of power.
**Prueba del equilibrio de fuerza** Una prueba para determinar si todos los cilindros producen la misma cantidad de potencia de salida. En una situación ideal, todos los cilindros producirían exactamente la misma cantidad de poder.

**Power tools** Tools that use other forces than that generated from the body. They can use compressed air, electricity, or hydraulic pressure to generate and multiply force.
**Herramientas de motor** La herramientas que usan otras fuerzas que las producidas por el cuerpo. Pueden usar el aire bajo presión, la electricidad, o la presión hidráulica para engendrar y multiplicar la fuerza.

**Preignition** Defect that is the result of spark occurring too soon.
**Autoencendido** Un defecto que resulta de una chispa que ocurre demasiado temprano.

**Profilometer** A tool capable of electrically sensing the distances between peaks to determine the finish of a cut.
**Perfilómetro** Una herramiento capaz de detectar electrónicamente las distancias entre dos puntos altos para determinar cuando terminar un corte.

**Pushrod** A connecting link between the lifter and rocker arm. Engines designed with the camshaft located in the block use pushrods to transfer motion from the lifters to the rocker arms.
**Varilla de presión** Una conexión entre la levantaválvulas y el balancín empuja válvulas. Los motores diseñados con el árbol de levas en el bloque usan las varillas de presión para transferir el movimiento de la levantaválvulas a los balancines.

## Q

**Quenching** The cooling of gases as a result of compressing them into a thin area. The quench area has a few thousandths of an inch clearance between the piston and combustion chamber. Placing the crown of the piston this close to the cooler cylinder head prevents the gases in this area from igniting prematurely.
**Amortiguamiento** El enfriamiento de los gases que resulta de su compresión en una área pequeña. El área de amortiguamiento es una holgura de algunas milésimas partes de una pulgada entre el pistón y la cámara de combustión. Al poner el punto superior de la cabeza del pistón tan cerca a la cabeza del cilindro enfriada previene que los gases en esta área se enciendan prematuramente.

## R

**Radiator** A heat exchanger used to transfer heat from the engine to the air passing through it.
**Radiador** Un cambiador de calor que sirve para transferir el calor del motor al aire que lo atraviesa.

**Ream** Process of accurately finishing a hole with a rotating fluted tool.
**Escriar** El proceso de acabar un taladro precisamente con una herramienta acanalada giratoria.

**Reciprocating** An up-and-down or back-and-forth motion.
**Alternativo** Un movimiento oscilante de arriba a abajo o de un lado a otro.

**Reed valve** A one-way check valve. The reed opens to allow the air/fuel mixture to enter from one direction, while closing to prevent movement in the other direction.
**Válvula de lengüeta** Una válvula de una vía. La lengüeta se abre para permitir entrar la mezcla de aire/combustible de una dirección, mientras que se cierre para prevenir el movimiento de la otra dirección.

**Resource Conservation and Recovery Act (RCRA)** Law that makes users of hazardous materials responsible for the material from the time it becomes a waste until disposal is complete.
**Acta de Conservación y Recobro de Recursos (RCRA)** Un ley que hace responsable a los que usan las materiales peligrosas desde el tiempo que se convierte en un producto residual hasta que se haya completado su disposición.

**Retarded camshaft** A camshaft design that has the exhaust valves open more than the intake valves at TDC.
**Arbol de levas retrasado** Un diseño del árbol de levas cuyas válvulas de escape abren más que las válvulas de admisión en el PMS.

**Ridge reamer** A cutting tool used to remove the ridge at the top of the cylinder.
**Escariador de reborde** Una herramienta de cortar que sirve para quitar el reborde en la parte superior del cilindro.

**Ring noise** A noise caused by worn rings or cylinders. Other causes include broken piston ring lands and too little tension of the ring against the cylinder wall.
**Ruido de los anillos** Un ruido causado por los anillos o cilindros desgastados. Otra causas incluyen las mesetas de pistones rotas o una tensión insuficiente del anillo contra el muro del cilindro.

**Ring seating** The process of lapping the rings against the cylinder wall, accomplished by the movement of the piston in the cylinder.
**Asiento del anillo** El proceso de asentar los anillos a pulso con el muro del cilindro que se lleva acabo por el movimiento del pistón dentro del cilindro.

**Rocker arm** Pivots that transfer the motion of the pushrods or followers to the valve stem.
**Balancín** Un punto pivote que transfiere el movimiento de las levantaválvulas o de los seguidores al vástago de la válvula.

**Rocker arm ratio** A mathematical comparison of rocker arm dimensions. The rocker arm ratio compares the center-to-valve-stem measurement against the center-to-pushrod measurement.
**Indice del balancín** Una comparación de las dimensiones del balancín. El índice del balancín compara las dimensiones del balancín del centro-al-vástago de la válvula con las dimensiones del centro-al-levantaválvulas.

**Rod beaming** The process of polishing the beams of the rods to prevent stress risers. This is done by blending the casting seam on the sides of the rods.
**Pulido de las varillas** El proceso de pulir los resaltos de las varillas para prevenir la deformación de colada. Esto se lleva acabo puliendo la mazarota en los lados de las varillas.

**Rod length to stroke ratio** A mathematical comparison between the length of the connecting rod and the length of the engine's stroke. It is determined by dividing the connecting rod length by the stroke.
**Indice de longitud de la biela a la carrera** Una comparación matiemática entre la longitud de la biela y la longitud de la carrera del motor. Se determine dividiendo la longitud de la biela por la carrera.

**Rotary valve** A valve that rotates to cover and uncover the intake port. A rotary valve is usually designed as a flat disc that is driven from the crankshaft.
**Válvula rotativa** Una válvula que gira para cubrir y descubrir la puerta de admisión. Suelen ser diseñadas como un disco plano impulsado por el cigüeñal.

## S

**Saddle** The portion of the crankcase bore that holds the bearing half in place.
**Asiento del cojinete (silleta)** La parte del taladro del cigüeñal que mantiene en su lugar a la mitad con el cojinete.

**SAE** Society of Automotive Engineers.
**SAE** Asociación de Ingenieros Automotrices.

**Safety goggles** Safety devices that provide eye protection from all sides. Goggles fit against the face and forehead to seal off the eyes from outside elements.
**Gafas de seguridad** Proporciona la protección a los ojos de todos lados siendo que quedan apretados contra la cara y el frente para formar un sello para los ojos contra los elementos exteriores.

**Scale** The distance of the marks from each other on a measuring tool.
**Escala** La distancia entre las marcas en una herramienta de medir.

**Score** A scratch, ridge, or groove marring a finish surface.
**Raya** Un rasguño, un arruga, o una muesca que echa a perder una superficie.

**Scuffing** Scraping and heavy wear between two surfaces.
**Erosión** El rozamiento y desgaste fuerte entre dos superficies.

**Seal** Component used to seal between a stationary part and a moving one.
**Junta** Un componente que se emplea para sellar entre una parte fija y una que mueva.

**Sealant** A special liquid material commonly used to fill irregularities between the gasket and its mating surface. Some sealants are designed to be used in place of a gasket.
**Compuesto obturador** Un material líquido especial que normalmente se emlpea para rellenar las irregularidades entre un empaque y su superficie de contacto. Algunos compuestos son diseñados de uso sin empaque.

**Seat** A surface upon which another part rests.
**Asiento** Una superficie sobre la cual queda otra parte.

**Seat pressure** Term that indicates spring tension with the spring at installed height. Seat pressure is also known as valve closed pressure.
**Presión del asiento** Un término que indica la tensión del resorte cuando esta en su altura de instalación. La presión del asiento tambien se conoce como la presión con la válvula cerrada.

**Seat runout (concentricity)** A measure of how circular the valve seat is in relation to the valve guide.
**Excentricidad del asiento (concentricidad)** Una medida de lo circular del asiento de la válvula con relación a la guía de la válvula.

**Second-order vibration** Vibration that occurs twice per revolution.
**Vibración de segunda orden** Una vibración que ocurre dos veces por revolución.

**Seize** When one surface moving upon another causes scratches. The metal transfer can become severe enough to cause the moving component to stop.
**Rayar** Cuando los movimientos de una suprficie sobre otra causan las rayas. La transferencia del metal puede ser tan severa que para al componente en movimiento.

**Service sleeve (speedy-sleeve)** A metal sleeve that is pressed over a damaged sealing area to provide a new, smooth surface for the seal lip.

**Camisa de servicio (manguito rápido)** Un camisa de metal que se coloca sobre una área de sello dañada que provee una superficie nueva y lisa para el borde del sello.

**Short** An electrical defect that allows electrical current to bypass its normal path.

**Cortocircuito** Un defecto eléctrico que permite que el corriente eléctrico sobrepasa su rumbo normal.

**Shot** Round beads that are used for cleaning when etching of the metal is not desired.

**Granalla** Las bolitas que se emplean en la limpieza cuando el grabado del metal no es deseado.

**Shot-peening** A tempering process that uses shot under pressure to tighten the outside surface of the metal. Shot-peening is used to help strengthen the component and reduce chances of stress or surface cracks.

**Chorreo con granalla** Un proceso de templado que usa la granalla bajo presión para estrechar la superficie exterior del metal. El chorreo con granalla fortalece al componente y disminuye la aparencia del fatiga o de grietas en la superficie.

**Silicon** A nonmetallic element that can be doped to provide good lubrication properties. The melting point of silicon is 2,570°F (1,410°C).

**Silicio** Un elemento no metálico que se puede agregar para proporcionar las propriedades buenas de la lubricación. El silicio se funde hacia 2,570°F (1,410°C).

**Single pattern camshaft** A camshaft design in which the lobes are ground the same on both opening and closing ramps.

**Arbol de levas de un solo modelo** Un diseño de árbol de levas en el cual los lóbulos se rectifican de la misma manera en ambos lados de apertura y cerradura.

**Sizing point** The location the manufacturer designates for measuring the diameter of the piston to determine clearance.

**Punto de calibración** El lugar indicado por el fabricante para efectuar las medidas del diámetro del pistón para determinar la holgura.

**Sleeving** The process of boring the cylinder to accept a sleeve.

**Preparar para camisa** El proceso de taladrar un cilindro para aceptar una camisa.

**Slipper skirt** A piston skirt ground to provide additional clearance between the piston and the counterweights of the crankshaft. Without this recessed area, the piston would contact the crankshaft when shorter connecting rods are used.

**Faldilla deslizante** Una faldilla del pistón rectificada para proveer una holgura adicional entre el pistón y los contrapesos del cigüeñal. Sin esta área rebajada, el pistón podría rozar contra el cigüeñal cuando se emplean las bielas más cortas.

**Small end** Term that refers to the end of the connecting rod that accepts the piston pin.

**Extremo pequeño** Un término que refiere a la extremidad de la biela que accepta el eje del pistón.

**Small-hole gauge** An instrument used to measure holes or bores that are smaller than a telescoping gauge can measure.

**Calibre de taladros chicos** Un instrumento que se emplea para medir los agujeros o taladros que son demasiado pequeños para medirse con un calibrador telescópico.

**Soak tanks** Cleaning tanks equipped with a large basket which holds the parts while they are submerged into a caustic solution or detergent. Some soak tanks are equipped with an agitation system.

**Tanques (cubos) de remojo** Los tanques equipados con un capacho para sostener las partes que se sumergen en una solución cáustica o en el detergente. Algunos tanques tienen un sistema de agitación.

**Solid lifters (mechanical lifters)** Components that provide a rigid connection between the camshaft and the valves.

**Levantaválvulas macizas (o mecánicas)** Los componentes que proveen una conexión rígida entre el árbol de levas y las válvulas.

**Split overlap** Term describing when the intake and exhaust valves are equally open at TDC.

**Apertura compratida** Un término que describe que las válvulas de admisión y escape estan igualmente abiertas en el PMS.

**Spread** Descriptive term applied when the diameter at the outside parting edges of a bearing shell exceeds the inside diameter of the mating housing bore.

**Aplastamiento** Un término descriptivo que se aplica cuando el diámetro de los bordes exteriores del casquillo de un cojinete es más grande que el diámetro interior del taladro en el superficie de contacto del cárter.

**Spring free length** The height the spring stands when not loaded.

**Longitud libre del resorte** La altura del resorte cuando no tiene carga.

**Spring squareness** Refers to how true to vertical the entire spring is.

**Escuadrado del resorte** Refiere a si la posición del resorte entero esta en línea recta al vertical.

**Squish area** The area of the combustion chamber where the piston is very close to the cylinder head. The air/fuel mixture is rapidly pushed out of this area as the piston approaches TDC, causing turbulence and forcing the mixture toward the spark plug. The squish area can also double as the quench area.

**Area de compresión** El área de la cámara de combustión en donde el pistón esta muy cerca a la cabeza del cilindro. La mezcla del aire/combustible se expulsa rapidamente de esta área al aproximarse el pistón al PMS, causando una turbulencia y empujando la mezcla hacia la bujía. El área de compresión tambien puede servir de área de extinción.

**Steam cleaners** A pressure washer that uses a soap solution, heated under pressure to a temperature higher than its normal boiling point. The superheated solution boils once it leaves the nozzle as it shoots against the object being cleaned.
**Limpiadoras de vapor** Una limpiadora a presión que utiliza una solución de jabón, calentado bajo presión a una temperatura más elevada de su punto de ebullición. La solución sobrecalentada hierve al salir de la boquilla projectada hacia el objeto para limpiar.

**Stellite** A hard facing material made from a cobalt-based material with a high chromium content.
**Estelita** Una material de recarga compuesta de una material de base cobáltico con un contenido muy alto del cromo.

**Still timing** The process of adjusting base ignition timing without the engine running.
**Regulación sin marcha** El proceso de ajustar el avance del encendido fundamental sin que esté en marcha el motor.

**Stone dressing** Dressing a stone refers to using a diamond tool to clean and restore the stone's surface.
**Reacondicionar la muela** Para recondicionar una muela se emplea una herramienta de diamante para limpiar y rectificar la agudeza de la superficie de la muela.

**Stratified charge combustion chamber** Combustion chamber design that uses a two-stage combustion process to provide for lower emissions and good fuel economy.
**Cámara de combustión de carga estratificada** Un diseño de cámara de combustión que utilisa un proceso de combustión de dos etapas para proporcionar las emisiones más bajas y un rendimiento bueno.

**Stress risers** Defects in the component resulting in a weakness of the metal. The defect tends to decrease the tensile strength of the metal, and stress applied to the area of the defect tends to cause breakage.
**Deformación de colada** Los defectos en el componente que causan una debilidad del metal. El defecto suele disminuir la resistencia a la tracción del metal y al aplicar una carga en el área del defecto muchas veces causa la quebradura.

**Stroke** The distance traveled by the piston from TDC to BDC.
**Carrera** La distancia que viaja el pistón dle PMS al PMI.

**Surface-to-volume ratio** A mathematical comparison between the surface area of the combustion chamber and the volume of the combustion chamber. The greater the surface area, the more area the mixture can cling to, and mixture that clings to the metal will not burn completely because the metal cools it. Typical surface-to-volume ratio is 7.5:1.
**Indice superficie-volumen** Una comparación matemática entre el área de la superficie de la cámara de combustión y el volumen de la cámara de combustión. Lo mayor el área de la superficie, lo mayor el área en donde puede pegarse la mezcla y la mezcla pegada al metal no se quemará completamente puesto que la enfría el metal. El índice del superficie-volumen es el 7.5:1.

**Sweep grinding** A crankshaft grinding method using a stone that is swept back and forth across the journal surface.
**Rectificado de barrido** Un método de rectificación del cigüeñal empleando una muela que mueve de un lado al otro encima de la superficie de muñón.

**Swirl chamber** Combustion chamber that creates an airflow in a horizontal direction.
**Cámara vórtice** Cámara de combustión que crea un flujo de aire de una dirección horizontal.

**Synchronization timing** Timing design used on vehicles with computer-controlled fuel-injection systems. A pickup in the distributor is used to synchronize the crankshaft position to the camshaft position.
**Regulación (tiempo) sincronizado** Un diseño de regulación empleado en los vehículos equipados con sistemas de inyección de combustible de control computerizado. Un captador en el distribuidor sirve para sincronizar la posición del cigüeñal con la posición del árbol de levas.

## T

**Telescoping gauge** A precision tool used in conjunction with outside micrometers to measure the inside diameter of a hole. Telescoping gauges are sometimes called snap gauges.
**Calibrador telescópico** Una herramienta de precisión que se emplea junta con los micrómetros exteriores para medir los diámetros interiores de un agujero. Tambien se llaman calibradores de brocha.

**Tensile strength** The metal's resistance to be pulled apart.
**Resistencia a la tracción** La resistencia de un metal a ser estirado.

**Thermal cleaners (pyrolytic ovens)** A parts cleaner that uses high temperatures to bake the grease and grime into ash.
**Limpiadores térmicos (hornos pirolíticos)** Una limpiadora de partes que usa las temperaturas elevadas para convertir la grasa y el lodo en cenizas.

**Thermodynamics** The study of relationship between heat energy and mechanical energy.
**Termodinámica** El estudio de la relación entre la energía del calor y la energía mecánica.

**Thermostat** A control device that allows the engine to reach normal operating temperatures quickly and maintains the desired temperatures.
**Termostato** Un dispositivo de control que permite que le motor llegue rápidamente a las temperaturas de funcionamiento normales y mantenga las temperaturas deseadas.

**Thread depth** The height of the thread of a bolt from its base to the top of its peak.
**Profundidad de la rosca** La longitud de la rosca de un perno desde su fondo a la parte superior.

**Throating** Machining process using a 60-degree stone to narrow the contact surface.
**Rectificación** Un proceso de rebajar a máquina empleando una muela de 60 grados para disminuir la superficie de contacto.

**Throw** The distance from the center of the crankshaft main bearing to the center of the connecting rod journal.
**Codo del cigüeñal** La distancia del centro del muñón principal del cigüeñal al centro del muñón de la biela.

**Throw-off** The quantity of oil that escapes at the end of the bearings and lubricates adjacent engine parts while the engine is running.
**Expulsión** La cantidad del aceite que escapa de las extremidades de los cojinetes y lubrifica las partes contiguas del motor al estar en marcha el motor.

**Thrust bearing** A double-flanged bearing used to prevent the crankshaft from sliding back and forth.
**Cojinete de empuje** Un cojinete con doble brida que sirve para prevenir que el cigüeñal desliza de un lado a otro.

**Timing** The process of identifying when an event is to occur.
**Sincronización** El proceso de identificar cuando debe ocurir un evento.

**Timing chain** The chain that drives the camshaft off of the crankshaft.
**Cadena de sincronización** La cadena procedente del cigüeñal que acciona al árbol de levas.

**Tolerance** A permissible variation between the two extremes of a specification or dimension.
**Tolerancia** Una variación que se permite entre los dos extremos de una especificación o una dimensión.

**Top dead center (TDC)** Term used to indicate the piston is at the very top of its stroke.
**Punto muerto superior (PMS)** El término que indica que el pistón esta en la cima de su carrera.

**Topping** Refers to the use of the 30-degree stone to lower the contact surface on the valve face.
**Despunte** Refiere al uso de una muela de 30 grados para rebajar al superficie de contacto en la cara de la válvula.

**Torque** Twisting force applied to a shaft or bolt.
**Torsión** Una fuerza de torsión aplicada en un eje o en un perno.

**Torque convertor** A series of components that work together to reduce slip and multiply torque. It provides a fluid coupling between the engine and the automatic transmission.
**Convertidor del par** Un serie de componentes que trabajan juntos para disminuir el deslizamiento y multiplicar el par. Provee un acoplamiento flúido entre el motor y la transmisión automática.

**Torque plates** Metal blocks about 2 inches thick that are bolted to the cylinder block at the cylinder head mating surface to prevent twisting during honing and boring operations.
**Chapas de torsión** Los bloques de metal midiendo unas dos pulgadas de grueso empernados al monoblock en la superficie de contacto de la cabeza del cilindro que previenen que se retuerce durante las opraciones de esmerilado y taladreo.

**Torque wrench** Wrench that measures the amount of twisting force applied to a fastener.
**Llave de torsión** Una llave que sirve para medir la cantidad de la fuerza de torsión que se aplica a una fijación.

**Torsional vibration** Vibration that is the result of the crankshaft twisting and snapping back during each revolution.
**Vibración torsional** Una vibración que resulta cuando el cigüeñal se tuerce y regresa repentinamente en cada revolución.

**Transverse-mounted engine** Engine placement in which the block faces from side to side instead of front to back within the vehicle.
**Motor de motaje transversal** La ubicación del motor en la cual la longitud del monoblock queda de un lado a otro dentro del vehículo en vez de estar colocado de frente a atrás.

**Tumble port** Intake port design that creates an airflow in a vertical direction.
**Puerta de remolino** Un diseño de la puerta de entrada que crea un flujo del aire en una dirección vertical.

**Turbulence** A rapid movement of the air/fuel mixture. Turbulence results in improved combustion because of better mixing.
**Turbulencia** Un movimiento rápido de la mezcla del aire/combustible. La turbulencia resulta en una combustión mejorada porque se mezcla mejor.

## U

**Undersize bearing** Bearing with the same outside diameter as standard bearings but constructed of thicker bearing material in order to fit an undersize crankshaft journal.
**Cojinete de dimensión inferior** Un cojinete que tiene el diámetro exterior del mismo tamaño que los cojinetes normales pero que se ha fabricado de una material más gruesa para que puede quedar en un muñón de cigüeñal más pequeño.

## V

**Vacuum** A pressure lower than atmospheric pressure. Vacuum in the engine is created when the volume of the cylinder above the piston is increased. This results in the atmospheric pressure pushing the air/fuel mixture into the area of lowered pressure above the piston.
**Vacío** Una presión más baja que la presión atmosférica. El vacío en un motor se crea al aumentar el volumen del cilindro arriba del pistón. Esto resulta en que la presión atmosférica empuja la mezcla air/combustible dentro del área arriba del pistón.

**Vacuum testing** Testing that determines the engine's ability to provide sufficient pressure differentials to allow the induction of the air/fuel mixture into the cylinder. Results of vacuum testing indicate internal engine condition, fuel delivery abilities, ignition system condition, and valve timing.

**Prueba del vacío** Una prueba que determina la habilidad del motor en proveer las diferenciales de presión adecuadas que permiten la inducción de la mezcla aire/combustible al cilindro. Los resultados indicarán la condición interna del motor, las habilidades de entregar el combustible, la condición del sistema de encendido y la sincronización de las válvulas.

**Valley pans** A pan located under the intake manifold in the valley of a "V"-type engine used to prevent the formation of deposits on the underside of the intake manifold.
**Cárter en V** Un cárter ubicado en la parte inferior del múltiple de admisión en la parte más baja de un motor tipo "V" que previene la formación de los depósitos en la parte inferior del múltiple de admisión.

**Valve** Device that controls the flow of gases into and out of the engine cylinder.
**Válvula** Un dispositivo que controla el flujo de los gases entrando y saliendo del cilindro del motor.

**Valve cover gaskets** Components that seal the connection between the valve cover and cylinder head. The valve cover gasket is not subject to pressures, but must be able to seal hot, thinning oil.
**Juntas de la tapa de válvula** Los componentes que sellan la conexión entre la tapa de las válvulas y la cabeza del cilindro. La junta de la tapa de válvula no se sujeta a la presión, pero si tiene que sellar el aceite caliente muy fluido.

**Valve cupping** A deformation of the valve head caused by heat and combustion pressures.
**Embutición de la válvula** Una deformación de la cabeza de la válvula causada por el calor y las presiones de la combustión.

**Valve float** A condition that allows the valve to remain open longer than it is intended. Valve float is the effect of inertia on the valve.
**Flotación de la válvula** Una condición que permite que queda abierta la válvula por más tiempo de lo designado. Es el efecto que tiene la inercia en la válvula.

**Valve guide** A part of the cylinder head that supports and guides the valve stem.
**Guía de válvula** Una parte de la cabeza del cilindro que apoya y guía al vástago de la válvula.

**Valve guide bore gauge** Instument that provides a quick measurement of the valve guide. It can also be used to measure taper and out-of-round.
**Verificador del calibrador para la guía de válvula** Un instrumento que provee una medida rápida de la guía de la válvula. Tambien puede servir para medir lo cónico y lo ovulado.

**Valve guide clearance** Measurement of the difference between the valve stem diameter and the guide bore diameter.
**Holgura de la guía de válvula** La medida de la diferencia entre el diámetro del vástago de la válvula y el diámetro del taladro de la guía.

**Valve overhang** The area of the face between the seat contact and the margin.
**Desborde de la válvula** El área de la cara entre el contacto del asiento y el margen.

**Valve overlap** The length of time, measured in degrees of crankshaft revolution, that the intake and exhaust valves of the same combustion chamber are open simultaneously.
**Períodos de abertura de las válvulas** El período del tiempo, que se mide en los grados de las revoluciones del cigüeñal, en el cual las válvulas de admisión y de escape de la misma cámara de combustión estan abiertas simultáneamente.

**Valve seat** Machined surface of the cylinder head that provides the matting surface for the valve face. The valve seat can be either machined into the cylinder head or a separate component that is pressed into the cylinder head.
**Asiento de la válvula** La superficie acabada a máquina de la cabeza del cilindro que provee una superficie de contacto para la cara de la válvula. El asiento puede ser labrado a máquina en la cabeza del cilindro o puede ser un componente aparte para prensar en la cabeza de la válvula.

**Valve seat recession** The loss of metal from the valve seat which causes the seat to recede into the cylinder head.
**Encastre del asiento de la válvula** La pérdida del metal del asiento de la válvula, lo que causa que el asiento retroceda de la cabeza del cilindro.

**Valve seat runout gauge** Instrument that provides a quick measurement of the valve seat concentricity.
**Calibrador de la excentricidad del asiento de la válvula** Un instrumento que provee una medida rápida de la concentricidad del asiento de la válvula.

**Valve spring** A coil of specially constructed metal used to force the valve closed, providing a positive seal between the valve face and seat.
**Resorte de válvula** Un rollo de metal de construcción especial que provee la fuerza para mantener cerrada la válvula, así proveyendo en sello positivo entre la cara de la válvula y el asiento.

**Valve spring tension tester** Device used to measure the open and closed valve spring pressures.
**Probador de tensión del resorte de válvula** Un dispositivo que se emplea en medir las presiones de los resortes de las válvulas en la posición abierta o cerrada.

**Valve train** The series of components that work together to open and close the valves.
**Tren de válvulas** Un serie de componentes que trabajan juntos para abrir y cerrar las válvulas.

**Vernier calipers** Calipers that use a vernier scale which allows for measurement to a precision of ten to twenty-five times as fine as the base scale.
**Pie de rey** Los calibres que emplean una escala vernier permitiendo una medida precisa de diez a veinticinco veces más finas que la escala fundamental.

**Vibration dampener** *See* harmonic balancer.
**Amortiguador de vibraciones** Vea equilibrador armónico.

**Viscosity** The measure of oil thickness.
**Viscosidad** La medida de lo espeso de un aceite.

**Volumetric efficiency** A measurement of the amount of air/fuel mixture that actually enters the combustion chamber compared to the amount that could be drawn in.
**Rendimiento volumétrico** Una medida de la cantidad de la mezcla del aire/combustible que actualmente entra en la cámara de combustión comparada con la cantidad que podría entrar.

## W

**Wear-in** The required time needed for the ring to conform to the shape of the cylinder bore.
**Tiempo de estreno** El tiempo que se requiere para el anillo se conforme a la forma del taladro del cilindro.

**Wedge chamber** Combustion chamber employing a wedge design used to create turbulence to mix the air and fuel.
**Cámara biselada** Una cámara de combustión que emplea un diseño biselado para crear una turbulencia que mezcla el aire y el combustible.

**Wet compression testing** Compression test done after adding a small amount of oil to the cylinder. Wet compression testing is performed if the cylinder fails the dry compression test.
**Prueba de compresión húmedo** La prueba de compresión que se lleva acabo añadiendo una pequeña cantidad del aceite al cilindro. Se efectúa si el cilindro reprueba la prueba de compresión en seco.

**Wet sleeves** Replacement cylinder sleeves that are surrounded by engine coolant.
**Camisas húmedas** Las camisas de repuesta del cilindro que se rodean por el fluido refrigerante del motor.

# INDEX

## A



## B



## C

**D**



**E**

## F



## G



## H

## I



## J



## K



## L



## M



## N



## O

## P



## Q



## R



## S

## T



## U



## V

## W



## Z